A TREATISE ON PROBABILITY

A TREATISE ON PROBABILITY

JOHN MAYNARD KEYNES

GYAN

Published by

GYAN

Gyan Publishing House
5, Ansari Road
Daryaganj, New Delhi-110002
Phone: 011-47034999, 9811692060
E-mail: books@gyanbooks.com

Distribution Network
gyanbooks.com
India, USA, Canada, UK, Australia, France

ISBN: 978-81-212-8630-5 (PB)
First Published, 1921

2nd Impression 2023

Printed at: Gyan Press, Delhi.

A TREATISE ON PROBABILITY
Author: JOHN MAYNARD KEYNES

A TREATISE ON PROBABILITY

A TREATISE ON PROBABILITY

BY

JOHN MAYNARD KEYNES

FELLOW OF KING'S COLLEGE, CAMBRIDGE

MACMILLAN AND CO., LIMITED
ST. MARTIN'S STREET, LONDON
1921

PREFACE

THE subject matter of this book was first broached in the brain of Leibniz, who, in the dissertation, written in his twenty-third year, on the mode of electing the kings of Poland, conceived of Probability as a branch of Logic. A few years before, "un problème," in the words of Poisson, "proposé à un austère janséniste par un homme du monde, a été l'origine du calcul des probabilités." In the intervening centuries the algebraical exercises, in which the Chevalier de la Méré interested Pascal, have so far predominated in the learned world over the profounder enquiries of the philosopher into those processes of human faculty which, by determining reasonable preference, guide our choice, that Probability is oftener reckoned with Mathematics than with Logic. There is much here, therefore, which is novel, and, being novel, unsifted, inaccurate, or deficient. I propound my systematic conception of this subject for criticism and enlargement at the hand of others, doubtful whether I myself am likely to get much further, by waiting longer, with a work, which, beginning as a Fellowship Dissertation, and interrupted by the war, has already extended over many years.

It may be perceived that I have been much influenced by W. E. Johnson, G. E. Moore, and Bertrand Russell, that is to say, by Cambridge, which, with great debts to the writers of Continental Europe, yet continues in direct succession the English tradition of Locke and Berkeley and Hume, of Mill and Sidgwick, who, in spite of their divergences of doctrine, are united in a preference for what is matter of fact, and have conceived their subject as a branch rather of science than of the creative imagination, prose writers, hoping to be understood.

J. M. KEYNES.

KING'S COLLEGE, CAMBRIDGE,
May 1, 1920.

CONTENTS

PART I

FUNDAMENTAL IDEAS

CHAPTER I

PAGE

THE MEANING OF PROBABILITY 3

CHAPTER II

PROBABILITY IN RELATION TO THE THEORY OF KNOWLEDGE . 10

CHAPTER III

THE MEASUREMENT OF PROBABILITIES 20

CHAPTER IV

THE PRINCIPLE OF INDIFFERENCE 41

CHAPTER V

OTHER METHODS OF DETERMINING PROBABILITIES . . 65

CHAPTER VI

THE WEIGHT OF ARGUMENTS 71

CHAPTER VII

PAGE

Historical Retrospect 79

CHAPTER VIII

The Frequency Theory of Probability . . . 92

CHAPTER IX

The Constructive Theory of Part I. summarised . . 111

PART II

FUNDAMENTAL THEOREMS

CHAPTER X

Introductory 115

CHAPTER XI

The Theory of Groups, with special reference to Logical Consistence, Inference, and Logical Priority . . 123

CHAPTER XII

The Definitions and Axioms of Inference and Probability 133

CHAPTER XIII

The Fundamental Theorems of Necessary Inference. . 139

CHAPTER XIV

The Fundamental Theorems of Probable Inference . 144

CHAPTER XV

PAGE

NUMERICAL MEASUREMENT AND APPROXIMATION OF PROBABILITIES 158

CHAPTER XVI

OBSERVATIONS ON THE THEOREMS OF CHAPTER XIV., AND THEIR DEVELOPMENTS, INCLUDING TESTIMONY . . . 164

CHAPTER XVII

SOME PROBLEMS IN INVERSE PROBABILITY, INCLUDING AVERAGES 186

PART III

INDUCTION AND ANALOGY

CHAPTER XVIII

INTRODUCTION 217

CHAPTER XIX

THE NATURE OF ARGUMENT BY ANALOGY 222

CHAPTER XX

THE VALUE OF MULTIPLICATION OF INSTANCES, OR PURE INDUCTION . 233

CHAPTER XXI

THE NATURE OF INDUCTIVE ARGUMENT CONTINUED . . 242

CHAPTER XXII

THE JUSTIFICATION OF THESE METHODS . . . 251

CHAPTER XXIII

SOME HISTORICAL NOTES ON INDUCTION . . . 265

NOTES ON PART III. 274

PART IV

SOME PHILOSOPHICAL APPLICATIONS OF PROBABILITY

CHAPTER XXIV

PAGE

The Meanings of Objective Chance, and of Randomness . 281

CHAPTER XXV

Some Problems arising out of the Discussion of Chance . 293

CHAPTER XXVI

The Application of Probability to Conduct . . 307

PART V

THE FOUNDATIONS OF STATISTICAL INFERENCE

CHAPTER XXVII

The Nature of Statistical Inference . . . 327

CHAPTER XXVIII

The Law of Great Numbers 332

CHAPTER XXIX

The Use of *à priori* Probabilities for the Prediction of Statistical Frequency—the Theorems of Bernoulli, Poisson, and Tchebycheff 337

CHAPTER XXX

The Mathematical use of Statistical Frequencies for the Determination of Probability *à posteriori*—the Methods of Laplace 367

CHAPTER XXXI

PAGE

The Inversion of Bernoulli's Theorem 384

CHAPTER XXXII

The Inductive Use of Statistical Frequencies for the Determination of Probability *à posteriori*—the Methods of Lexis 391

CHAPTER XXXIII

Outline of a Constructive Theory 406

BIBLIOGRAPHY 429

INDEX 459

PART I

FUNDAMENTAL IDEAS

CHAPTER I

THE MEANING OF PROBABILITY

"J'ai dit plus d'une fois qu'il faudrait une nouvelle espèce de logique, qui traiteroit des degrés de Probabilité."—LEIBNIZ.

1. PART of our knowledge we obtain direct; and part by argument. The Theory of Probability is concerned with that part which we obtain by argument, and it treats of the different degrees in which the results so obtained are conclusive or inconclusive.

In most branches of academic logic, such as the theory of the syllogism or the geometry of ideal space, all the arguments aim at demonstrative certainty. They claim to be *conclusive.* But many other arguments are rational and claim some weight without pretending to be certain. In Metaphysics, in Science, and in Conduct, most of the arguments, upon which we habitually base our rational beliefs, are admitted to be inconclusive in a greater or less degree. Thus for a philosophical treatment of these branches of knowledge, the study of probability is required.

The course which the history of thought has led Logic to follow has encouraged the view that doubtful arguments are not within its scope. But in the actual exercise of reason we do not wait on certainty, or deem it irrational to depend on a doubtful argument. If logic investigates the general principles of valid thought, the study of arguments, to which it is rational to attach *some* weight, is as much a part of it as the study of those which are demonstrative.

2. The terms *certain* and *probable* describe the various degrees of rational belief about a proposition which different amounts of knowledge authorise us to entertain. All propositions are true or false, but the knowledge we have of them depends on our circumstances; and while it is often convenient to speak of

propositions as certain or probable, this expresses strictly a relationship in which they stand to a *corpus* of knowledge, actual or hypothetical, and not a characteristic of the propositions in themselves. A proposition is capable at the same time of varying degrees of this relationship, depending upon the knowledge to which it is related, so that it is without significance to call a proposition probable unless we specify the knowledge to which we are relating it.

To this extent, therefore, probability may be called subjective. But in the sense important to logic, probability is not subjective. It is not, that is to say, subject to human caprice. A proposition is not probable because we think it so. When once the facts are given which determine our knowledge, what is probable or improbable in these circumstances has been fixed objectively, and is independent of our opinion. The Theory of Probability is logical, therefore, because it is concerned with the degree of belief which it is *rational* to entertain in given conditions, and not merely with the actual beliefs of particular individuals, which may or may not be rational.

Given the body of direct knowledge which constitutes our ultimate premisses, this theory tells us what further rational beliefs, certain or probable, can be derived by valid argument from our direct knowledge. This involves purely logical relations between the propositions which embody our direct knowledge and the propositions about which we seek indirect knowledge. What particular propositions we select as the premisses of *our* argument naturally depends on subjective factors peculiar to ourselves; but the relations, in which other propositions stand to these, and which entitle us to probable beliefs, are objective and logical.

3. Let our premisses consist of any set of propositions h, and our conclusion consist of any set of propositions a, then, if a knowledge of h justifies a rational belief in a of degree α, we say that there is a *probability-relation* of degree α between a and h.[1]

In ordinary speech we often describe the *conclusion* as being doubtful, uncertain, or only probable. But, strictly, these terms ought to be applied, either to the degree of our *rational belief* in the conclusion, or to the relation or argument between two sets of propositions, knowledge of which would afford grounds for a corresponding degree of rational belief.[2]

[1] This will be written $a/h=\alpha$.

[2] See also Chapter II. § 5.

4. With the term "event," which has taken hitherto so important a place in the phraseology of the subject, I shall dispense altogether.[1] Writers on Probability have generally dealt with what they term the "happening" of "events." In the problems which they first studied this did not involve much departure from common usage. But these expressions are now used in a way which is vague and unambiguous; and it will be more than a verbal improvement to discuss the truth and the probability of *propositions* instead of the occurrence and the probability of *events*.[2]

5. These general ideas are not likely to provoke much criticism. In the ordinary course of thought and argument, we are constantly assuming that knowledge of one statement, while not *proving* the truth of a second, yields nevertheless *some ground* for believing it. We assert that we *ought* on the evidence to prefer such and such a belief. We claim rational grounds for assertions which are not conclusively demonstrated. We allow, in fact, that statements may be unproved, without, for that reason, being unfounded. And it does not seem on reflection that the information we convey by these expressions is wholly subjective. When we argue that Darwin gives valid grounds for our accepting his theory of natural selection, we do not simply mean that we are psychologically inclined to agree with him; it is certain that we also intend to convey our belief that we are acting rationally in regarding his theory as probable. We believe that there is some real objective relation between Darwin's evidence and his conclusions, which is independent of the mere fact of our belief, and which is just as real and objective, though of a different degree, as that which would exist if the argument were as demonstrative as a syllogism. We are claiming, in fact, to cognise correctly a logical connection between one set of propositions which we call our evidence and which we suppose ourselves to know, and another set which we call our conclusions, and to which we attach more or less weight

[1] Except in those chapters (Chap. XVII., for example) where I am dealing chiefly with the work of others.

[2] The first writer I know of to notice this was Ancillon in *Doutes sur les bases du calcul des probabilités* (1794): "Dire qu'un fait passé, présent ou à venir est probable, c'est dire qu'une proposition est probable." The point was emphasised by Boole, *Laws of Thought*, pp. 7 and 167. See also Czuber, *Wahrscheinlichkeitsrechnung*, vol. i. p. 5, and Stumpf, *Über den Begriff der mathematischen Wahrscheinlichkeit.*

according to the grounds supplied by the first. It is this type of objective relation between sets of propositions—the type which we claim to be correctly perceiving when we make such assertions as these—to which the reader's attention must be directed.

6. It is not straining the use of words to speak of this as the relation of probability. It is true that mathematicians have employed the term in a narrower sense; for they have often confined it to the limited class of instances in which the relation is adapted to an algebraical treatment. But in common usage the word has never received this limitation.

Students of probability in the sense which is meant by the authors of typical treatises on *Wahrscheinlichkeitsrechnung* or *Calcul des probabilités*, will find that I do eventually reach topics with which they are familiar. But in making a serious attempt to deal with the fundamental difficulties with which all students of mathematical probabilities have met and which are notoriously unsolved, we must begin at the beginning (or almost at the beginning) and treat our subject widely. As soon as mathematical probability ceases to be the merest algebra or pretends to guide our decisions, it immediately meets with problems against which its own weapons are quite powerless. And even if we wish later on to use probability in a narrow sense, it will be well to know first what it means in the widest.

7. Between two sets of propositions, therefore, there exists a relation, in virtue of which, if we know the first, we can attach to the latter some degree of rational belief. This relation is the subject-matter of the logic of probability.

A great deal of confusion and error has arisen out of a failure to take due account of this *relational* aspect of probability. From the premisses "a implies b" and "a is true," we can conclude something about b—namely that b is true—which does not involve a. But, if a is so related to b, that a knowledge of it renders a probable belief in b rational, we cannot conclude anything whatever about b which has not reference to a; and it is not true that every set of self-consistent premisses which includes a has this same relation to b. It is as useless, therefore, to say "b is probable" as it would be to say "b is equal," or "b is greater than," and as unwarranted to conclude that, because a makes b probable, therefore a and c together make b

probable, as to argue that because *a* is less than *b*, therefore *a* and *c* together are less than *b*.

Thus, when in ordinary speech we name some opinion as probable without further qualification, the phrase is generally elliptical. We mean that it is probable when certain considerations, implicitly or explicitly present to our minds at the moment, are taken into account. We use the word for the sake of shortness, just as we speak of a place as being three miles distant, when we mean three miles distant from where we are then situated, or from some starting-point to which we tacitly refer. No proposition is in itself either probable or improbable, just as no place can be intrinsically distant; and the probability of the same statement varies with the evidence presented, which is, as it were, its origin of reference. We may fix our attention on our own knowledge and, treating this as our origin, consider the probabilities of all other suppositions,—according to the usual practice which leads to the elliptical form of common speech; or we may, equally well, fix it on a proposed conclusion and consider what degree of probability this would derive from various sets of assumptions, which might constitute the *corpus* of knowledge of ourselves or others, or which are merely hypotheses.

Reflection will show that this account harmonises with familiar experience. There is nothing novel in the supposition that the probability of a theory turns upon the evidence by which it is supported; and it is common to assert that an opinion was probable on the evidence at first to hand, but on further information was untenable. As our knowledge or our hypothesis changes, our conclusions have new probabilities, not in themselves, but relatively to these new premisses. New logical relations have now become important, namely those between the conclusions which we are investigating and our new assumptions; but the old relations between the conclusions and the former assumptions still exist and are just as real as these new ones. It would be as absurd to deny that an opinion *was* probable, when at a later stage certain objections have come to light, as to deny, when we have reached our destination, that it was ever three miles distant; and the opinion still *is* probable in relation to the old hypotheses, just as the destination is still three miles distant from our starting-point.

8. A *definition* of probability is not possible, unless it contents us to define degrees of the probability-relation by reference to degrees of rational belief. We cannot analyse the probability-relation in terms of simpler ideas. As soon as we have passed from the logic of implication and the categories of truth and falsehood to the logic of probability and the categories of knowledge, ignorance, and rational belief, we are paying attention to a new logical relation in which, although it is logical, we were not previously interested, and which cannot be explained or defined in terms of our previous notions.

This opinion is, from the nature of the case, incapable of positive proof. The presumption in its favour must arise partly out of our failure to find a definition, and partly because the notion presents itself to the mind as something new and independent. If the statement that an opinion was probable on the evidence at first to hand, but became untenable on further information, is not solely concerned with psychological belief, I do not know how the element of logical doubt is to be defined, or how its substance is to be stated, in terms of the other indefinables of formal logic. The attempts at definition, which have been made hitherto, will be criticised in later chapters. I do not believe that any of them accurately represent that particular logical relation which we have in our minds when we speak of the probability of an argument.

In the great majority of cases the term "probable" seems to be used consistently by different persons to describe the same concept. Differences of opinion have not been due, I think, to a radical ambiguity of language. In any case a desire to reduce the indefinables of logic can easily be carried too far. Even if a definition is discoverable in the end, there is no harm in postponing it until our enquiry into the object of definition is far advanced. In the case of "probability" the object before the mind is so familiar that the danger of misdescribing its qualities through lack of a definition is less than if it were a highly abstract entity far removed from the normal channels of thought.

9. This chapter has served briefly to indicate, though not to define, the subject matter of the book. Its object has been to emphasise the existence of *a logical relation between two sets of propositions* in cases where it is not possible to argue demonstratively from one to the other. This is a contention

of a most fundamental character. It is not entirely novel, but has seldom received due emphasis, is often overlooked, and sometimes denied. The view, that probability arises out of the existence of a specific relation between premiss and conclusion, depends for its acceptance upon a reflective judgment on the true character of the concept. It will be our object to discuss, under the title of Probability, the principal properties of this relation. First, however, we must digress in order to consider briefly what we mean by *knowledge*, *rational belief*, and *argument*.

CHAPTER II

PROBABILITY IN RELATION TO THE THEORY OF KNOWLEDGE

1. I do not wish to become involved in questions of epistemology to which I do not know the answer; and I am anxious to reach as soon as possible the particular part of philosophy or logic which is the subject of this book. But some explanation is necessary if the reader is to be put in a position to understand the point of view from which the author sets out; I will, therefore, expand some part of what has been outlined or assumed in the first chapter.

2. There is, first of all, the distinction between that part of our belief which is rational and that part which is not. If a man believes something for a reason which is preposterous or for no reason at all, and what he believes turns out to be true for some reason not known to him, he cannot be said to believe it *rationally*, although he believes it and it is in fact true. On the other hand, a man may rationally believe a proposition to be *probable*, when it is in fact false. The distinction between rational belief and mere belief, therefore, is not the same as the distinction between true beliefs and false beliefs. The highest degree of rational belief, which is termed *certain* rational belief, corresponds to *knowledge*. We may be said to know a thing when we have a certain rational belief in it, and *vice versa*. For reasons which will appear from our account of probable degrees of rational belief in the following paragraph, it is preferable to regard *knowledge* as fundamental and to define *rational belief* by reference to it.

3. We come next to the distinction between that part of our rational belief which is certain and that part which is only probable. Belief, whether rational or not, is capable of degree. The highest degree of rational belief, or rational certainty of

belief, and its relation to knowledge have been introduced above. What, however, is the relation to knowledge of *probable* degrees of rational belief?

The proposition (*say*, q) that we *know* in this case is not the same as the proposition (*say*, p) in which we have a probable degree (*say*, a) of rational belief. If the evidence upon which we base our belief is h, then what we *know*, namely q, is that the proposition p bears the probability-relation of degree a to the set of propositions h; and this knowledge of ours justifies us in a rational belief of degree a in the proposition p. It will be convenient to call propositions such as p, which do not contain assertions about probability-relations, "primary propositions"; and propositions such as q, which assert the existence of a probability-relation, "secondary propositions."[1]

4. Thus knowledge of a proposition always corresponds to certainty of rational belief in it and at the same time to actual truth in the proposition itself. We cannot know a proposition unless it is in fact true. A probable degree of rational belief in a proposition, on the other hand, arises out of knowledge of some corresponding secondary proposition. A man may rationally believe a proposition to be probable when it is in fact false, if the secondary proposition on which he depends is true and certain; while a man cannot rationally believe a proposition to be probable even when it is in fact true, if the secondary proposition on which he depends is not true. Thus rational belief of whatever degree can only arise out of knowledge, although the knowledge may be of a proposition secondary, in the above sense, to the proposition in which the rational degree of belief is entertained.

5. At this point it is desirable to colligate the three senses in which the term *probability* has been so far employed. In its most fundamental sense, I think, it refers to the logical relation between two sets of propositions, which in § 4 of Chapter I. I have termed the probability-relation. It is with this that I shall be mainly concerned in the greater part of this Treatise. Derivative from this sense, we have the sense in which, as above, the term *probable* is applied to the degrees of rational belief arising out of knowledge of secondary propositions which assert the

[1] This classification of "primary" and "secondary" propositions was suggested to me by Mr. W. E. Johnson.

existence of probability-relations in the fundamental logical sense. Further it is often convenient, and not necessarily misleading, to apply the term *probable* to the proposition which is the object of the probable degree of rational belief, and which bears the probability-relation in question to the propositions comprising the evidence.

6. I turn now to the distinction between direct and indirect knowledge—between that part of our rational belief which we know directly and that part which we know by argument.

We start from things, of various classes, with which we have, what I choose to call without reference to other uses of this term, *direct acquaintance*. Acquaintance with such things does not in itself constitute knowledge, although knowledge arises out of acquaintance with them. The most important classes of things with which we have direct acquaintance are our own sensations, which we may be said to *experience*, the ideas or meanings, about which we have thoughts and which we may be said to *understand*, and facts or characteristics or relations of sense-data or meanings, which we may be said to *perceive* ;—experience, understanding, and perception being three forms of direct acquaintance.

The objects of knowledge and belief—as opposed to the objects of direct acquaintance which I term sensations, meanings, and perceptions—I shall term *propositions*.

Now our knowledge of propositions seems to be obtained in two ways: directly, as the result of contemplating the objects of acquaintance; and indirectly, *by argument*, through perceiving the probability-relation of the proposition, about which we seek knowledge, to other propositions. In the second case, at any rate at first, what we know is not the proposition itself but a secondary proposition involving it. When we know a secondary proposition involving the proposition p as subject, we may be said to have indirect knowledge *about* p.

Indirect knowledge about p may in suitable conditions lead to rational belief in p of an appropriate degree. If this degree is that of certainty, then we have not merely indirect knowledge *about* p, but indirect knowledge *of* p.

7. Let us take examples of direct knowledge. From acquaintance with a sensation of yellow I can pass directly to a knowledge of the proposition "I have a sensation of yellow." From acquaintance with a sensation of yellow and with the meanings of "yellow," "colour," "existence," I may be able to pass to a direct knowledge of the propositions "I understand the meaning of yellow," "my sensation of yellow exists," "yellow is a colour." Thus, by some mental process of which it is difficult to give an account, we are able to pass from direct acquaintance with things to a knowledge of propositions about the things of which we have sensations or understand the meaning.

Next, by the contemplation of propositions of which we have direct knowledge, we are able to pass indirectly to knowledge of or about other propositions. The mental process by which we pass from direct knowledge to indirect knowledge is in some cases and in some degree capable of analysis. We pass from a knowledge of the proposition *a* to a knowledge about the proposition *b* by perceiving a logical relation between them. With this logical relation we have direct acquaintance. The logic of knowledge is mainly occupied with a study of the logical relations, direct acquaintance with which permits direct knowledge of the secondary proposition asserting the probability-relation, and so to indirect knowledge about, and in some cases of, the primary proposition.

It is not always possible, however, to analyse the mental process in the case of indirect knowledge, or to say by the perception of *what* logical relation we have passed from the knowledge of one proposition to knowledge about another. But although in some cases we *seem* to pass directly from one proposition to another, I am inclined to believe that in all legitimate transitions of this kind some logical relation of the proper kind must exist between the propositions, even when we are not explicitly aware of it. In any case, whenever we pass to knowledge about one proposition by the contemplation of it in relation to another proposition of which we have knowledge—even when the process is unanalysable—I call it an *argument*. The knowledge, such as we have in ordinary thought by passing from one proposition to another without being able to say what logical relations, if any, we have perceived between them, may be termed uncompleted knowledge. And knowledge, which results from a distinct apprehension of the relevant logical relations, may be termed knowledge proper.

8. In this way, therefore, I distinguish between direct and indirect knowledge, between that part of our rational belief which is based on direct knowledge and that part which is based on argument. About what *kinds* of things we are capable of knowing propositions directly, it is not easy to say. About our own existence, our own sense-data, some logical ideas, and some logical relations, it is usually agreed that we have direct knowledge. Of the law of gravity, of the appearance of the other side of the moon, of the cure for phthisis, of the contents of Bradshaw, it is usually agreed that we do *not* have direct knowledge. But many questions are in doubt. Of *which* logical ideas and relations we have direct acquaintance, as to whether we can ever know directly the existence of *other people*, and as to when we are knowing propositions about sense-data directly and when we are interpreting them—it is not possible to give a clear answer. Moreover, there is another and peculiar kind of derivative knowledge—by memory.

At a given moment there is a great deal of our knowledge which we know neither directly nor by argument—we remember it. We may remember it as knowledge, but forget how we originally knew it. What we once knew and now consciously remember, can fairly be called knowledge. But it is not easy to draw the line between conscious memory, unconscious memory or habit, and pure instinct or irrational associations of ideas (acquired or inherited)—the last of which cannot fairly be called knowledge, for unlike the first two it did not even arise (in us at least) out of knowledge. Especially in such a case as that of what our eyes tell us, it is difficult to distinguish between the different ways in which our beliefs have arisen. We cannot always tell, therefore, what is remembered knowledge and what is not knowledge at all; and when knowledge is remembered, we do not always remember at the same time whether, originally, it was direct or indirect.

Although it is with knowledge by argument that I shall be mainly concerned in this book there is one kind of direct knowledge, namely of secondary propositions, with which I cannot help but be involved. In the case of every argument, it is only directly that we can know the secondary proposition which makes the argument itself valid and rational. When we know something by argument this must be through direct acquaintance with some logical relation between the conclusion and the premiss.

In *all* knowledge, therefore, there is some direct element; and logic can never be made purely mechanical. All it can do is so to arrange the reasoning that the logical relations, which have to be perceived directly, are made explicit and are of a simple kind.

9. It must be added that the term *certainty* is sometimes used in a merely psychological sense to describe a state of mind without reference to the logical grounds of the belief. With this sense I am not concerned. It is also used to describe the highest degree of rational belief; and this is the sense relevant to our present purpose. The peculiarity of certainty is that knowledge of a secondary proposition involving certainty, together with knowledge of what stands in this secondary proposition in the position of evidence, leads to *knowledge of*, and not merely *about*, the corresponding primary proposition. Knowledge, on the other hand, of a secondary proposition involving a degree of probability lower than certainty, together with knowledge of the premiss of the secondary proposition, leads only to a *rational belief of the appropriate degree* in the primary proposition. The knowledge present in this latter case I have called knowledge *about* the primary proposition or conclusion of the argument, as distinct from knowledge *of* it.

Of probability we can say no more than that it is a lower degree of rational belief than certainty; and we may say, if we like, that it deals with degrees of certainty.[1] Or we may make probability the more fundamental of the two and regard certainty as a special case of probability, as being, in fact, the *maximum probability*. Speaking somewhat loosely we may say that, if our premisses make the conclusion certain, then it *follows* from the premisses; and if they make it very probable, then it very nearly follows from them.

It is sometimes useful to use the term "impossibility" as the negative correlative of "certainty," although the former sometimes has a different set of associations. If a is certain, then the contradictory of a is impossible. If a knowledge of a makes b certain, then a knowledge of a makes the contradictory

[1] This view has often been taken, *e.g.*, by Bernoulli and, incidentally, by Laplace; also by Fries (see Czuher, *Entwicklung*, p. 12). The view, occasionally held, that probability is concerned with degrees of truth, arises out of a confusion between certainty and truth. Perhaps the Aristotelian doctrine that future events are neither true nor false arose in this way.

of b impossible. Thus a proposition is impossible with respect to a given premiss, if it is disproved by the premiss; and the relation of impossibility is the relation of minimum probability.[1]

10. We have distinguished between rational belief and irrational belief and also between rational beliefs which are certain in degree and those which are only probable. Knowledge has been distinguished according as it is direct or indirect, according as it is of primary or secondary propositions, and according as it is *of* or merely *about* its object.

In order that we may have a rational belief in a proposition p of the degree of certainty, it is necessary that one of two conditions should be fulfilled—(i.) that we know p directly; or (ii.) that we know a set of propositions h, and also know some secondary proposition q asserting a certainty-relation between p and h. In the latter case h may include secondary as well as primary propositions, but it is a necessary condition that all the propositions h should be *known*. In order that we may have rational belief in p of a lower degree of probability than certainty, it is necessary that we know a set of propositions h, and also know some secondary proposition q asserting a probability-relation between p and h.

In the above account one possibility has been ruled out. It is assumed that we cannot have a rational belief in p of a degree less than certainty except through knowing a secondary proposition of the prescribed type. Such belief can only arise, that is to say, by means of the perception of some probability-relation. To employ a common use of terms (though one inconsistent with the use adopted above), I have assumed that all direct knowledge is certain. All knowledge, that is to say, which is obtained in a manner strictly direct by contemplation of the objects of acquaintance and without any admixture whatever of argument and the contemplation of the logical bearing of any other knowledge on this, corresponds to *certain* rational belief and not to a merely probable degree of rational belief. It is true that there do *seem* to be degrees of knowledge and rational belief, when the source of

[1] Necessity and Impossibility, in the senses in which these terms are used in the theory of Modality, seem to correspond to the relations of Certainty and Impossibility in the theory of probability, the other modals, which comprise the intermediate degrees of possibility, corresponding to the intermediate degrees of probability. Almost up to the end of the seventeenth century the traditional treatment of modals is, in fact, a primitive attempt to bring the relations of probability within the scope of formal logic.

the belief is solely in acquaintance, as there are when its source is in argument. But I think that this appearance arises partly out of the difficulty of distinguishing direct from indirect knowledge, and partly out of a confusion between *probable* knowledge and *vague* knowledge. I cannot attempt here to analyse the meaning of vague knowledge. It is certainly not the same thing as knowledge proper, whether certain or probable, and it does not seem likely that it is susceptible of strict logical treatment. At any rate I do not know how to deal with it, and in spite of its importance I will not complicate a difficult subject by endeavouring to treat adequately the theory of vague knowledge.

I assume then that only true propositions can be known, that the term "probable knowledge" ought to be replaced by the term "probable degree of rational belief," and that a probable degree of rational belief cannot arise directly but only as the result of an argument, out of the knowledge, that is to say, of a secondary proposition asserting some logical probability-relation in which the object of the belief stands to some known proposition. With arguments, if they exist, the *ultimate* premisses of which are known in some other manner than that described above, such as might be called "probable knowledge," my theory is not adequate to deal without modification.[1]

For the objects of certain belief which is based on direct knowledge, as opposed to certain belief arising indirectly, there is a well-established expression; propositions, in which our rational belief is both certain and direct, are said to be *self-evident.*

11. In conclusion, the relativity of knowledge to the individual may be briefly touched on. Some part of knowledge—knowledge of our own existence or of our own sensations—is clearly relative to individual experience. We cannot speak of knowledge absolutely—only of the knowledge of a particular person. Other parts of knowledge—knowledge of the axioms of logic, for example—may seem more objective. But we must admit, I think, that this too is relative to the constitution of the human mind, and that the constitution of the human mind may vary in some degree from man to man. What is self-evident to me and what

[1] I do not mean to imply, however, at any rate at present, that the ultimate premisses of an argument need always be *primary* propositions.

I really know, may be only a probable belief to you, or may form no part of your rational beliefs at all. And this may be true not only of such things as *my* existence, but of some logical axioms also. Some men—indeed it is obviously the case—may have a greater power of logical intuition than others. Further, the difference between some kinds of propositions over which human intuition seems to have power, and some over which it has none, may depend wholly upon the constitution of our minds and have no significance for a perfectly objective logic. We can no more assume that all true secondary propositions are or ought to be universally known than that all true primary propositions are known. The perceptions of some relations of probability may be outside the powers of some or all of us.

What we know and what probability we can attribute to our rational beliefs is, therefore, subjective in the sense of being relative to the individual. But given the body of premisses which our subjective powers and circumstances supply to us, and given the kinds of logical relations, upon which arguments can be based and which we have the capacity to perceive, the conclusions, which it is rational for us to draw, stand to these premisses in an objective and wholly logical relation. Our logic is concerned with drawing conclusions by a series of steps of certain specified kinds from a *limited* body of premisses.

With these brief indications as to the relation of Probability, as I understand it, to the Theory of Knowledge, I pass from problems of ultimate analysis and definition, which are not the primary subject matter of this book, to the logical theory and superstructure, which occupies an intermediate position between the ultimate problems and the applications of the theory, whether such applications take a generalised mathematical form or a concrete and particular one. For this purpose it would only encumber the exposition, without adding to its clearness or its accuracy, if I were to employ the perfectly exact terminology and minute refinements of language, which are necessary for the avoidance of error in very fundamental enquiries. While taking pains, therefore, to avoid any divergence between the substance of this chapter and of those which succeed it, and to employ only such periphrases as could be translated, *if desired*, into perfectly exact language, I shall not cut myself off from the convenient, but looser, expressions, which have been habitually employed by previous writers and have the advantage of being, in a general way at least, immediately intelligible to the reader.[1]

[1] This question, which faces all contemporary writers on logical and philosophical subjects, is in my opinion much more a question of *style*—and therefore to be settled on the same sort of considerations as other such questions—than is generally supposed. There are occasions for very exact methods of statement, such as are employed in Mr. Russell's *Principia Mathematica*. But there are advantages also in writing the English of Hume. Mr. Moore has developed in *Principia Ethica* an intermediate style which in his hands has force and beauty. But those writers, who strain after exaggerated precision without going the whole hog with Mr. Russell, are sometimes merely pedantic. They lose the reader's attention, and the repetitious complication of their phrases eludes his comprehension, without their really attaining, to compensate, a complete precision. Confusion of thought is not always best avoided by technical and unaccustomed expressions, to which the mind has no immediate reaction of understanding; it is possible, under cover of a careful formalism, to make statements, which, if expressed in plain language, the mind would immediately repudiate. There is much to be said, therefore, in favour of understanding the substance of what you are saying *all the time*, and of never reducing the substantives of your argument to the mental status of an x or y.

CHAPTER III

THE MEASUREMENT OF PROBABILITIES

1. I HAVE spoken of probability as being concerned with *degrees* of rational belief. This phrase implies that it is in some sense quantitative and perhaps capable of measurement. The theory of probable arguments must be much occupied, therefore, with *comparisons* of the respective weights which attach to different arguments. With this question we will now concern ourselves.

It has been assumed hitherto as a matter of course that probability is, in the full and literal sense of the word, measurable. I shall have to limit, not extend, the popular doctrine. But, keeping my own theories in the background for the moment, I will begin by discussing some existing opinions on the subject.

2. It has been sometimes supposed that a numerical comparison between the degrees of any pair of probabilities is not only conceivable but is actually within our power. Bentham, for instance, in his *Rationale of Judicial Evidence*,[1] proposed a scale on which witnesses might mark the degree of their certainty; and others have suggested seriously a 'barometer of probability.'[2]

That such comparison is *theoretically possible*, whether or not we are actually competent in every case to make the comparison, has been the generally accepted opinion. The following quotation[3] puts this point of view very well:

"I do not see on what ground it can be doubted that every

[1] Book i. chap vi. (referred to by Venn).

[2] The reader may be reminded of Gibbon's proposal that:—"A Theological Barometer might be formed, of which the Cardinal (Baronius) and our countryman, Dr. Middleton, should constitute the opposite and remote extremities, as the former sunk to the lowest degree of credulity, which was compatible with learning, and the latter rose to the highest pitch of scepticism, in any wise consistent with Religion."

[3] W. F. Donkin, *Phil. Mag.*, 1851. He is replying to an article by J. D. Forbes (*Phil. Mag.*, Aug. 1849) which had cast doubt upon this opinion.

definite state of belief concerning a proposed hypothesis is in itself capable of being represented by a numerical expression, however difficult or impracticable it may be to ascertain its actual value. It would be very difficult to estimate in numbers the *vis viva* of all of the particles of a human body at any instant; but no one doubts that it is capable of numerical expression. I mention this because I am not sure that Professor Forbes has distinguished the difficulty of *ascertaining numbers* in certain cases from a supposed difficulty of *expression by means of numbers*. The former difficulty is real, but merely relative to our knowledge and skill; the latter, if real, would be absolute and inherent in the subject-matter, which I conceive is not the case."

De Morgan held the same opinion on the ground that, wherever we have differences of degree, numerical comparison *must* be theoretically possible.[1] He assumes, that is to say, that all probabilities can be placed in an *order* of magnitude, and argues from this that they must be measurable. Philosophers, however, who are mathematicians, would no longer agree that, even if the premiss is sound, the conclusion follows from it. Objects can be arranged in an order, which we can reasonably call one of degree or magnitude, without its being possible to conceive a system of measurement of the differences between the individuals.

This opinion may also have been held by others, if not by De Morgan, in part because of the narrow associations which Probability has had for them. The Calculus of Probability has received far more attention than its logic, and mathematicians, under no compulsion to deal with the whole of the subject, have naturally confined their attention to those special cases, the existence of which will be demonstrated at a later stage, where algebraical representation is possible. Probability has become associated, therefore, in the minds of theorists with those problems in which we are presented with a number of exclusive and exhaustive alternatives of equal probability; and the principles, which are readily applicable in such circumstances, have been supposed, without much further enquiry, to possess general validity.

3. It is also the case that theories of probability have been

[1] "Whenever the terms greater and less can be applied, there twice, thrice, etc., can be conceived, though not perhaps measured by us."—"Theory of Probabilities," *Encyclopaedia Metropolitana*, p. 395. He is a little more guarded in his *Formal Logic*, pp. 174, 175; but arrives at the same conclusion so far as probability is concerned.

propounded and widely accepted, according to which its numerical character is necessarily involved in the definition. It is often said, for instance, that probability is the ratio of the number of "favourable cases" to the total number of "cases." If this definition is accurate, it follows that every probability can be properly represented by a number and in fact *is* a number; for a ratio is not a quantity at all. In the case also of definitions based upon statistical frequency, there must be by definition a numerical ratio corresponding to every probability. These definitions and the theories based on them will be discussed in Chapter VIII.; they are connected with fundamental differences of opinion with which it is not necessary to burden the present argument.

4. If we pass from the opinions of theorists to the experience of practical men, it might perhaps be held that a presumption in favour of the numerical valuation of all probabilities can be based on the practice of underwriters and the willingness of Lloyd's to insure against practically any risk. Underwriters are actually willing, it might be urged, to name a numerical measure in every case, and to back their opinion with money. But this practice shows no more than that many probabilities are greater or less than some numerical measure, not that they themselves are numerically definite. It is sufficient for the underwriter if the premium he names *exceeds* the probable risk. But, apart from this, I doubt whether in extreme cases the process of thought, through which he goes before naming a premium, is wholly rational and determinate; or that two equally intelligent brokers acting on the same evidence would always arrive at the same result. In the case, for instance, of insurances effected before a Budget, the figures quoted must be partly arbitrary. There is in them an element of caprice, and the broker's state of mind, when he quotes a figure, is like a bookmaker's when he names odds. Whilst he may be able to make sure of a profit, on the principles of the bookmaker, yet the individual figures that make up the book are, within certain limits, arbitrary. He may be almost certain, that is to say, that there will not be new taxes on more than one of the articles tea, sugar, and whisky; there may be an opinion abroad, reasonable or unreasonable, that the likelihood is in the order—whisky, tea, sugar; and he may, therefore, be able to effect insurances for equal amounts in each

at 30 per cent, 40 per cent, and 45 per cent. He has thus made sure of a profit of 15 per cent, however absurd and arbitrary his quotations may be. It is not necessary for the success of underwriting on these lines that the probabilities of these new taxes are really measurable by the figures $\frac{3}{10}$, $\frac{4}{10}$, and $\frac{45}{100}$; it is sufficient that there should be merchants willing to insure at these rates. These merchants, moreover, may be wise to insure even if the quotations are partly arbitrary; for they may run the risk of insolvency unless their possible loss is thus limited. That the transaction is in principle one of bookmaking is shown by the fact that, if there is a specially large demand for insurance against one of the possibilities, the rate rises;—the probability has not changed, but the "book" is in danger of being upset. A Presidential election in the United States supplies a more precise example. On August 23, 1912, 60 per cent was quoted at Lloyd's to pay a total loss should Dr. Woodrow Wilson be elected, 30 per cent should Mr. Taft be elected, and 20 per cent should Mr. Roosevelt be elected. A broker, who could effect insurances in equal amounts against the election of each candidate, would be certain at these rates of a profit of 10 per cent. Subsequent modifications of these terms would largely depend upon the number of applicants for each kind of policy. Is it possible to maintain that these figures in any way represent reasoned numerical estimates of probability?

In some insurances the arbitrary element seems even greater. Consider, for instance, the reinsurance rates for the *Waratah*, a vessel which disappeared in South African waters. The lapse of time made rates rise; the departure of ships in search of her made them fall; some nameless wreckage is found and they rise; it is remembered that in similar circumstances thirty years ago a vessel floated, helpless but not seriously damaged, for two months, and they fall. Can it be pretended that the figures which were quoted from day to day—75 per cent, 83 per cent, 78 per cent—were rationally determinate, or that the actual figure was not within wide limits arbitrary and due to the caprice of individuals? In fact underwriters themselves distinguish between risks which are properly insurable, either because their probability can be estimated between comparatively narrow numerical limits or because it is possible to make a "book" which covers all possibilities, and other risks which cannot be dealt with in this way and which cannot form the basis of a regular business of insurance,—although an occasional gamble may be indulged in. I believe, therefore, that the practice of underwriters weakens rather than supports the contention that all probabilities can be measured and estimated numerically.

5. Another set of practical men, the lawyers, have been more subtle in this matter than the philosophers.[1] A distinction, interesting for our present purpose, between probabilities, which can be estimated within somewhat narrow limits, and those which cannot, has arisen in a series of judicial decisions respecting damages. The following extract[2] from the *Times Law Reports* seems to me to deal very clearly in a mixture of popular and legal phraseology, with the logical point at issue:

This was an action brought by a breeder of racehorses to recover damages for breach of a contract. The contract was that Cyllene, a racehorse owned by the defendant, should in the season of the year 1909 serve one of the plaintiff's brood mares. In the summer of 1908 the defendant, without the consent of the plaintiff, sold Cyllene for £30,000 to go to South America. The plaintiff claimed a sum equal to the average profit he had made through having a mare served by Cyllene during the past four years. During those four years he had had four colts which had sold at £3300. Upon that basis his loss came to 700 guineas.

Mr. Justice Jelf said that he was desirous, if he properly could, to find some mode of legally making the defendant compensate the plaintiff; but the question of damages presented formidable and, to his mind, insuperable difficulties. The damages, if any, recoverable here must be either the estimated loss of profit or else nominal damages. The estimate could only be based on a succession of contingencies. Thus it was assumed that (*inter alia*) Cyllene would be alive and well at the time of the intended service; that the mare sent would be well bred and not barren; that she would not slip her foal; and that the foal would be born alive and healthy. In a case of this kind he could only

[1] Leibniz notes the subtle distinctions made by Jurisconsults between degrees of probability; and in the preface to a work, projected but unfinished, which was to have been entitled *Ad stateram juris de gradibus probationum et probabilitatum* he recommends them as models of logic in contingent questions (Couturat, *Logique de Leibniz*, p. 240).

[2] I have considerably compressed the original report (Sapwell *v.* Bass).

rely on the weighing of chances; and the law generally regarded damages which depended on the weighing of chances as too remote, and therefore irrecoverable. It was drawing the line between an estimate of damage based on probabilities, as in "Simpson *v.* L. and N.W. Railway Co." (1, Q.B.D., 274), where Cockburn, C.J., said: "To some extent, no doubt, the damage must be a matter of speculation, but that is no reason for not awarding any damages at all," and a claim for damages of a totally problematical character. He (Mr. Justice Jelf) thought the present case was well over the line. Having referred to "Mayne on Damages" (8th ed., p. 70), he pointed out that in "Watson *v.* Ambergah Railway Co." (15, Jur., 448) Patteson, J., seemed to think that the chance of a prize might be taken into account in estimating the damages for breach of a contract to send a machine for loading barges by railway too late for a show; but Erle, J., appeared to think such damage was too remote. In his Lordship's view the chance of winning a prize was not of sufficiently ascertainable value at the time the contract was made to be within the contemplation of the parties. Further, in the present case, the contingencies were far more numerous and uncertain. He would enter judgment for the plaintiff for nominal damages, which were all he was entitled to. They would be assessed at 1s.

One other similar case may be quoted in further elucidation of the same point, and because it also illustrates another point—the importance of making clear the assumptions relative to which the probability is calculated. This case [1] arose out of an offer of a Beauty Prize [2] by the *Daily Express*. Out of 6000 photographs submitted, a number were to be selected and published in the newspaper in the following manner:

The United Kingdom was to be divided into districts and the photographs of the selected candidates living in each district were to be submitted to the readers of the paper in the district, who were to select by their votes those whom they considered the most beautiful, and a Mr. Seymour Hicks was then to make an appointment with the 50 ladies obtaining the greatest number of votes and himself select 12 of them. The plaintiff, who came

[1] Chaplin *v.* Hicks (1911).

[2] The prize was to be a theatrical engagement and, according to the article, the probability of subsequent marriage into the peerage.

out head of one of the districts, submitted that she had not been given a reasonable opportunity of keeping an appointment, that she had thereby lost the value of her chance of one of the 12 prizes, and claimed damages accordingly. The jury found that the defendant had not taken reasonable means to give the plaintiff an opportunity of presenting herself for selection, and assessed the damages, provided they were capable of assessment, at £100, the question of the possibility of assessment being postponed. This was argued before Mr. Justice Pickford, and subsequently in the Court of Appeal before Lord Justices Vaughan Williams, Fletcher Moulton, and Farwell. Two questions arose —relative to what evidence ought the probability to be calculated, and was it numerically measurable? Counsel for the defendant contended that, "if the value of the plaintiff's chance was to be considered, it must be the value as it stood at the beginning of the competition, not as it stood after she had been selected as one of the 50. As 6000 photographs had been sent in, and there was also the personal taste of the defendant as final arbiter to be considered, the value of the chance of success was really incalculable." The first contention that she ought to be considered as one of 6000 not as one of 50 was plainly preposterous and did not hoodwink the court. But the other point, the personal taste of the arbiter, presented more difficulty. In estimating the chance, ought the Court to receive and take account of evidence respecting the arbiter's preferences in types of beauty? Mr. Justice Pickford, without illuminating the question, held that the damages were capable of estimation. Lord Justice Vaughan Williams in giving judgment in the Court of Appeal argued as follows:

As he understood it, there were some 50 competitors, and there were 12 prizes of equal value, so that the average chance of success was about one in four. It was then said that the questions which might arise in the minds of the persons who had to give the decisions were so numerous that it was impossible to apply the doctrine of averages. He did not agree. Then it was said that if precision and certainty were impossible in any case it would be right to describe the damages as unassessable. He agreed that there might be damages so unassessable that the doctrine of averages was not possible of application because the figures necessary to be applied were not forthcoming. Several cases were to be found in the reports where it had been so held, but he denied the proposition that because precision and certainty had not been arrived at, the jury had no function or duty to determine the damages. . . . He (the Lord Justice) denied that the mere fact that you could not assess with precision and certainty relieved a wrongdoer from paying damages for his breach of duty. He would not lay down that in every case it could be left to the jury to assess the damages; there were cases where the loss was so dependent on the mere unrestricted volition of another person that it was impossible to arrive at any assessable loss from the breach. It was true that there was no market here; the right to compete was personal and could not be transferred. He could not admit that a competitor who found herself one of 50 could have gone into the market and sold her right to compete. At the same time the jury might reasonably have asked themselves the question whether, if there was a right to compete, it could have been transferred, and at what price. Under these circumstances he thought the matter was one for the jury.

The attitude of the Lord Justice is clear. The plaintiff had evidently suffered damage, and justice required that she should be compensated. But it was equally evident, that, relative to the completest information available and account being taken of the arbiter's personal taste, the probability could be by no means estimated with numerical precision. Further, it was impossible to say how much weight ought to be attached to the fact that the plaintiff had been *head* of her district (there were *fewer* than 50 districts); yet it was plain that it made her chance *better* than the chances of those of the 50 left in, who were not head of their districts. Let rough justice be done, therefore. Let the case be simplified by ignoring some part of the evidence. The "doctrine of averages" is then applicable, or, in other words, the plaintiff's loss may be assessed at twelve-fiftieths of the value of the prize.[1]

6. How does the matter stand, then? Whether or not such a thing is theoretically conceivable, no exercise of the practical judgment is possible, by which a numerical value can actually be given to the probability of every argument. So far from

[1] The jury in assessing the damages at £100, however, cannot have argued so subtly as this; for the average value of a prize (I have omitted the details bearing on their value) could not have been fairly estimated so high as £400.

our being able to measure them, it is not even clear that we are always able to place them in an order of magnitude. Nor has any theoretical rule for their evaluation ever been suggested.

The doubt, in view of these facts, whether any two probabilities are in every case even theoretically capable of comparison in terms of numbers, has not, however, received serious consideration. There seems to me to be exceedingly strong reasons for entertaining the doubt. Let us examine a few more instances.

7. Consider an induction or a generalisation. It is usually held that each additional instance increases the generalisation's probability. A conclusion, which is based on three experiments in which the unessential conditions are varied, is more trustworthy than if it were based on two. But what reason or principle can be adduced for attributing a numerical measure to the increase ? [1]

Or, to take another class of instances, we may sometimes have some reason for supposing that one object belongs to a certain category if it has points of similarity to other known members of the category (*e.g.* if we are considering whether a certain picture should be ascribed to a certain painter), and the greater the similarity the greater the probability of our conclusion. But we cannot in these cases *measure* the increase ; we can say that the presence of certain peculiar marks in a picture increases the probability that the artist of whom those marks are known to be characteristic painted it, but we cannot say that the presence of these marks makes it two or three or any other number of times more probable than it would have been without them. We can say that one thing is more like a second object than it is like a third ; but there will very seldom be any meaning in saying that it is twice as like. Probability is, so far as measurement is concerned, closely analogous to similarity.[2]

[1] It is true that Laplace and others (even amongst contemporary writers) have believed that the probability of an induction is measurable by means of a formula known as the *rule of succession*, according to which the probability of an induction based on n instances is $\frac{n+1}{n+2}$. Those who have been convinced by the reasoning employed to establish this rule must be asked to postpone judgment until it has been examined in Chapter XXX. But we may point out here the absurdity of supposing that the odds are 2 to 1 in favour of a generalisation based on a single instance—a conclusion which this formula would seem to justify.

[2] There are very few writers on probability who have explicitly admitted that probabilities, though in some sense quantitative, may be incapable of

Or consider the ordinary circumstances of life. We are out for a walk—what is the probability that we shall reach home alive? Has this always a numerical measure? If a thunderstorm bursts upon us, the probability is less than it was before; but is it changed by some definite numerical amount? There might, of course, be data which would make these probabilities numerically comparable; it might be argued that a knowledge of the statistics of death by lightning would make such a comparison possible. But if such information is not included within the knowledge to which the probability is referred, this fact is not relevant to the probability actually in question and cannot affect its value. In some cases, moreover, where general statistics are available, the numerical probability which might be derived from them is inapplicable because of the presence of additional knowledge with regard to the particular case. Gibbon calculated his prospects of life from the volumes of vital statistics and the calculations of actuaries. But if a doctor had been called to his assistance the nice precision of these calculations would have become useless; Gibbon's prospects would have been better or worse than before, but he would no longer have been able to calculate to within a day or week the period for which he then possessed an even chance of survival.

In these instances we can, perhaps, arrange the probabilities in an order of magnitude and assert that the new datum strengthens or weakens the argument, although there is no basis for an estimate *how much* stronger or weaker the new argument is than the old. But in another class of instances is it even possible to arrange the probabilities in an *order* of magnitude, or to say that one is the greater and the other less?

8. Consider three sets of experiments, each directed towards establishing a generalisation. The first set is more numerous;

---

numerical comparison. Edgeworth, "Philosophy of Chance" (*Mind*, 1884, p. 225), admitted that "there may well be important quantitative, although not numerical, estimates" of probabilities. Goldschmidt (*Wahrscheinlichkeitsrechnung*, p. 43) may also be cited as holding a somewhat similar opinion. He maintains that a lack of comparability in the grounds often stands in the way of the measurability of the probable in ordinary usage, and that there are not necessarily good reasons for measuring the value of one argument against that of another. On the other hand, a numerical statement for the degree of the probable, although generally impossible, is not in itself contradictory to the notion; and of three statements, relating to the same circumstances, we can well say that one is more probable than another, and that one is the most probable of the three.

in the second set the irrelevant conditions have been more carefully varied; in the third case the generalisation in view is wider in scope than in the others. Which of these generalisations is on such evidence the most probable? There is, surely, no answer; there is neither equality nor inequality between them. We cannot always weigh the analogy against the induction, or the scope of the generalisation against the bulk of the evidence in support of it. If we have *more* grounds than before, comparison is possible; but, if the grounds in the two cases are quite different, even a comparison of more and less, let alone numerical measurement, may be impossible.

This leads up to a contention, which I have heard supported, that, although not all measurements and not all comparisons of probability are within our power, yet we can say in the case of every argument whether it is *more* or *less* likely than not. Is our expectation of rain, when we start out for a walk, always *more* likely than not, or *less* likely than not, or *as* likely as not? I am prepared to argue that on some occasions *none* of these alternatives hold, and that it will be an arbitrary matter to decide for or against the umbrella. If the barometer is high, but the clouds are black, it is not always rational that one should prevail over the other in our minds, or even that we should balance them,—though it will be rational to allow caprice to determine us and to waste no time on the debate.

9. Some cases, therefore, there certainly are in which no rational basis has been discovered for numerical comparison. It is not the case here that the method of calculation, prescribed by theory, is beyond our powers or too laborious for actual application. *No* method of calculation, however impracticable, has been suggested. Nor have we any *prima facie* indications of the existence of a common unit to which the magnitudes of all probabilities are naturally referrible. A degree of probability is not composed of some homogeneous material, and is not apparently divisible into parts of like character with one another. An assertion, that the magnitude of a given probability is in a numerical ratio to the magnitude of every other, seems, therefore, unless it is based on one of the current *definitions* of probability, with which I shall deal separately in later chapters, to be altogether devoid of the kind of support, which can usually be supplied in the case of quantities of which

the mensurability is not open to denial. It will be worth while, however, to pursue the argument a little further.

10. There appear to be four alternatives. Either in some cases there is no probability at all; or probabilities do not all belong to a single set of magnitudes measurable in terms of a common unit; or these measures always exist, but in many cases are, and *must remain*, unknown; or probabilities do belong to such a set and their measures are *capable* of being determined by us, although we are not always able so to determine them in practice.

11. Laplace and his followers excluded the first two alternatives. They argued that every conclusion has its place in the numerical range of probabilities from 0 to 1, *if only we knew it*, and they developed their theory of *unknown* probabilities.

In dealing with this contention, we must be clear as to what we mean by saying that a probability is *unknown*. Do we mean unknown through lack of skill in arguing from given evidence, or unknown through lack of evidence? The first is alone admissible, for new evidence would give us a new probability, not a fuller knowledge of the old one; we have not discovered the probability of a statement on given evidence, by determining its probability in relation to quite different evidence. We must not allow the theory of unknown probabilities to gain plausibility from the second sense. A relation of probability does not yield us, as a rule, information of much value, unless it invests the conclusion with a probability which lies between narrow numerical limits. In ordinary practice, therefore, we do not always regard ourselves as *knowing* the probability of a conclusion, unless we can estimate it numerically. We are apt, that is to say, to restrict the use of the expression *probable* to these numerical examples, and to allege in other cases that the probability is unknown. We might say, for example, that we do not know, when we go on a railway journey, the probability of death in a railway accident, unless we are told the statistics of accidents in former years; or that we do not know our chances in a lottery, unless we are told the number of the tickets. But it must be clear upon reflection that if we use the term in this sense,—which is no doubt a perfectly legitimate sense,—we ought to say that in the case of some arguments a relation of probability does not exist, and not that it is unknown. For it is not *this* probability that we have discovered, when the accession of new evidence makes it possible to frame a numerical estimate.

Possibly this theory of unknown probabilities may also gain strength from our practice of estimating arguments, which, as I maintain, have *no* numerical value, by reference to those that have. We frame two ideal arguments, that is to say, in which the general character of the evidence largely resembles what is actually within our knowledge, but which is so constituted as to yield a numerical value, and we judge that the probability of the actual argument lies between these two. Since our standards, therefore, are referred to numerical measures in many cases where actual measurement is impossible, and since the probability lies *between* two numerical measures, we come to believe that it must also, if only we knew it, possess such a measure itself.

12. To say, then, that a probability is unknown ought to mean that it is unknown to us through our lack of skill in arguing from given evidence. The evidence justifies a certain degree of knowledge, but the weakness of our reasoning power prevents our knowing what this degree is. At the best, in such cases, we only know *vaguely* with what degree of probability the premisses invest the conclusion. That probabilities can be unknown in this sense or known with less distinctness than the argument justifies, is clearly the case. We can through stupidity fail to make any estimate of a probability at all, just as we may through the same cause estimate a probability wrongly. As soon as we distinguish between the degree of belief which it is rational to entertain and the degree of belief actually entertained, we have in effect admitted that the true probability is *not* known to everybody.

But this admission must not be allowed to carry us too far. Probability is, *vide* Chapter II. (§ 12), relative in a sense to the principles of *human* reason. The degree of probability, which it is rational for *us* to entertain, does not presume perfect logical insight, and is relative in part to the secondary propositions which we in fact know; and it is not dependent upon whether more perfect logical insight is or is not conceivable. It is the degree of probability to which those logical processes lead, of which our minds are capable; or, in the language of Chapter II., which those secondary propositions justify, which we in fact know. If we do not take this view of probability, if we do not limit it in this way and make it, to this extent, relative to human powers, we are altogether adrift in the unknown; for we cannot ever know what degree of probability would be justified by the perception of logical relations which we are, and must always be, incapable of comprehending.

13. Those who have maintained that, where we cannot assign a numerical probability, this is not because there is none, but simply because we do not know it, have really meant, I feel sure, that with some addition to our knowledge a numerical value would be assignable, that is to say that our conclusions would have a numerical probability relative to *slightly different* premisses. Unless, therefore, the reader clings to the opinion that, in every one of the instances I have cited in the earlier paragraphs of this chapter, it is theoretically possible on *that* evidence to assign a numerical value to the probability, we are left with the first two of the alternatives of § 10, which were as follows: either in some cases there is no probability at all; or probabilities do not all belong to a single set of magnitudes measurable in terms of a common unit. It would be difficult to maintain that there is *no* logical relation whatever between our premiss and our conclusion in those cases where we cannot assign a numerical value to the probability; and if this is so, it is really a question of whether the logical relation has characteristics, other than mensurability, of a kind to justify us in calling it a probability-relation. Which of the two we favour is, therefore, partly a matter of definition. We might, that is to say, pick out from probabilities (in the widest sense) a set, if there is one, all of which are measurable in terms of a common unit, and call the members of this set, and them only, probabilities (in the narrow sense). To restrict the term 'probability' in this way would be, I think, very inconvenient. For it is possible, as I shall show, to find *several* sets, the members of each of which are measurable in terms of a unit common to all the members of that set; so that it would be in some degree arbitrary [1] which we chose. Further, the distinction between probabilities, which would be thus measurable and those which would not, is not fundamental.

At any rate I aim here at dealing with probability in its

[1] Not altogether; for it would be natural to select the set to which the relation of certainty belongs.

widest sense, and am averse to confining its scope to a limited type of argument. If the opinion that not all probabilities can be measured seems paradoxical, it may be due to this divergence from a usage which the reader may expect. Common usage, even if it involves, as a rule, a flavour of numerical measurement, does not *consistently* exclude those probabilities which are incapable of it. The confused attempts, which have been made, to deal with numerically indeterminate probabilities under the title of unknown probabilities, show how difficult it is to confine the discussion within the intended limits, if the original definition is too narrow.

14. I maintain, then, in what follows, that there are some pairs of probabilities between the members of which *no* comparison of magnitude is possible; that we can say, nevertheless, of some pairs of relations of probability that the one is greater and the other less, although it is not possible to measure the difference between them; and that in a very special type of case, to be dealt with later, a meaning can be given to a *numerical* comparison of magnitude. I think that the results of observation, of which examples have been given earlier in this chapter, are consistent with this account.

By saying that not all probabilities are measurable, I mean that it is not possible to say of every pair of conclusions, about which we have some knowledge, that the degree of our rational belief in one bears any numerical relation to the degree of our rational belief in the other; and by saying that not all probabilities are comparable in respect of more and less, I mean that it is not always possible to say that the degree of our rational belief in one conclusion is either equal to, greater than, or less than the degree of our belief in another.

We must now examine a philosophical theory of the quantitative properties of probability, which would explain and justify the conclusions, which reflection discovers, if the preceding discussion is correct, in the practice of ordinary argument. We must bear in mind that our theory must apply to all probabilities and not to a limited class only, and that, as we do not adopt a definition of probability which presupposes its numerical mensurability, we cannot directly argue from differences in degree to a numerical measurement of these differences. The problem is subtle and difficult, and the following solution is, therefore, proposed with hesitation; but I am strongly convinced that something resembling the conclusion here set forth is true.

15. The so-called magnitudes or degrees of knowledge or probability, in virtue of which one is greater and another less, really arise out of an *order* in which it is possible to place them. Certainty, impossibility, and a probability, which has an intermediate value, for example, constitute an ordered series in which the probability lies *between* certainty and impossibility. In the same way there may exist a second probability which lies *between* certainty and the first probability. When, therefore, we say that one probability is greater than another, this precisely means that the degree of our rational belief in the first case lies *between* certainty and the degree of our rational belief in the second case.

On this theory it is easy to see why comparisons of more and less are not always possible. They exist between two probabilities, only when they and certainty all lie on the same ordered series. But if more than one distinct series of probabilities exist, then it is clear that only those, which belong to the *same* series, can be compared. If the attribute 'greater' as applied to one of two terms arises solely out of the relative order of the terms in a series, then comparisons of greater and less must always be possible between terms which are members of the same series, and can never be possible between two terms which are not members of the same series. Some probabilities are not comparable in respect of more and less, because there exists more than one path, so to speak, between proof and disproof, between certainty and impossibility; and neither of two probabilities, which lie on independent paths, bears to the other and to certainty the relation of 'between' which is necessary for quantitative comparison.

If we are comparing the probabilities of two arguments, where the conclusion is the same in both and the evidence of one exceeds the evidence of the other by the inclusion of some fact which is favourably relevant, in such a case a relation seems clearly to exist between the two in virtue of which one lies *nearer* to certainty than the other. Several types of argument can be instanced in which the existence of such a relation is equally apparent. But we cannot assume its presence in every case or in comparing in respect of more and less the probabilities of every pair of arguments.

16. Analogous instances are by no means rare, in which, by a convenient looseness, the phraseology of quantity is misapplied in the same manner as in the case of probability. The simplest example is that of colour. When we describe the colour of one object as bluer than that of another, or say that it has more green in it, we do not mean that there are quantities blue and green of which the object's colour possesses more or less; we mean that the colour has a certain position in an order of colours and that it is nearer some standard colour than is the colour with which we compare it.

Another example is afforded by the cardinal numbers. We say that the number three is greater than the number two, but we do not mean that these numbers are quantities one of which possesses a greater magnitude than the other. The one is greater than the other by reason of its position in the order of numbers; it is further distant from the origin zero. One number is greater than another if the second number lies *between* zero and the first.

But the closest analogy is that of similarity. When we say of three objects A, B, and C that B is more like A than C is, we mean, not that there is any respect in which B is in itself quantitatively greater than C, but that, if the three objects are placed in an order of similarity, B is nearer to A than C is. There are also, as in the case of probability, *different* orders of similarity. For instance, a book bound in blue morocco is more like a book bound in red morocco than if it were bound in blue calf; and a book bound in red calf is more like the book in red morocco than if it were in blue calf. But there may be no comparison between the degree of similarity which exists between books bound in red morocco and blue morocco, and that which exists between books bound in red morocco and red calf. This illustration deserves special attention, as the analogy between orders of similarity and probability is so great that its apprehension will greatly assist that of the ideas I wish to convey. We say that one argument is more probable than another (*i.e.* nearer to certainty) in the same kind of way as we can describe one object as more like than another to a standard object of comparison.

17. Nothing has been said up to this point which bears on the question whether probabilities are ever capable of *numerical* comparison. It is true of some types of ordered series that

there are measurable relations of distance between their members as well as order, and that the relation of one of its members to an 'origin' can be numerically compared with the relation of another member to the same origin. But the legitimacy of such comparisons must be matter for special enquiry in each case.

It will not be possible to explain in detail how and in what sense a meaning can sometimes be given to the numerical measurement of probabilities until Part II. is reached. But this chapter will be more complete if I indicate briefly the conclusions at which we shall arrive later. It will be shown that a process of compounding probabilities can be defined with such properties that it can be conveniently called a process of *addition*. It will sometimes be the case, therefore, that we can say that one probability C is equal to the *sum* of two other probabilities A and B, *i.e.* $C = A + B$. If in such a case A and B are equal, then we may write this $C = 2A$ and say that C is double A. Similarly if $D = C + A$, we may write $D = 3A$, and so on. We can attach a meaning, therefore, to the equation $P = n.A$, where P and A are relations of probability, and n is a number. The relation of certainty has been commonly taken as the unit of such conventional measurements. Hence if P represents certainty, we should say, in ordinary language, that the magnitude of the probability A is $\frac{1}{n}$. It will be shown also that we can define a process, applicable to probabilities, which has the properties of arithmetical multiplication. Where numerical measurement is possible, we can in consequence perform algebraical operations of considerable complexity. The attention, out of proportion to their real importance, which has been paid, on account of the opportunities of mathematical manipulation which they afford, to the limited class of numerical probabilities, seems to be a part explanation of the belief, which it is the principal object of this chapter to prove erroneous, that *all* probabilities must belong to it.

18. We must look, then, at the quantitative characteristics of probability in the following way. Some sets of probabilities we can place in an ordered series, in which we can say of any pair that one is nearer than the other to certainty,—that the argument in one case is nearer proof than in the other, and that there is more reason for one conclusion than for the other. But we can only build up these ordered series in special cases. If we are given two distinct arguments, there is no general presumption that their two probabilities and certainty can be placed in an order. The burden of establishing the existence of such an order lies on us in each separate case. An endeavour will be made later to explain in a systematic way how and in what circumstances such orders can be established. The argument for the theory here proposed will then be strengthened. For the present it has been shown to be agreeable to common sense to suppose that an order exists in some cases and not in others.

19. Some of the principal properties of ordered series of probabilities are as follows:

(i.) Every probability lies on a path between impossibility and certainty; it is always true to say of a degree of probability, which is not identical either with impossibility or with certainty, that it lies *between* them. Thus certainty, impossibility and *any* other degree of probability form an ordered series. This is the same thing as to say that every argument amounts to proof, or disproof, or occupies an intermediate position.

(ii.) A path or series, composed of degrees of probability, is not in general compact. It is not necessarily true, that is to say, that any pair of probabilities in the same series have a probability between them.

(iii.) The same degree of probability can lie on more than one path (*i.e.* can belong to more than one series). Hence, if B lies between A and C, and also lies between A′ and C′, it does not follow that of A and A′ either lies between the other and certainty. The fact, that the same probability can belong to more than one distinct series, has its analogy in the case of similarity.

(iv.) If ABC forms an ordered series, B lying between A and C, and BCD forms an ordered series, C lying between B and D, then ABCD forms an ordered series, B lying between A and D.

20. The different series of probabilities and their mutual relations can be most easily pictured by means of a diagram. Let us represent an ordered series by points lying upon a path, all the

points on a given path belonging to the same series. It follows from (i.) that the points O and I, representing the relations of impossibility and certainty, lie on every path, and that all paths lie wholly between these points. It follows from (iv.) that the same point can lie on more than one path. It is possible, therefore, for paths to intersect and cross. It follows from (iv.) that the probability represented by a given point is greater than that represented by any other point which can be reached by passing along a path with a motion constantly towards the point of impossibility, and less than that represented by any point which can be reached by moving along a path towards the point of certainty. As there are independent paths there will be some pairs of points representing relations of probability such that we cannot reach one by moving from the other along a path always in the same direction.

These properties are illustrated in the annexed diagram. O represents impossibility, I certainty, and A a numerically measurable probability intermediate between O and I; U, V, W, X, Y, Z are non-numerical probabilities, of which, however, V is less than the numerical probability A, and is also less than W, X, and Y. X and Y are both greater than W, and greater than V, but are not comparable with one another, or with A. V and Z are both less than W, X, and Y, but are not comparable with one another; U is not quantitatively comparable with any of the probabilities V, W, X, Y, Z. Probabilities which are numerically comparable will all belong to one series, and the path of this series, which we may call the numerical path or strand, will be represented by OAI.

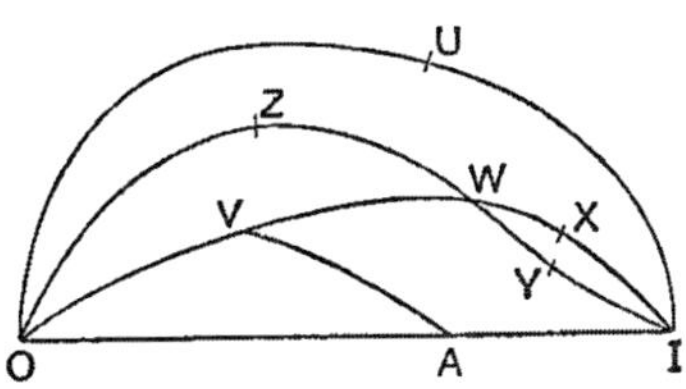

21. The chief results which have been reached so far are collected together below, and expressed with precision :—

(i.) There are amongst degrees of probability or rational belief various sets, each set composing an ordered series. These series are ordered by virtue of a relation of 'between.' If B is 'between' A and C, ABC form a series.

(ii.) There are two degrees of probability O and I *between* which *all* other probabilities lie. If, that is to say, A is a probability, OAI form a series. O represents impossibility and I certainty.

(iii.) If A lies between O and B, we may write this $\widehat{AB}$, so that $\widehat{OA}$ and $\widehat{AI}$ are true for all probabilities.

(iv.) If $\widehat{AB}$, the probability B is said to be greater than the probability A, and this can be expressed by $B > A$.

(v.) If the conclusion a bears the relation of probability P to the premiss h, or if, in other words, the hypothesis h invests the conclusion a with probability P, this may be written aPh. It may also be written $a/h = P$.

This latter expression, which proves to be the more useful of the two for most purposes, is of fundamental importance. If aPh and $a'Ph'$, *i.e.* if the probability of a relative to h is the same as the probability of a' relative to h', this may be written $a/h = a'/h'$. The value of the symbol a/h, which represents what is called by other writers 'the probability of a,' lies in the fact that it contains explicit reference to the *data* to which the probability relates the conclusion, and avoids the numerous errors which have arisen out of the omission of this reference.

CHAPTER IV

THE PRINCIPLE OF INDIFFERENCE

ABSOLUTE. 'Sure, Sir, this is not very reasonable, to summon my affection for a lady I know nothing of.'
SIR ANTHONY. 'I am sure, Sir, 'tis more unreasonable in you to object to a lady you know nothing of.' [1]

1. IN the last chapter it was assumed that in some cases the probabilities of two arguments may be *equal*. It was also argued that there are other cases in which one probability is, in some sense, greater than another. But so far there has been nothing to show *how* we are to know when two probabilities are equal or unequal. The recognition of equality, when it exists, will be dealt with in this chapter, and the recognition of inequality in the next. An historical account of the various theories about this problem, which have been held from time to time, will be given in Chapter VII.

2. The determination of equality between probabilities has received hitherto much more attention than the determination of inequality. This has been due to the stress which has been laid on the mathematical side of the subject. In order that numerical measurement may be possible, we must be given a number of *equally* probable alternatives. The discovery of a rule, by which equiprobability could be established, was, therefore, essential. A rule, adequate to the purpose, introduced by James Bernoulli, who was the real founder of mathematical probability,[2] has been widely adopted, generally under the title of *The Principle of Non-Sufficient Reason*, down to the present time. This description is clumsy and unsatisfactory, and, if it is justifiable to break away from tradition, I prefer to call it *The Principle of Indifference*.

[1] Quoted by Mr. Bosanquet with reference to the Principle of Non-Sufficient Reason.

[2] See also Chap. VII.

The Principle of Indifference asserts that if there is no *known* reason for predicating of our subject one rather than another of several alternatives, then relatively to such knowledge the assertions of each of these alternatives have an *equal* probability. Thus *equal* probabilities must be assigned to each of several arguments, if there is an absence of positive ground for assigning *unequal* ones.

This rule, as it stands, may lead to paradoxical and even contradictory conclusions. I propose to criticise it in detail, and then to consider whether any valid modification of it is discoverable. For several of the criticisms which follow I am much indebted to Von Kries's *Die Principien der Wahrscheinlichkeit.*[1]

3. If every probability was necessarily either greater than, equal to, or less than any other, the Principle of Indifference would be plausible. For if the evidence affords no ground for attributing unequal probabilities to the alternative predications, it seems to follow that they must be equal. If, on the other hand, there need be neither equality nor inequality between probabilities, this method of reasoning fails. Apart, however, from this objection, which is based on the arguments of Chapter III., the plausibility of the principle will be most easily shaken by an exhibition of the contradictions which it involves. These fall under three or four distinct heads. In §§ 4-9 my criticism will be purely destructive, and I shall not attempt in these paragraphs to indicate my own way out of the difficulties.

4. Consider a proposition, about the subject of which we know only the meaning, and about the truth of which, as applied to this subject, we possess no external relevant evidence. It has been held that there are here two exhaustive and exclusive alternatives—the truth of the proposition and the truth of its contradictory—while our knowledge of the subject affords no ground for preferring one to the other. Thus if a and $\bar{a}$ are contradictories, about the subject of which we have no outside knowledge, it is inferred that the probability of each is $\frac{1}{2}$.[2] In

[1] Published in 1886. A brief account of Von Kries's principal conclusions will be given on p. 87. A useful summary of his book will be found in a review by Meinong, published in the *Göttingische gelehrte Anzeigen* for 1890 (pp. 56-75).

[2] Cf. (*e.g.*) the well-known passage in Jevons's *Principles of Science*, vol. i. p. 243, in which he assigns the probability $\frac{1}{2}$ to the proposition "A Platythliptic Coefficient is positive." Jevons points out, by way of proof, that no other

the same way the probabilities of two other propositions, b and c, having the same subject as a, may be each $\frac{1}{2}$. But without having any evidence bearing on the subject of these propositions we may know that the predicates are contraries amongst themselves, and, therefore, exclusive alternatives—a supposition which leads by means of the same principle to values inconsistent with those just obtained. If, for instance, having no evidence relevant to the colour of this book, we could conclude that $\frac{1}{2}$ is the probability of 'This book is red,' we could conclude equally that the probability of each of the propositions 'This book is black' and 'This book is blue' is also $\frac{1}{2}$. So that we are faced with the impossible case of *three* exclusive alternatives all as likely as not. A defender of the Principle of Indifference might rejoin that we are assuming knowledge of the proposition: 'Two different colours cannot be predicated of the same subject at the same time'; and that, if we know this, it constitutes relevant outside evidence. But such evidence is about the predicate, not about the subject. Thus the defender of the Principle will be driven on, either to confine it to cases where we know nothing about either the subject or the predicate, which would be to emasculate it for all practical purposes, or else to revise and amplify it, which is what we propose to do ourselves.

The difficulty cannot be met by saying that we must know and take account of the *number* of possible contraries. For the number of contraries to any proposition on any evidence is always infinite; $\bar{a}b$ is contrary to a for all values of b. The same point can be put in a form which does not involve contraries or contradictories. For example, $a/h=\frac{1}{2}$ and $ab/h=\frac{1}{2}$, if h is

probability could reasonably be given. This, of course, involves the assumption that every proposition must have some numerical probability. Such a contention was first criticised, so far as I am aware, by Bishop Terrot in the *Edin. Phil. Trans.* for 1856. It was deliberately rejected by Boole in his last published work on probability: "It is a plain consequence," he says (*Edin. Phil. Trans.* vol. xxi. p. 624), "of the logical theory of probabilities, that the state of expectation which accompanies entire ignorance of an event is properly represented, not by the fraction $\frac{1}{2}$, but by the indefinite form $\frac{0}{0}$." Jevons's particular example, however, is also open to the objection that we do not even know the *meaning* of the subject of the proposition. Would he maintain that there is any sense in saying that for those who know no Arabic the probability of every statement expressed in Arabic is even? How far has he been influenced in the choice of his example by known characteristics of the predicate 'positive'? Would he have assigned the probability $\frac{1}{2}$ to the proposition 'A Platythliptic Coefficient is a perfect cube'? What about the proposition 'A Platythliptic Coefficient is allogeneous'?

irrelevant both to a and to b, in the sense required by the crude Principle of Indifference.[1] It follows from this that, if a is true, b must be true also. If it follows from the absence of positive *data* that 'A is a red book' has a probability of $\frac{1}{2}$, and that the probability of 'A is red' is also $\frac{1}{2}$, then we may deduce that, if A is red, it must certainly be a book.

We may take it, then, that the probability of a proposition, about the subject of which we have no extraneous evidence, is *not* necessarily $\frac{1}{2}$. Whether or not this conclusion discredits the Principle of Indifference, it is important on its own account, and will help later on to confute some famous conclusions of Laplace's school.

5. Objection can now be made in a somewhat different shape. Let us suppose as before that there is no positive evidence relating to the subjects of the propositions under examination which would lead us to discriminate in any way between certain alternative predicates. If, to take an example, we have no information whatever as to the area or population of the countries of the world, a man is as likely to be an inhabitant of Great Britain as of France, there being no reason to prefer one alternative to the other.[2] He is also as likely to be an inhabitant of Ireland as of France. And on the same principle he is as likely to be an inhabitant of the British Isles as of France. And yet these conclusions are plainly inconsistent. For our first two propositions together yield the conclusion that he is twice as likely to be an inhabitant of the British Isles as of France.

Unless we argue, as I do not think we can, that the knowledge that the British Isles are composed of Great Britain and Ireland is a ground for supposing that a man is more likely to inhabit them than France, there is no way out of the contradiction. It is not plausible to maintain, when we are considering the relative populations of different areas, that the number of *names* of sub-divisions which are within our knowledge, is, in the absence of any evidence as to their size, a piece of relevant evidence.

At any rate, many other similar examples could be invented,

[1] a/h stands for 'the probability of a on hypothesis h.'

[2] This example raises a difficulty similar to that raised by Von Kries's example of the meteor. Stumpf has propounded an invalid solution of Von Kries's difficulty. Against the example proposed here, Stumpf's solution has less plausibility than against Von Kries's.

which would require a special explanation in each case; for the above is an instance of a perfectly general difficulty. The possible alternatives may be *a*, *b*, *c*, and *d*, and there may be no means of discriminating between them; but equally there may be no means of discriminating between (*a* or *b*), *c*, and *d*. This difficulty could be made striking in a variety of ways, but it will be better to criticise the principle further from a somewhat different side.

6. Consider the specific volume of a given substance.[1] Let us suppose that we know the specific volume to lie between 1 and 3, but that we have no information as to whereabouts in this interval its exact value is to be found. The Principle of Indifference would allow us to assume that it is as likely to lie between 1 and 2 as between 2 and 3; for there is no reason for supposing that it lies in one interval rather than in the other. But now consider the specific density. The specific density is the reciprocal of the specific volume, so that if the latter is v the former is $\frac{1}{v}$. Our *data* remaining as before, we know that the specific density must lie between 1 and $\frac{1}{3}$, and, by the same use of the Principle of Indifference as before, that it is as likely to lie between 1 and $\frac{2}{3}$ as between $\frac{2}{3}$ and $\frac{1}{3}$. But the specific volume being a determinate function of the specific density, if the latter lies between 1 and $\frac{2}{3}$, the former lies between 1 and $1\frac{1}{2}$, and if the latter lies between $\frac{2}{3}$ and $\frac{1}{3}$, the former lies between $1\frac{1}{2}$ and 3. It follows, therefore, that the specific volume is as likely to lie between 1 and $1\frac{1}{2}$ as between $1\frac{1}{2}$ and 3; whereas we have already proved, relatively to precisely the same *data*, that it is as likely to lie between 1 and 2 as between 2 and 3. Moreover, any other function of the specific volume would have suited our purpose equally well, and by a suitable choice of this function we might have proved in a similar manner that any division whatever of the interval 1 to 3 yields sub-intervals of equal probability. Specific volume and specific density are simply alternative methods of measuring the *same* objective quantity; and there are many methods which might be adopted, each yielding on the application of the Principle of Indifference a different probability for a given objective variation in the quantity.[2]

[1] This example is taken from Von Kries, *op. cit.* p. 24. Von Kries does not seem to me to explain correctly how the contradiction arises.

[2] A. Nitsche ("Die Dimensionen der Wahrscheinlichkeit und die Evidenz der Ungewissheit," *Vierteljahrsschr. f. wissensch. Philos.* vol. xvi. p. 29, 1892), in

The arbitrary nature of particular methods of measurement of this and of many other physical quantities is easily explained. The objective quality measured may not, strictly speaking, possess numerical quantitativeness, although it has the properties necessary for measurement by means of correlation with numbers. The values which it can assume may be capable of being ranged in an order, and it will sometimes happen that the series which is thus formed is *continuous*, so that a value can always be found whose order in the series is between any two selected values; but it does not follow from this that there is any meaning in the assertion that one value is *twice* another value. The relations of continuous order can exist between the terms of a series of values, without the relations of numerical quantitativeness necessarily existing also, and in such cases we can adopt a largely arbitrary measure of the successive terms, which yields results which may be satisfactory for many purposes, those, for instance, of mathematical physics, though not for those of probability. This method is to select some other series of quantities or numbers, each of the terms of which corresponds in order to one and only one of the terms of the series which we wish to measure. For instance, the series of characteristics, differing in degree, which are measured by specific volume, have this relation to the series of numerical ratios between the volumes of equal masses of the substances, the specific volumes of which are in question, and of water. They have it also to the corresponding ratios which give rise to the measure of specific density. But these only yield conventional measurements, and the numbers with which we correlate the

criticising Von Kries, argues that the alternatives to which the principle must be applied are the smallest *physically* distinguishable intervals, and that the probability of the specific volume's lying within a certain range of values turns on the *number* of such distinguishable intervals in the range. This procedure might conceivably provide the correct method of computation, but it does not therefore restore the credit of the Principle of Indifference. For it is argued, not that the results of applying the principle are always wrong, but that it does not lead unambiguously to the correct procedure. If we do not know the number of distinguishable intervals we have *no* reason for supposing that the specific volume lies between 1 and 2 rather than 2 and 3, and the principle can therefore be applied as it has been applied above. And even if we do know the number and reckon intervals as equal which contain an equal number of 'physically distinguishable' parts, is it certain that this does not simply provide us with a new system of measurement, which has the same conventional basis as the methods of specific volume and specific density, and is no more the one correct measure than these are?

terms which we wish to measure can be selected in a variety of ways. It follows that equal intervals between the numbers which represent the ratios do not necessarily correspond to equal intervals between the qualities under measurement; for these numerical differences depend upon which convention of measurement we have selected.

7. A somewhat analogous difficulty arises in connection with the problems of what is known as 'geometrical' or 'local' probability.[1] In these problems we are concerned with the position of a point or infinitesimal area or volume within a continuum.[2] The number of cases here is indefinite, but the Principle of Indifference has been held to justify the supposition that equal lengths or areas or volumes of the continuum are, in the absence of discriminating evidence, equally likely to contain the point. It has long been known that this assumption leads in numerous cases to contradictory conclusions. If, for instance, two points A and A′ are taken at random on the surface of a sphere, and we seek the probability that the lesser of the two arcs of the great circle AA′ is less than a, we get one result by assuming that the probability of a point's lying on a given portion of the sphere's surface is proportional to the area of that portion, and another result by assuming that, if a point lies on a given great circle, the probability of its lying on a given arc of that circle is proportional to the length of the arc, each of these assumptions being equally justified by the Principle of Indifference.

Or consider the following problem: if a chord in a circle is drawn at random, what is the probability that it will be less than the side of the inscribed equilateral triangle. One can argue:—

(*a*) It is indifferent at what point one end of the chord lies. If we suppose this end fixed, the direction is then

[1] The best accounts of this subject are to be found in Czuber, *Geometrische Wahrscheinlichkeiten und Mittelwerte*; Czuber, *Wahrscheinlichkeitsrechnung*, vol. i. pp. 75-109; Crofton, *Encycl. Brit.* (9th edit.), article 'Probability'; Borel, *Éléments de la théorie des probabilités*, chaps. vi.-viii.; a few other references are given in the following pages, and a number of discussions of individual problems will be found in the mathematical volumes of the *Educational Times*. The interest of the subject is primarily mathematical, and no discussion of its principal problems will be attempted here.

[2] As Czuber points out (*Wahrscheinlichkeitsrechnung*, vol. i. p. 84), all problems, whether geometrical or arithmetical, which deal with a continuum and with non-enumerable aggregates, are commonly discussed under the name of 'geometrical probability.' See also Lämmel, *Untersuchungen*.

chosen at random. In this case the answer is easily shown to be $\frac{2}{3}$.

(b) It is indifferent in what direction we suppose the chord to lie. Beginning with this apparently not less justifiable assumption, we find that the answer is $\frac{1}{2}$.

(c) To choose a chord at random, one must choose its middle point at random. If the chord is to be less than the side of the inscribed equilateral triangle, the middle point must be at a greater distance from the centre than half the radius. But the area at a greater distance than this is $\frac{3}{4}$ of the whole. Hence our answer is $\frac{3}{4}$. [1]

In general, if x and $f(x)$ are both continuous variables, varying always in the same or in the opposite sense, and x must lie between a and b, then the probability that x lies between c and d, where $a<c<d<b$, seems to be $\frac{d-c}{b-a}$, and the probability that $f(x)$ lies between $f(c)$ and $f(d)$ to be $\frac{f(d)-f(c)}{f(b)-f(a)}$. These expressions, which represent the probabilities of necessarily concordant conclusions, are not, as they ought to be, equal.[2]

8. More than one attempt has been made to separate the cases in which the Principle of Indifference can be legitimately applied to examples of geometrical probability from those in which it cannot. M. Borel argues that the mathematician can *define* the geometrical probability that a point M lies on a certain segment PQ of AD as proportional to the length of the segment, but that this definition is *conventional* until its consequences have been confirmed *à posteriori* by their conformity with the results of empirical observation. He points out that in actual cases there are generally some considerations present which lead us to prefer one of the possible assumptions to the others. Whether or not this is so, the proposed procedure amounts to an abandonment of the Principle of Indifference as a valid criterion, and leaves our choice undetermined when further evidence is not forthcoming.

M. Poincaré, who also held that judgments of equiprobability in such cases depend upon a 'convention,' endeavoured to mini-

[1] Bertrand, *Calcul des probabilités*, p. 5.

[2] See (*e.g.*) Borel, *Éléments de la théorie des probabilités*, p. 85.

mise the importance of the arbitrary element by showing that, under certain conditions, the result is independent of the particular convention which is chosen. Instead of assuming that the point is equally likely to lie in every infinitesimal interval dx we may represent the probability of its lying in this interval by the function $\phi(x)dx$. M. Poincaré showed that, in the game of *rouge et noir*, for instance, where we have a number of compartments arranged in a circle coloured alternately black and white, if we can assume that $\phi(x)$ is a regular function, continuous and with continuous differential coefficients, then, whatever the particular form of the function, the probability of black is approximately equal to that of white.[1]

Whether or not investigations on these lines prove to have a practical value, they have not, I think, any theoretical importance. If, as I maintain, the probability $\phi(x)$ is not necessarily numerical, it is not a generally justifiable assumption to take its continuity for granted. We have, in the particular example quoted, a number of alternatives, half of which lead to black and half to white ; the assumption of continuity amounts to the assumption that for every white alternative there is a black alternative whose probability is very nearly equal to that of the white. Naturally in such a case we can get an approximately equal probability for the whites as a whole and for the blacks as a whole, without assuming equal probability for each alternative individually. But this fact has no bearing on the theoretical difficulties which we are discussing.

M. Bertrand is so much impressed by the contradictions of geometrical probability that he wishes to exclude all examples in which the number of alternatives is *infinite*.[2] It will be argued in the sequel that something resembling this is true. The discussion of this question will be resumed in §§ **21-25**.

9. There is yet another group of cases, distinct in character from those considered so far, in which the principle does not seem to provide us with unambiguous guidance. The typical example is that of an urn containing black and white balls in an

[1] Poincaré, *Calcul des probabilités*, pp. 126 *et seq.*

[2] Bertrand, *Calcul des probabilités*, p. 4 : "L'infini n'est pas un nombre ; on ne doit pas, sans explication, l'introduire dans les raisonnements. La précision illusoire des mots pourrait faire naître des contradictions. Choisir au hasard, entre un nombre infini de cas possibles, n'est pas une indication suffisante."

unknown proportion.[1] The Principle of Indifference can be claimed to support the most usual hypothesis, namely, that all possible numerical *ratios* of black and white are equally probable. But we might equally well assume that all possible *constitutions* [2] of the system of balls are equally probable, so that each individual ball is assumed equally likely to be black or white. It would follow from this that an approximately equal number of black and white balls is more probable than a large excess of one colour. On this hypothesis, moreover, the drawing of one ball and the resulting knowledge of its colour leaves unaltered the probabilities of the various possible constitutions of the rest of the bag; whereas on the first hypothesis knowledge of the colour of one ball, drawn and not replaced, manifestly alters the probability of the colour of the next ball to be drawn. Either of these hypotheses seems to satisfy the Principle of Indifference, and a believer in the absolute validity of the principle will doubtless adopt that one which enters his mind first.[3]

The same point is very clearly illustrated by an example which I take from Von Kries. Two cards, chosen from different packs, are placed face downwards on the table; one is taken up and found to be of a black suit: what is the chance that the other is black also? One would naturally reply that the chance is even. But this is based on the supposition, relatively unpopular with writers on the subject, that every 'constitution' is equally probable, *i.e.* that each individual card is as likely to be black as red. If we prefer this assumption, we must relin-

[1] The difficulty in question was first pointed out by Boole, *Laws of Thought*, pp. 369-370. After discussing the Law of Succession, Boole proceeds to show that "there are other hypotheses, as strictly involving the principle of the 'equal distribution of knowledge or ignorance' which would also conduct to conflicting results." See also Von Kries, *op. cit.* pp. 31-34, 59, and Stumpf, *Über den Begriff der mathematischen Wahrscheinlichkeit*, Bavarian Academy, 1892, pp. 64-68.

[2] If A and B are two balls, A white, B black, and A black, B white, are different 'constitutions.' But if we consider different numerical ratios, these two cases are indistinguishable, and count as one only.

[3] C. S. Peirce in his *Theory of Probable Inference* (Johns Hopkins *Studies in Logic*), pp. 172, 173, argues that the 'constitution' hypothesis is alone valid, on the ground that, of the two hypotheses, only this one is consistent with itself. I agree with his conclusion, and shall give at the close of the chapter the fundamental considerations which lead to the rejection of the 'ratio' hypothesis. Stumpf points out that the probability of drawing a white ball is, in any case, $\frac{1}{2}$. This is true; but the probability of a second white clearly depends upon which of the two hypotheses has been preferred. Nitsche (*loc. cit.* p. 31) seems to miss the point of the difficulty in the same way.

quish the text-book theory that the drawing of a black ball from an urn, containing black and white balls in unknown proportions, affects our knowledge as to the proportion of black and white amongst the remaining balls.

The alternative—or text-book—theory assumes that there are three equal possibilities—one of each colour, both black, both red. If both cards are black, we are twice as likely to turn up a black card than if only one is black. *After* we have turned up a black, the probability that the other is black is, therefore, twice as great as the probability that it is red. The chance of the second's being black is therefore $\frac{2}{3}$.[1] The Principle of Indifference has nothing to say against either solution. Until some further criterion has been proposed we seem compelled to agree with Poincaré that a preference for either hypothesis is wholly arbitrary.

10. Such, then, are the kinds of result to which an unguarded use of the Principle of Indifference may lead us. The difficulties, to which attention has been drawn, have been noticed before; but the discredit has not been emphatically thrown on the original source of error. Yet the principle certainly remains as a *negative* criterion; two propositions cannot be equally probable, so long as there *is* any ground for discriminating between them. The principle is a necessary, but not, as it seems, a sufficient condition.

The enunciation of some sufficient rule is certainly essential if we are to make any progress in the subject. But the difficulty of discovering a correct principle is considerable. This difficulty is partly responsible, I think, for the doubts which philosophers and many others have often felt regarding any practical application of the Calculus. Many candid persons, when confronted with the results of Probability, feel a strong sense of the uncertainty of the logical basis upon which it seems to rest. It is difficult to find an intelligible account of the meaning of 'probability,' or of how we are ever to determine the probability of any particular proposition; and yet treatises on the subject profess to arrive at complicated results of the greatest precision and the most profound practical importance.

The incautious methods and exaggerated claims of the school of Laplace have undoubtedly contributed towards the existence of these sentiments. But the general scepticism, which I believe

[1] This is Poisson's solution, *Recherches*, p. 96.

to be much more widely spread than the literature of the subject admits, is more fundamental. In this matter Hume need not have felt "affrighted and confounded with that forelorn solitude, in which I am placed in my philosophy," or have fancied himself "some strange uncouth monster, who not being able to mingle and unite in society, has been expell'd all human commerce, and left utterly abandon'd and disconsolate." In his views on probability, he stands for the plain man against the sophisms and ingenuities of "metaphysicians, logicians, mathematicians, and even theologians."

Yet such scepticism goes too far. The judgments of probability, upon which we depend for almost all our beliefs in matters of experience, undoubtedly depend on a strong psychological propensity in us to consider objects in a particular light. But this is no ground for supposing that they are nothing more than "lively imaginations." The same is true of the judgments in virtue of which we assent to other logical arguments; and yet in such cases we believe that there may be present some element of objective validity, transcending the psychological impulsion, with which primarily we are presented. So also in the case of probability, we may believe that our judgments can penetrate into the real world, even though their credentials are subjective.

11. We must now inquire how far it is possible to rehabilitate the Principle of Indifference or find a substitute for it. There are several distinct difficulties which need attention in a discussion of the problems raised in the preceding paragraphs. Our first object must be to make the Principle itself more precise by disclosing how far its application is mechanical and how far it involves an appeal to logical intuition.

12. Without compromising the objective character of relations of probability, we must nevertheless admit that there is little likelihood of our discovering a method of recognising particular probabilities, without any assistance whatever from intuition or direct judgment. Inasmuch as it is always assumed that we can sometimes judge directly that a conclusion *follows from* a premiss, it is no great extension of this assumption to suppose that we can sometimes recognise that a conclusion *partially follows from*, or stands in a relation of probability to, a premiss. Moreover, the failure to explain or define 'probability' in terms of other logical notions, creates a presumption that particular relations of probability must be, in the first instance, directly recognised as such, and cannot be evolved by rule out of *data* which themselves contain no statements of probability.

On the other hand, although we cannot exclude every element of direct judgment, these judgments may be limited and controlled, perhaps, by logical rules and principles which possess a general application. While we may possess a faculty of direct recognition of many relations of probability, as in the case of many other logical relations, yet some may be much more easily recognisable than others. The object of a logical system of probability is to enable us to know the relations, which cannot be easily perceived, by means of other relations which we can recognise more distinctly—to convert, in fact, vague knowledge into more distinct knowledge.[1]

13. Let us seek to distinguish between the element of direct judgment and the element of mechanical rule in the Principle of Indifference. The enunciation of this principle, as it is ordinarily expressed, cloaks, but does not avoid, the former element. It is in part a formula and in part an appeal to direct inspection; but in addition to the obscurity and ambiguity of the formula, the appeal to intuition is not as explicit as it should be. The principle states that 'there must be no known reason for preferring one of a set of alternatives to any other.' What does this mean? What are 'reasons,' and how are we to know whether they do or do not justify us in preferring one alternative to another? I do not know any discussion of Probability in which this question has been so much as asked. If, for example, we are considering the probability of drawing a black ball from an urn containing balls which are

[1] As it is the aim of trigonometry to determine the position of an object, which is in a sense visible, not by a direct observation of it, but by observing some other object together with certain relations, so an indirect method of this kind is the aim of all logical system. If the truth of *some* propositions, and the validity of *some* arguments, could not be recognised directly, we could make no progress. We may have, moreover, some power of direct recognition where it is not necessary in our logical system that we should make use of it. In these cases the method of logical proof increases the certainty of knowledge, which we might be able to possess in a more doubtful manner without it. In other cases, that, for instance, of a complicated mathematical theorem, it enables us to know propositions to be true, which are altogether beyond the reach of our direct insight; just as we can often obtain knowledge about the position of a partially visible or even invisible object by starting with observations of other objects.

black and white, we assume that the difference of *colour* between the balls is not a reason for preferring either alternative. But how do we know this, unless by a judgment that, on the evidence in hand, our knowledge of the colours of the balls is *irrelevant* to the probability in question? We know of *some* respects in which the alternatives differ; but we judge that a knowledge of *these* differences is not relevant. If, on the other hand, we were taking the balls out of the urn with a magnet, and knew that the black balls were of iron and the white of tin, we might regard the fact, that a ball was iron and not tin, as very important in determining the probability of its being drawn. Before, then, we can begin to apply the Principle of Indifference, we must have made a number of direct judgments to the effect that the probabilities under consideration are unaffected by the inclusion in the evidence of certain particular details. We have no right to say of any known difference between the two alternatives that it is 'no reason' for preferring one of them, unless we have judged that a knowledge of this difference is irrelevant to the probability in question.

14. A brief digression is now necessary, in order to introduce some new terms. There are in general two principal types of probabilities, the magnitudes of which we seek to compare,—those in which the evidence is the same and the conclusions different, and those in which the evidence is different but the conclusion the same. Other types of comparison may be required, but these two are by far the commonest. In the first we compare the likelihood of two conclusions on given evidence; in the second we consider what difference a change of evidence makes to the likelihood of a given conclusion. In symbolic language we may wish to compare x/h with y/h, or x/h with x/h_1h. We may call the first type judgments of *preference*, or, when there is equality between x/h and y/h, of *indifference*; and the second type we may call judgments of *relevance*, or, when there is equality between x/h and x/h_1h, of *irrelevance*. In the first we consider whether or not x is to be preferred to y on evidence h; in the second we consider whether the addition of h_1 to evidence h is relevant to x.

The Principle of Indifference endeavours to formulate a rule which will justify judgments of *indifference*. But the rule that there must be no ground for preferring one alternative to another, involves, if it is to be a guiding rule at all, and not a *petitio principii*, an appeal to judgments of *irrelevance*.

The simplest definition of Irrelevance is as follows: h_1 is irrelevant to x on evidence h, if the probability of x on evidence hh_1 is the same as its probability on evidence h.[1] But for a reason which will appear in Chapter VI., a stricter and more complicated definition, as follows, is theoretically preferable: h_1 is irrelevant to x on evidence h, if there is no proposition, inferrible from h_1h but not from h, such that its addition to evidence h affects the probability of x.[2] Any proposition which is irrelevant in the strict sense is, of course, also irrelevant in the simpler sense; but if we were to adopt the simpler definition, it would sometimes occur that a part of evidence would be relevant, which taken as a whole was irrelevant. The more elaborate definition by avoiding this proves in the sequel more convenient. If the condition $x/h_1h = x/h$ alone is satisfied, we may say that the evidence h_1 is 'irrelevant as a whole.'[3]

It will be convenient to define also two other phrases. h_1 and h_2 are independent and complementary parts of the evidence, if between them they make up h and neither can be inferred from the other. If x is the conclusion, and h_1 and h_2 are independent and complementary parts of the evidence, then h_1 is relevant if the addition of it to h_2 affects the probability of x.[4]

Some propositions regarding irrelevance will be proved in Part II. If $\bar{h}_1$ is the contradictory of h_1 and $x/h_1h = x/h$, then $x/\bar{h}_1h = x/h$. Thus the contradictory of irrelevant evidence is also irrelevant. Also, if $x/yh = x/h$, it follows that $y/xh = y/h$. Hence if, on initial evidence h, y is irrelevant to x, then, on the same initial evidence, x is irrelevant to y, *i.e.* if in a given state of knowledge one occurrence has no bearing on another, then equally the second has no bearing on the first.

15. This distinction enables us to formulate the Principle of Indifference at any rate more precisely. There must be no *relevant* evidence relating to one alternative, unless there is *corresponding* evidence relating to the other; our relevant

[1] That is to say, h_1 is irrelevant to x/h if $x/h_1h = x/h$.

[2] That is to say, h_1 is irrelevant to x/h, if there is no proposition h'_1 such that $h'_1/h_1h = 1$, $h'_1/h \neq 1$, and $x/h'_1h \neq x/h$.

[3] Where no misunderstanding can arise, the qualification 'as a whole' will be sometimes omitted.

[4] *I.e* (in symbolism) h_1 and h_2 are independent and complementary parts of h if $h_1h_2 = h$, $h_1/h_2 \neq 1$, and $h_2/h_1 \neq 1$. Also h_1 is relevant if $x/h \neq x/h_2$.

evidence, that is to say, must be symmetrical with regard to the alternatives, and must be applicable to each in the same manner. This is the rule at which the Principle of Indifference somewhat obscurely aims. We must first determine what parts of our evidence are relevant on the whole by a series of judgments of relevance, not easily reduced to rule, of the type described above. If this relevant evidence is *of the same form* for both alternatives, then the Principle authorises a judgment of indifference.

16. This rule can be expressed more precisely in symbolic language. Let us assume, to begin with, that the alternative conclusions are expressible in the forms $\phi(a)$ and $\phi(b)$, where $\phi(x)$ is a propositional function.[1] The difference between them, that is to say, can be represented in terms of a single variable.

The Principle of Indifference is applicable to the alternatives $\phi(a)$ and $\phi(b)$, when the evidence h is so constituted that, if $f(a)$ is an independent part of h (see § **14**) which is relevant to $\phi(a)$, and does not contain any independent parts which are irrelevant to $\phi(a)$, then h includes $f(b)$ also.

The rule can be extended by successive steps to cases in which we have more than one variable. We can, if the necessary conditions are fulfilled, successively compare the probabilities of $\phi(a_1a_2)$ and $\phi(b_1a_2)$, and of $\phi(b_1a_2)$ and $\phi(b_1b_2)$, and establish equality between $\phi(a_1a_2)$ and $\phi(b_1b_2)$.

This elucidation is suited to most of the cases to which the Principle of Indifference is ordinarily applied. Thus in the favourite examples in which balls are drawn from urns, we can infer from our evidence no relevant proposition about white balls, such that we cannot infer a corresponding proposition about black balls. Most of the examples, to which the mathematical theory of chances has been applied, and which depend upon the Principle of Indifference, can be arranged, I think, in the forms which the rule requires as formulated above.

17. We can now clear up the difficulties which arose over the group of cases dealt with in § **9**, the typical example of which was the problem of the urn containing black and white balls in an unknown proportion. This more precise enunciation of the Principle enables us to show that of the two solutions the equiprobability of each 'constitution' is alone legitimate, and the

[1] If $\phi(a)$, $\phi(b)$, etc., are propositions, and x is a variable, capable of taking the values a, b, etc., then $\phi(x)$ is a propositional function.

equiprobability of each numerical ratio erroneous. Let us write the alternative 'The proportion of black balls is x'$\equiv\phi(x)$, and the datum 'There are n balls in the bag, with regard to none of which it is known whether they are black or white'$\equiv h$. On the 'ratio' hypothesis it is argued that the Principle of Indifference justifies the judgment of indifference, $\phi(x)/h = \phi(y)/h$. In order that this may be valid, it must be possible to state the relevant evidence in the form $f(x)$ $f(y)$. But this is not the case. If $x=\frac{1}{2}$ and $y=\frac{1}{4}$, we have relevant knowledge about the way in which a proportion of black balls of one half can arise, which is not identical with our knowledge of the way in which a proportion of one quarter can arise. If there are four balls, A, B, C, D, one half are black, if A, B or A, C or A, D or B, C or B, D or C, D are black; and one quarter are black, if A or B or C or D are black. These propositions are not identical in form, and only by a false judgment of irrelevance can we ignore them. On the 'constitution' hypothesis, however, where A, B black and A, C black are treated as distinct alternatives, this want of symmetry in our relevant evidence cannot arise.

18. We can also deal with the point which was illustrated by the difficulty raised in § 4. We considered there the probabilities of a and its contradictory $\bar{a}$ when there is no external evidence relevant to either. What exactly do we mean by saying that there is *no* relevant evidence? Is the addition of the word *external* significant? If a represents a particular proposition, we must know something about it, namely, its *meaning*. May not the apprehension of its meaning afford us some relevant evidence? If so, such evidence must not be excluded. If, then, we say that there is *no* relevant evidence, we must mean no evidence beyond what arises from the mere apprehension of the meaning of the symbol a. If we attach *no* meaning to the symbol, it is useless to discuss the value of the probability; for the probability, which belongs to a proposition as an object of knowledge, not as a form of words, cannot in such a case exist.

What exactly does the symbol a stand for in the above? Does it stand for any proposition of which we know no more than that it is a proposition? Or does it stand for a particular proposition which we understand but of which we know no more than is involved in understanding it? In the former case we cannot extend our result to a proposition of which we know even the meaning; for we should then know *more* than that it is a proposition; and in the latter case we cannot say what the probability of a is as compared with that of its contradictory, until we know *what particular proposition* it stands for; for, as we have seen, the proposition itself may supply relevant evidence.

This suggests that a source of much confusion may lie in the use of symbols and the notion of variables in probability. In the logic of implication, which deals not with probability but with truth, what is true of a variable must be equally true of all instances of the variable. In Probability, on the other hand, we must be on our guard wherever a variable occurs. In Implication we may conclude that ψ is true of anything of which ϕ is true. In Probability we may conclude no more than that ψ is probable of anything of which we *only* know that ϕ is true of it. If x stands for anything of which $\phi(x)$ is true, as soon as we substitute in probability any particular value, whose meaning we know, for x, the value of the probability may be affected; for knowledge, which was irrelevant before, may now become relevant. Take the following example: Does $\phi(a)/\psi(a) = \phi(b)/\psi(b)$? That is to say, is the probability of ϕ's being true of a, given only that ψ is true of a, equal to the probability of ϕ's being true of b, given only that ψ is true of b? If this simply means that the probability of an object's satisfying ϕ about which nothing is known except that it satisfies ψ is equal to ditto ditto, the equation is an identity. For in this case $\phi(a)/\psi(a)$ *means* the same as $\phi(b)/\psi(b)$, *i.e.* we know *nothing* about x and y except that they satisfy ψ, and there is nothing whatever by which we can distinguish a from b. But if a and b represent specific entities, which we can distinguish, then the equality does not necessarily hold. If, for instance, $\phi(x)$ stands for 'x is Socrates,' then it is plainly false that $\phi(a)/\psi(a) = \phi(b)/\psi(b)$, where a stands for Socrates and b does not.

19. Bearing this danger in mind, we can now give further precision to the enunciation of the Principle of Indifference given in § 16. Our knowledge of the meaning of a must be taken account of *so far as it is relevant*; and the Principle is only satisfied if we have corresponding knowledge about the meaning of b. Thus $\phi(a)/h = \phi(b)/h$ may be true for one pair of values a, b, and not true for another pair of values a', b'.

This makes it possible to explain in part the contradiction discussed in § 4. Even if it were true that the probability of a is $\frac{1}{2}$, when we know nothing except that a is a proposition, it does not follow that the probability of 'This book is red' is $\frac{1}{2}$, when we know the meanings of 'book' and 'red,' even if we know no more than this. Knowledge arising directly out of acquaintance with the meaning of 'red' may be sufficient to enable us to infer that 'red' and 'not-red' are not satisfactory alternatives to which to apply the Principle of Indifference. How this may come about will be discussed in §§ 20, 21.

But the contradictions are not yet really solved; for some of the difficulties discussed in § 4 can arise even when we know no more of a and b than that they are *different* propositions. In fact, although we have now stated more clearly than before how the Principle should be enunciated, it is not yet possible to explain or to avoid all the contradictions to which it led us in §§ 4 to 7. For this purpose we must proceed to a further qualification.

20. The examples, in which the Principle of Indifference broke down, had a great deal in common. We broke up the field of possibility, as we may term it, into a number of areas by a series of disjunctive judgments. But the alternative areas were not *ultimate*. They were capable of further subdivision into other areas *similar in kind* to the former. The paradoxes and contradictions arose, in each case, when the alternatives, which the Principle of Indifference treated as equivalent, actually contained or might contain a different or an indefinite number of more elementary units.

In the type of cases in which the Principle of Indifference seemed to permit the assertion that, in the absence of relevant evidence, a proposition is as likely as its contradictory, its contradictory is not an ultimate and indivisible alternative (in the sense to be explained in § 21 below), even if the proposition itself satisfies this condition. For its contradictory can be disjunctively resolved into an indefinite number of sets of contraries to the proposition. It was out of this that our difficulties first arose. 'This book is not red' includes amongst others the alternatives 'This book is black' and 'This book is blue.' It is not, therefore, an ultimate alternative.

In the same way the contradiction of § 5 arose out of the possibility of splitting the alternatives 'He inhabits the British

Isles' into the sub-alternatives 'He inhabits Ireland or he inhabits Great Britain.' And in the third type of case, to which the example of specific volume and density belongs, the alternative 'v lies in the interval 1 to 2' can be broken up into the sub-alternatives 'v lies in the interval 1 to $1\frac{1}{2}$ or $1\frac{1}{2}$ to 2.'

21. This, then, seems to point the way to the qualification of which we are in search. We must enunciate some formal rule which will exclude those cases, in which one of the alternatives involved is itself a disjunction of sub-alternatives *of the same form*. For this purpose the following condition is proposed.

Let the alternatives, the equiprobability of which we seek to establish by means of the Principle of Indifference, be $\phi(a_1)$, $\phi(a_2)$ . . . $\phi(a_r)$,[1] and let the evidence be h. Then it is a necessary condition for the application of the principle, that these should be, relatively to the evidence, *indivisible* alternatives of the form $\phi(x)$. We may define a divisible alternative in the following manner:

An alternative $\phi(a_r)$ is *divisible* if

$$\text{(i.)}\ [\phi(a_r) \equiv \phi(a_{r'}) + \phi(a_{r''})]/h = 1,$$
$$\text{(ii.)}\ \phi(a_{r'})\ .\ \phi(a_{r''})/h = o,$$
$$\text{(iii.)}\ \phi(a_{r'})/h \neq o \text{ and } \phi(a_{r''})/h \neq o$$

The condition that the sub-alternatives must be *of the same form* as the original alternatives, *i.e.* expressible by means of the same propositional function $\phi(x)$, deserves attention. It might be the case that the original alternatives had nothing substantial in common; *i.e.* $\phi(x) \equiv x$ is the only propositional function common to all of them, the alternatives being $a_1, a_2, \ldots, a_r$. In these circumstances the condition in question cannot be satisfied. For the proposition a_r can always be resolved into the disjunction $a_r b + a_r \bar{b}$, where b is any proposition and $\bar{b}$ its contradictory. If, on the other hand, the alternatives which we are comparing can be expressed in the forms $\phi(a_1)$ and $\phi(a_2)$, where the function $\phi(x)$ is distinct from x, it is not necessarily the case that either of these can be resolved into a disjunctive combination of terms which can be expressed in their turn in the same form.

Dispensing with symbolism, we can express these conditions as follows: Our knowledge must not enable us to split up the

[1] The more complicated cases in which the propositional function, of which the alternatives are instances, involves more than one variable (see § 16), can be dealt with in a similar manner *mutatis mutandis*.

alternative $\phi(a_r)$ into a disjunction of two sub-alternatives, (i.) which are themselves expressible in the same form ϕ, (ii.) which are mutually exclusive, and (iii.) which, on the evidence, are possible.

In short, the Principle of Indifference is not applicable to a pair of alternatives, if we know that either of them is capable of being further split up into a pair of possible but incompatible alternatives of the same form as the original pair.

22. This rule commends itself to common sense. If we know that the two alternatives are compounded of a different number or of an indefinite number of sub-alternatives which are in other respects similar, so far as our evidence goes to the original alternatives, then this is a relevant fact of which we must take account. And as it affects the two alternatives in differing and unsymmetrical ways, it breaks down the fundamental condition for the valid application of the Principle of Indifference.

Neither this consideration nor that discussed in §§ **18** and **19** substantially modify the Principle of Indifference as enunciated in § **16**. They have only served to make explicit what was always implicit in the Principle, by explaining the manner in which our knowledge of the *form and meaning* of the alternatives may be a relevant part of the evidence. The apparent contradictions arose from paying attention to what we may term the *extraneous* evidence only, to the neglect of such part of the evidence as bore upon the form and meaning of the alternatives.

23. The application of this result to the examples cited in § **18** is not difficult. It excludes the class of cases in which a proposition and its contradictory constitute the alternatives. For if b is the proposition and $\bar{b}$ its contradictory, we cannot find a propositional function $\phi(x)$ which will satisfy the necessary conditions. It deals also with the type of contradiction which arose in considering the probability that an individual taken at random was an inhabitant of a given region. If, on the other hand, the term 'country' is so defined that one country cannot include two countries, then an individual is, relatively to suitable hypotheses, as likely to be an inhabitant of one as of another. For the function $\phi(x)$, where $\phi(x) \equiv$ 'the individual is an inhabitant of country x,' satisfies the conditions. And it deals with the example of ranges of specific volume and specific density, because there is no range which does not contain within itself two similar ranges. As there are in this case no definite units by which we can define *equal* ranges, the device, which will be referred to in § 25 for dealing with geometrical probabilities, is not available.

24. It is worth while to add that the qualification of § 21 is fatal to the practical utility of the Principle of Indifference in those cases only in which it is possible to find *no* ultimate alternatives which satisfy the conditions. For if the original alternatives each comprise a definite number of indivisible and indifferent sub-alternatives, we can compute their probabilities. It is often the case, however, that we cannot by any process of finite subdivision arrive at indivisible sub-alternatives, or that, if we can, they are not on the evidence indifferent. In the examples given above, for instance, where $\phi(x) \equiv x$, or where x is a part of unspecified magnitude in a continuum, there are *no* indivisible sub-alternatives. The first type comprises all cases, amongst others, in which we weigh the probabilities of a proposition and its contradictory; and the second includes a great number of cases in which physical or geometrical quantities are involved.

25. We can now return to the numerous paradoxes which arise in the study of geometrical probability (see §§ 7, 8). The qualification of § 21 enables us, I think, to discover the source of the confusion. Our alternatives in these problems relate to certain areas or segments or arcs, and however small the elements are which we adopt as our alternatives, they are made up of yet smaller elements which would also serve as alternatives. Our rule, therefore, is not satisfied, and, as long as we enunciate them in this shape, we cannot employ the Principle of Indifference. But it is easy in most cases to discover another set of alternatives which do satisfy the condition, and which will often serve our purpose equally well. Suppose, for instance, that a point lies on a line of length $m.l.$, we may write the alternative 'the interval of length l on which the point lies is the xth interval of that length as we move along the line from left to right' $\equiv \phi(x)$; and the Principle of Indifference can then be applied safely to the m alternatives $\phi(1), \phi(2) \ldots \phi(m)$, the number m increasing as the length l of the intervals is diminished. There is no reason why l should not be of any definite length however small.

If we deal with the problems of geometrical probability in

this way, we shall avoid the contradictory conclusions, which arise from confusing together *distinct* elementary areas. In the problem, for instance, of the chord drawn at random in a circle, which is discussed in § 7, the chord is regarded, not as a one-dimensional line, but as the limit of an area, the shape of which is different in each of the variant solutions. In the first solution it is the limit of a triangle, the length of the base of which tends to zero; in the second solution it is the limit of a quadrilateral, two of the sides of which are parallel and at a distance apart which tends to zero; and in the third solution the area is defined by the limiting position of a central section of undefined shape. These distinct hypotheses lead inevitably to different results. If we were dealing with a strictly linear chord, the Principle of Indifference would yield us no result, as we could not enunciate the alternatives in the required form; and if the chord is an elementary area, we must know the shape of the area of which it is the limit. So long as we are careful to enunciate the alternatives in a form to which the Principle of Indifference can be applied unambiguously, we shall be prevented from confusing together distinct problems, and shall be able to reach conclusions in geometrical probability which are unambiguously valid.

The substance of this explanation can be put in a slightly different way by saying that it is not a matter of indifference in these cases in what manner we proceed to the limit. We must assign the probabilities *before* proceeding to the limit, which we can do unambiguously. But if the problem in hand does not stop at small finite lengths, areas, or volumes, and we have to proceed to the limit, then the final result depends upon the shape in which the body approaches the limit. Mathematicians will recognise an analogy between this case and the determination of potential at points *within* a conductor. Its value depends upon the shape of the area which in the limit represents the point.

26. The positive contributions of this chapter to the determination of valid judgments of equiprobability are two. In the first place we have stated the Principle of Indifference in a more accurate form, by displaying its necessary dependence upon judgments of relevance and so bringing out the hidden element of direct judgment or intuition, which it has always involved. It has been shown that the Principle lays down a rule by which direct judgments of relevance and irrelevance can lead on to judgments of preference and indifference. In the second place, some types of consideration, which are in fact relevant, but which are in danger of being overlooked, have been brought into prominence. By this means it has been possible to avoid the various types of doubtful and contradictory conclusions to which the Principle seemed to lead, so long as we applied it without due qualification.

CHAPTER V

OTHER METHODS OF DETERMINING PROBABILITIES

1. THE recognition of the fact, that not all probabilities are numerical, limits the scope of the Principle of Indifference. It has always been agreed that a numerical measure can actually be obtained in those cases only in which a reduction to a set of exclusive and exhaustive *equiprobable* alternatives is practicable. Our previous conclusion that numerical measurement is often impossible agrees very well, therefore, with the argument of the preceding chapter that the rules, in virtue of which we can assert equiprobability, are somewhat limited in their field of application.

But the recognition of this same fact makes it more necessary to discuss the principles which will justify comparisons of more and less between probabilities, where numerical measurement is theoretically, as well as practically, impossible. We must, for the reasons given in the preceding chapter, rely in the last resort on direct judgment. The object of the following rules and principles is to reduce the judgments of preference and relevance, which we are compelled to make, to a few relatively simple types.[1]

2. We will enquire first in what circumstances we can expect a comparison of more and less to be theoretically possible. I am inclined to think that this is a matter about which, rather unexpectedly perhaps, we are able to lay down definite rules. We are able, I think, always to compare a pair of probabilities which are

(i.) of the type ab/h and a/h,

or (ii.) of the type a/hh_1 and a/h,

provided the additional evidence h_1 contains only one independent piece of relevant information.

[1] Parts of Chap. XV. are closely connected with the topics of the following paragraphs, and the discussion which is commenced here is concluded there.

(i.) The propositions of Part II. will enable us to prove that

$$ab/h < a/h \text{ unless } b/ah = 1;$$

that is to say, the probability of our conclusion is diminished by the addition to it of something, which on the hypothesis of our argument cannot be inferred from it. This proposition will be self-evident to the reader. The rule, that the probability of two propositions jointly is, in general, less than that of either of them separately, includes the rule that the attribution of a more specialised concept is less probable than the attribution of a less specialised concept.

(ii.) This condition requires a little more explanation. It states that the probability a/hh_1 is always greater than, equal to, or less than the probability a/h, if h_1 contains no pair of complementary and independent parts[1] both relevant to a/h. If h_1 is favourable, $a/hh_1 > a/h$. Similarly, if h_2 is favourable to a/hh_1, $a/hh_1h_2 > a/hh_1$. The reverse holds if h_1 and h_2 are unfavourable. Thus we can compare a/hh' and a/h, in every case in which the relevant independent parts of the additional evidence h' are either all favourable, or all unfavourable. In cases in which our additional evidence is equivocal, part taken by itself being favourable and part unfavourable, comparison is not necessarily possible. In ordinary language we may assert that, according to our rule, the addition to our evidence of a single fact always has a definite bearing on our conclusion. It either leaves its probability unaffected and is irrelevant, or it has a definitely favourable or unfavourable bearing, being favourably or unfavourably relevant. It cannot affect the conclusion in an indefinite way, which allows no comparison between the two probabilities. But if the addition of one fact is favourable, and the addition of a second is unfavourable, it is not necessarily possible to compare the probability of our original argument with its probability when it has been modified by the addition of *both* the new facts.

Other comparisons are possible by a combination of these two principles with the Principle of Indifference. We may find, for instance, that $a/hh_1 > a/h$, that $a/h = b/h$, that $b/h > b/hh_2$, and that, therefore, $a/hh_1 > b/hh_2$. We have thus obtained a comparison between a pair of probabilities, which are not of the types discussed above, but without the introduction

[1] See Chap. IV. § 14 for the meaning of these terms.

of any fresh principle. We may denote comparisons of this type by (iii.).

3. Whether any comparisons are possible which do not fall within any of the categories (i.), (ii.), or (iii.), I do not feel certain. We undoubtedly make a number of direct comparisons which do not seem to be covered by them. We judge it more probable, for instance, that Caesar invaded Britain than that Romulus founded Rome. But even in such cases as this, where a reduction into the regular form is not obvious, it might prove possible if we could clearly analyse the real grounds of our judgment. We might argue in this instance that, whereas Romulus's founding of Rome rests solely on tradition, we have *in addition* evidence of another kind for Caesar's invasion of Britain, and that, in so far as our belief in Caesar's invasion rests on tradition, we have reasons of a precisely similar kind as for our belief in Romulus *without* the additional doubt involved in the maintenance of a tradition between the times of Romulus and Caesar. By some such analysis as this our judgment of comparison might be brought within the above categories.

The process of reaching a judgment of comparison in this way may be called 'schematisation.'[1] We take initially an ideal scheme which falls within the categories of comparison. Let us represent 'the historical tradition x has been handed down from a date many years previous to the time of Caesar' by $\psi_1(x)$; 'the historical tradition x has been handed down from the time of Caesar' by $\psi_2(x)$; 'the historical tradition x has extra-traditional support' by $\psi_3(x)$; and the two traditions, the Romulus tradition and the Caesar tradition respectively, by a and b. Then if our relevant evidence h were of the form $\psi_1(a)\psi_2(b)\psi_3(b)$, it is easily seen that the comparison $a/h < b/h$ could be justified on the lines laid down above.[2] A further judgment, that our actual evidence presented no relevant divergence from this schematic form, would then establish the practical conclusion. As I am not aware of any plausible judgment of comparison which we make in common practice, but which is clearly incapable of reduction to some schematic form, and as I see no logical basis for such a comparison, I feel justified in

[1] This phrase is used by Von Kries, *op. cit.* p. 179, in a somewhat similar connection.

[2] For $a/\psi_2(a) = b/\psi_2(b)$; $a/\psi_1(a) < a/\psi_2(a)$; $b/\psi_2(b) < b/\psi_2(b)\psi_3(b)$; $a/\psi_1(a) = a/h$; and $b/\psi_2(b)\psi_3(b) = b/h$.

doubting the *possibility* of comparing the probabilities of arguments dissimilar in form and incapable of schematic reduction. But the point must remain very doubtful until this part of the subject has received a more prolonged consideration.

4. Category (ii.) is very wide, and evidently covers a great variety of cases. If we are to establish general principles of argument and so avoid excessive dependence on direct individual judgments of relevance, we must discover some new and more particular principles included within it. Two of these—those of Analogy and of Induction—are excessively important, and will be the subject of Part III. of this book. In addition to these a few criteria will be examined and established in Chapter XIV., §§ 4 and 8 (49.1). We must be content here (pending the symbolic developments of Part II.) with the two observations following:

(1) The addition of new[1] evidence h_1 to a doubtful[2] argument a/h is *favourably* relevant, if either of the following conditions is fulfilled:—(*a*) if $a/h\bar{h}_1 = 0$; (*b*) if $a/hh_1 = 1$. Divested of symbolism, this merely amounts to a statement that a piece of evidence is favourable if, in conjunction with the previous evidence, it is either a necessary or a sufficient condition for the truth of our conclusion.

(2) It might plausibly be supposed that evidence would be favourable to our conclusion which is favourable to favourable evidence—*i.e.* that, if h_1 is favourable to x/h and x is favourable to a/h, h_1 is favourable to a/h. Whilst, however, this argument is frequently employed under conditions, which, if explicitly stated, would justify it, there are also conditions in which this is not so, so that it is not necessarily valid. For the very deceptive fallacy involved in the above supposition, Mr. Johnson has suggested to me the name of the *Fallacy of the Middle Term*. The general question—If h_1 is favourable to x/h and x is favourable to a/h, in what conditions is h_1 favourable to a/h?—will be examined in Chapter XIV. §§ 4 and 8 (49.1). In the meantime, the intuition of the reader towards the fallacy may be assisted by the following observations, which are due to Mr. Johnson:

Let x, x', x'' . . . be exclusive and exhaustive alternatives under datum h. Let h_1 and a be *concordant* in regard to *each* of

[1] h_1 is *new* evidence so long as $h_1/h \neq 1$.

[2] The argument is *doubtful* so long as a/h is neither certain nor impossible.

these alternatives: *i.e.* any hypothesis which is strengthened by h_1 will strengthen a, and any hypothesis which is weakened by h_1 will weaken a. It is obvious that, if h_1 strengthens *some* of the hypotheses x, x', x'' . . ., it will weaken *others*. This fact helps us to see why we cannot consider the concordance of h_1 and a in regard to one *single* alternative, but must be able to assert their concordance with regard to *every one* of the exclusive and exhaustive alternatives, including the particular one taken. But a further condition is needed, which (as we shall show) is obviously satisfied in two typical problems at least. This further condition is that, for *each* hypothesis x, x', x'' . . ., it shall hold that, were this hypothesis known to be true, the knowledge of h_1 would *not weaken* the probability of a.

These two conditions are *sufficient* to ensure that h_1 shall strengthen a (independently of knowledge of x, x', x'' . . .); and, in a sense, they appear to be *necessary*; for, unless they are satisfied, the dependence of h_1 upon a would be (so to speak) *accidental* as regards the 'middle terms,' (x, x', x'' . . .).

The necessity for reference to *all* the alternatives x, x', x'' . . . is analogous to the requirement of distribution of the middle term in ordinary syllogism. Thus, from premises "All P is x, all S is x," the conclusion that "S's are P" does not formally follow; but given "all P is x and all S is x'" it *does* follow that "no S are P", where x' is any contrary to x. The two conditions taken together would be analogous to the argument: all x S is P; all x' S is P; all x'' S is P; . . . therefore all S is P.

First Typical Problem.—An urn contains an unknown proportion of differently coloured balls. A ball is drawn and replaced. Then x, x', x'' . . . stand for the various possible proportions. Let h_1 mean "a white ball has been drawn"; and let a mean "a white ball will be again drawn." Then any hypothesis which is strengthened by h_1 will strengthen a; and any hypothesis which is weakened by h_1 will weaken a. Moreover, were any one of these hypotheses known to be true, the knowledge of h_1 would not weaken the probability of a. Hence, in the absence of definite knowledge as regards x, x', x'' . . ., the knowledge of h_1 would strengthen the probability of a.

Second Typical Problem.—Let a certain event have taken place; which may have been x, x', x'' or . . . Let h_1 mean that A reports so and so; and let a mean that B reports *similarly* or identically. The phrase *similarly* merely indicates that any hypothesis as to the actual fact, which would be strengthened by A's report, would be strengthened by B's report. Of course, even if the reports were verbally *identical*, A's evidence would not necessarily strengthen the hypothesis in an *equal* degree with B's; because A and B may be unequally expert or intelligent. Now, in such cases, we may further affirm (in general), that, were the actual nature of the event known, the knowledge of A's report on it *would not weaken* (though it also need not strengthen) the probability that B would give a *similar* report. Hence, in the absence of such knowledge, the knowledge of h_1 would strengthen the probability of a.

5. Before leaving this part of the argument we must emphasise the part played by direct judgment in the theory here presented. The rules for the determination of equality and inequality between probabilities all depend upon it at some point. This seems to me quite unavoidable. But I do not feel that we should regard it as a weakness. For we have seen that most, and perhaps all, cases can be determined by the application of general principles to one simple type of direct judgment. No more is asked of the intuitive power applied to particular cases than to determine whether a new piece of evidence tells, on the whole, for or against a given conclusion. The application of the rules involves no wider assumptions than those of other branches of logic.

While it is important, in establishing a control of direct judgment by general principles, not to conceal its presence, yet the fact that we ultimately depend upon an intuition need not lead us to suppose that our conclusions have, therefore, no basis in reason, or that they are as subjective in validity as they are in origin. It is reasonable to maintain with the logicians of the Port Royal that we may draw a conclusion which is truly probable by paying attention to all the circumstances which accompany the case, and we must admit with as little concern as possible Hume's taunt that "when we give the preference to one set of arguments above another, we do nothing but decide from our feeling concerning the superiority of their influence."

CHAPTER VI

THE WEIGHT OF ARGUMENTS

1. THE question to be raised in this chapter is somewhat novel; after much consideration I remain uncertain as to how much importance to attach to it. The magnitude of the probability of an argument, in the sense discussed in Chapter III., depends upon a balance between what may be termed the favourable and the unfavourable evidence; a new piece of evidence which leaves this balance unchanged, also leaves the probability of the argument unchanged. But it seems that there may be another respect in which some kind of quantitative comparison between arguments is possible. This comparison turns upon a balance, not between the favourable and the unfavourable evidence, but between the *absolute* amounts of relevant knowledge and of relevant ignorance respectively.

As the relevant evidence at our disposal increases, the magnitude of the probability of the argument may either decrease or increase, according as the new knowledge strengthens the unfavourable or the favourable evidence; but *something* seems to have increased in either case,—we have a more substantial basis upon which to rest our conclusion. I express this by saying that an accession of new evidence increases the *weight* of an argument. New evidence will sometimes decrease the probability of an argument, but it will always increase its 'weight.'

2. The measurement of evidential weight presents similar difficulties to those with which we met in the measurement of probability. Only in a restricted class of cases can we compare the weights of two arguments in respect of more and less. But this must always be possible where the conclusion of the two arguments is the same, and the relevant evidence in the one includes and exceeds the evidence in the other. If the new evidence

is 'irrelevant,' in the more precise of the two senses defined in § 14 of Chapter IV., the weight is left unchanged. If any part of the new evidence is relevant, then the value is increased.

The reason for our stricter definition of 'relevance' is now apparent. If we are to be able to treat 'weight' and 'relevance' as correlative terms, we must regard evidence as relevant, part of which is favourable and part unfavourable, even if, taken as a whole, it leaves the probability unchanged. With this definition, to say that a new piece of evidence is 'relevant' is the same thing as to say that it increases the 'weight' of the argument.

A proposition cannot be the subject of an argument, unless we at least attach some *meaning* to it, and this meaning, even if it only relates to the form of the proposition, may be relevant in some arguments relating to it. But there may be no other relevant evidence; and it is sometimes convenient to term the probability of such an argument an *à priori* probability. In this case the weight of the argument is at its lowest. Starting, therefore, with minimum weight, corresponding to *à priori* probability, the evidential weight of an argument rises, though its probability may either rise or fall, with every accession of relevant evidence.

3. Where the conclusions of two arguments are different, or where the evidence for the one does not overlap the evidence for the other, it will often be impossible to compare their weights, just as it may be impossible to compare their probabilities. Some rules of comparison, however, exist, and there seems to be a close, though not a complete, correspondence between the conditions under which pairs of arguments are comparable in respect of probability and of weight respectively. We found that there were three principal types in which comparison of probability was possible, other comparisons being based on a combination of these:—

(i.) Those based on the Principle of Indifference, subject to certain conditions, and of the form $\phi a/\psi a.h_1 = \phi b/\psi b.h_2$, where h_1 and h_2 are irrelevant to the arguments.

(ii.) $a/hh_1 \gtreqless a/h$, where h_1 is a single unit of information, containing no independent parts which are relevant.

(iii.) $ab/h \leq a/h$.

Let us represent the evidential weight of the argument, whose probability is a/h, by $V(a/h)$. Then, corresponding to the above, we find that the following comparisons of weight are possible :—

(i.) $V(\phi a/\psi a.h_1) = V(\phi b/\psi b.h_2)$, where h_1 and h_2 are irrelevant in the strict sense. Arguments, that is to say, to which the Principle of Indifference is applicable, have equal evidential weights.

(ii.) $V(a/hh_1) > V(a/h)$, unless h_1 is irrelevant, in which case $V(a/hh_1) = V(a/h)$. The restriction on the composition of h_1, which is necessary in the case of comparisons of magnitude, is not necessary in the case of weight.

There is, however, no rule for comparisons of weight corresponding to (iii.) above. It might be thought that $V(ab/h) < V(a/h)$, on the ground that the more complicated an argument is, relative to given premisses, the less is its evidential weight. But this is invalid. The argument ab/h is further off proof than was the argument a/h; but it is nearer disproof. For example, if $ab/h = 0$ and $a/h > 0$, then $V(ab/h) > V(a/h)$. In fact it would seem to be the case that the weight of the argument a/h is always equal to that of $\bar{a}/h$, where $\bar{a}$ is the contradictory of a; *i.e.*, $V(a/h) = V(\bar{a}/h)$. For an argument is always as near proving or disproving a proposition, as it is to disproving or proving its contradictory.

4. It may be pointed out that if $a/h = b/h$, it does not necessarily follow that $V(a/h) = V(b/h)$. It has been asserted already that if the first equality follows *directly* from a single application of the Principle of Indifference, the second equality also holds. But the first equality can exist in other cases also. If, for instance, a and b are members respectively of *different* sets of three equally probable exclusive and exhaustive alternatives, then $a/h = b/h$; but these arguments may have very different weights. If, however, a and b can each, relatively to h, be inferred from the other, *i.e.* if $a/bh = 1$ and $b/ah = 1$, then $V(a/h) = V(b/h)$. For in proving or disproving one, we are necessarily proving or disproving the other.

Further principles could, no doubt, be arrived at. The above can be combined to reach results in cases upon which unaided common-sense might feel itself unable to pronounce with confidence. Suppose, for instance, that we have three exclusive and exhaustive alternatives, a, b, and c, and that $a/h = b/h$ in virtue of the Principle of Indifference, then we have $V(a/h) = V(b/h)$ and $V(a/h) = V(\bar{a}/h)$, so that $V(b/h) = V(\bar{a}/h)$. It is also true, since $\bar{a}/(b+c)h=1$ and $(b+c)/\bar{a}h=1$, that $V(\bar{a}/h)=V((b+c)/h)$. Hence $V(b/h)=V((b+c)/h)$.

5. The preceding paragraphs will have made it clear that the weighing of the *amount* of evidence is quite a separate process from the *balancing* of the evidence for and against. In so far, however, as the question of weight has been discussed at all, attempts have been made, as a rule, to explain the former in terms of the latter. If $x/h_1h_2=\frac{2}{3}$ and $x/h_1=\frac{3}{4}$, it has sometimes been supposed that it is *more probable* that x/h_1h_2 really is $\frac{2}{3}$ than that x/h_1 really is $\frac{3}{4}$. According to this view, an increase in the amount of evidence strengthens the probability of the probability, or, as De Morgan would say, the presumption of the probability. A little reflection will show that such a theory is untenable. For the probability of x *on hypothesis* h_1 is independent of whether as a matter of fact x is or is not true, and if we find out subsequently that x is true, this does not make it false to say that on hypothesis h_1 the probability of x is $\frac{3}{4}$. Similarly the fact that x/h_1h_2 is $\frac{2}{3}$ does not impugn the conclusion that x/h_1 is $\frac{3}{4}$, and unless we have made a mistake in our judgment or our calculation on the evidence, the two probabilities *are* $\frac{2}{3}$ and $\frac{3}{4}$ respectively.

6. A second method, by which it might be thought, perhaps, that the question of weight has been treated, is the method of *probable error*. But while probable error is sometimes connected with weight, it is primarily concerned with quite a different question. 'Probable error,' it should be explained, is the name given, rather inconveniently perhaps, to an expression which arises when we consider the probability that a given quantity is measured by one of a number of different magnitudes. Our *data* may tell us that one of these magnitudes is the most probable measure of the quantity; but in some cases it will also tell us how probable each of the other possible magnitudes of the quantity is. In such cases we can determine the probability that the quantity will have a magnitude which does not differ from the most probable by more than a specified amount. The amount, which the difference between the actual value of the quantity and its most probable value is as likely as not to exceed, is the 'probable error.' In many practical questions the existence of a small probable error is of the greatest importance, if our conclusions are to prove valuable. The probability that

the quantity has any particular magnitude may be very small; but this may matter very little, if there is a high probability that it lies within a certain range.

Now it is obvious that the determination of probable error is intrinsically a different problem from the determination of weight. The method of probable error is simply a summation of a number of alternative and exclusive probabilities. If we say that the most probable magnitude is x and the probable error y, this is a way, convenient for many purposes, of summing up a number of probable conclusions regarding a variety of magnitudes other than x which, on the evidence, the quantity may possess. The connection between probable error and weight, such as it is, is due to the fact that in scientific problems a large probable error is not uncommonly due to a great lack of evidence, and that as the available evidence increases there is a tendency for the probable error to diminish. In these cases the probable error may conceivably be a good practical measure of the weight.

It is necessary, however, in a theoretical discussion, to point out that the connection is casual, and only exists in a limited class of cases. This is easily shown by an example. We may have *data* on which the probability of $x=5$ is $\frac{1}{3}$, of $x=6$ is $\frac{1}{4}$, of $x=7$ is $\frac{1}{5}$, of $x=8$ is $\frac{1}{6}$, and of $x=9$ is $\frac{1}{20}$. Additional evidence might show that x must either be 5 or 8 or 9, the probabilities of each of these conclusions being $\frac{7}{16}$, $\frac{5}{16}$, $\frac{4}{16}$. The evidential weight of the latter argument is greater than that of the former, but the probable error, so far from being diminished, has been increased. There is, in fact, no reason whatever for supposing that the probable error must necessarily diminish, as the weight of the argument is increased.

The typical case, in which there may be a *practical* connection between weight and probable error, may be illustrated by the two cases following of balls drawn from an urn. In each case we require the probability of drawing a white ball; in the first case we know that the urn contains black and white in equal proportions; in the second case the proportion of each colour is unknown, and each ball is as likely to be black as white. It is evident that in either case the probability of drawing a white ball is $\frac{1}{2}$, but that the weight of the argument in favour of this conclusion is greater in the first case. When we consider the most probable proportion in which balls will be drawn in the long run, if after each withdrawal they are replaced, the question of probable error enters in, and we find that the greater evidential weight of the argument on the first hypothesis is accompanied by the smaller probable error.

This conventionalised example is typical of many scientific problems. The more we know about any phenomenon, the less likely, as a rule, is our opinion to be modified by each additional item of experience. In such problems, therefore, an argument of high weight concerning some phenomenon is likely to be accompanied by a low probable error, when the character of a series of similar phenomena is under consideration.

7. Weight cannot, then, be explained in terms of probability. An argument of high weight is not 'more likely to be right' than one of low weight; for the probabilities of these arguments only state relations between premiss and conclusion, and these relations are stated with equal accuracy in either case. Nor is an argument of high weight one in which the probable error is small; for a small probable error only means that magnitudes in the neighbourhood of the most probable magnitude have a relatively high probability, and an increase of evidence does not necessarily involve an increase in these probabilities.

The conclusion, that the 'weight' and the 'probability' of an argument are independent properties, may possibly introduce a difficulty into the discussion of the application of probability to practice.[1] For in deciding on a course of action, it seems plausible to suppose that we ought to take account of the weight as well as the probability of different expectations. But it is difficult to think of any clear example of this, and I do not feel sure that the theory of 'evidential weight' has much practical significance.

Bernoulli's second maxim, that we must take into account all the information we have, amounts to an injunction that we should be guided by the probability of that argument, amongst those of which we know the premisses, of which the evidential weight is the greatest. But should not this be re-enforced by a further maxim, that we ought to make the weight of our arguments as great as possible by getting all the information we can?[2] It is

[1] See also Chapter XXVI. § 7.

[2] Cf. Locke, *Essay concerning Human Understanding*, book ii. chap. xxi. § 67: "He that judges without informing himself to the utmost that he is capable, cannot acquit himself of judging amiss."

difficult to see, however, to what point the strengthening of an argument's weight by increasing the evidence ought to be pushed. We may argue that, when our knowledge is slight but capable of increase, the course of action, which will, relative to such knowledge, probably produce the greatest amount of good, will often consist in the acquisition of more knowledge. But there clearly comes a point when it is no longer worth while to spend trouble, before acting, in the acquisition of further information, and there is no evident principle by which to determine *how far* we ought to carry our maxim of strengthening the weight of our argument. A little reflection will probably convince the reader that this is a very confusing problem.

8. The fundamental distinction of this chapter may be briefly repeated. One argument has more *weight* than another if it is based upon a greater amount of relevant evidence; but it is not always, or even generally, possible to say of two sets of propositions that one set embodies *more* evidence than the other. It has a greater *probability* than another if the balance in its favour, of what evidence there is, is greater than the balance in favour of the argument with which we compare it; but it is not always, or even generally, possible to say that the balance in the one case is greater than the balance in the other. The weight, to speak metaphorically, measures the *sum* of the favourable and unfavourable evidence, the probability measures the *difference*.

9. The phenomenon of 'weight' can be described from the point of view of other theories of probability than that which is adopted here. If we follow certain German logicians in regarding probability as being based on the disjunctive judgment, we may say that the weight is increased when the number of alternatives is reduced, although the ratio of the number of favourable to the number of unfavourable alternatives may not have been disturbed; or, to adopt the phraseology of another German school, we may say that the weight of the probability is increased, as the field of possibility is contracted.

The same distinction may be explained in the language of the frequency theory.[1] We should then say that the weight is increased if we are able to employ as the class of reference a class which is contained in the original class of reference.

10. The subject of this chapter has not usually been discussed

[1] See Chap. VIII.

by writers on probability, and I know of only two by whom the question has been explicitly raised:[1] Meinong, who threw out a suggestion at the conclusion of his review of Von Kries' "Principien," published in the *Göttingische gelehrte Anzeigen* for 1890 (see especially pp. 70-74), and A. Nitsche, who took up Meinong's suggestion in an article in the *Vierteljahrsschrift für wissenschaftliche Philosophie*, 1892, vol. xvi. pp. 20-35, entitled "Die Dimensionen der Wahrscheinlichkeit und die Evidenz der Ungewissheit."

Meinong, who does not develop the point in any detail, distinguishes probability and weight as 'Intensität' and 'Qualität,' and is inclined to regard them as two independent dimensions in which the judgment is free to move—they are the two dimensions of the 'Urteils-Continuum.' Nitsche regards the weight as being the measure of the reliability (Sicherheit) of the probability, and holds that the probability continually approximates to its true magnitude (reale Geltung) as the weight increases. His treatment is too brief for it to be possible to understand very clearly what he means, but his view seems to resemble the theory already discussed that an argument of high weight is 'more likely to be right' than one of low weight.

[1] There are also some remarks by Czuber (*Wahrscheinlichkeitsrechnung*, vol. i. p. 202) on the *Erkenntnisswert* of probabilities obtained by different methods, which may have been intended to have some bearing on it.

CHAPTER VII

HISTORICAL RETROSPECT

1. The characteristic features of our Philosophy of Probability must be determined by the solutions which we offer to the problems attacked in Chapters III. and IV. Whilst a great part of the logical calculus, which will be developed in Part II., would be applicable with slight modification to several distinct theories of the subject, the ultimate problems of establishing the premisses of the calculus bring into the light every fundamental difference of opinion.

These problems are often, for this reason perhaps, left on one side by writers whose interest chiefly lies in the more formal parts of the subject. But Probability is not yet on so sound a basis that the formal or mathematical side of it can be safely developed in isolation, and some attempts have naturally been made to solve the problem which Bishop Butler sets to the logician in the concluding words of the brief discussion on probability with which he prefaces the *Analogy*.[1]

In this chapter, therefore, we will review in their historical order the answers of Philosophy to the questions, how we know relations of probability, what ground we have for our judgments, and by what method we can advance our knowledge.

2. The natural man is disposed to the opinion that probability is essentially connected with the inductions of experience and, if he is a little more sophisticated, with the Laws of Causation

1 "It is not my design to inquire further into the nature, the foundation and measure of probability; or whence it proceeds that *likeness* should beget that presumption, opinion and full conviction, which the human mind is formed to receive from it, and which it does necessarily produce in every one; or to guard against the errors to which reasoning from analogy is liable. This belongs to the subject of logic, and is a part of that subject which has not yet been thoroughly considered."

and of the Uniformity of Nature. As Aristotle says, "the probable is that which usually happens." Events do not always occur in accordance with the expectations of experience; but the laws of experience afford us a good ground for supposing that they usually will. The occasional disappointment of these expectations prevents our predictions from being more than probable; but the ground of their probability must be sought in this experience, and in this experience only.

This is, in substance, the argument of the authors of the Port Royal Logic (1662), who were the first to deal with the logic of probability in the modern manner: "In order for me to judge of the truth of an event, and to be determined to believe it or not believe it, it is not necessary to consider it abstractly, and in itself, as we should consider a proposition in geometry; but it is necessary to pay attention to all the circumstances which accompany it, internal as well as external. I call internal circumstances those which belong to the fact itself, and external those which belong to the persons by whose testimony we are led to believe it. This being done, if all the circumstances are such that it never or rarely happens that the like circumstances are the concomitants of falsehood, our mind is led, naturally, to believe that it is true."[1] Locke follows the Port Royal Logicians very closely: "Probability is likeliness to be true. . . . The *grounds of it* are, in short, these two following. *First*, the conformity of anything with our own knowledge, observation, and experience. *Secondly*, the testimony of others, vouching their observation and experience";[2] and essentially the same opinion is maintained by Bishop Butler: "When we determine a thing to be probably true, suppose that an event has or will come to pass, it is from the mind's remarking in it a likeness to some other event, which we have observed has come to pass. And this observation forms, in numberless instances, a presumption, opinion, or full conviction that such event has or will come to pass."[3]

Against this view of the subject the criticisms of Hume were directed: "The idea of cause and effect is derived from *experience*, which informs us, that such particular objects, in all past

[1] Eng. Trans., p. 353.

[2] *An Essay concerning Human Understanding*, book iv. "Of Knowledge and Opinion."

[3] Introduction to the *Analogy*.

instances, have been constantly conjoined with each other. . . . According to this account of things . . . probability is founded on the presumption of a resemblance betwixt those objects, of which we have had experience, and those, of which we have had none; and therefore 'tis impossible this presumption can arise from probability."[1] "When we are accustomed to see two impressions conjoined together, the appearance or idea of the one immediately carries us to the idea of the other. . . . Thus all probable reasoning is nothing but a species of sensation. 'Tis not solely in poetry and music, we must follow our taste and sentiment, but likewise in philosophy. When I am convinced of any principle, 'tis only an idea, which strikes more strongly upon me. When I give the preference to one set of arguments above another, I do nothing but decide from my feeling concerning the superiority of their influence."[2] Hume, in fact, points out that, while it is true that past experience gives rise to a psychological anticipation of some events rather than of others, no ground has been given for the validity of this superior anticipation.

3. But in the meantime the subject had fallen into the hands of the mathematicians, and an entirely new method of approach was in course of development. It had become obvious that many of the judgments of probability which we in fact make do not depend upon past experience in a way which satisfies the canons laid down by the Port Royal Logicians or by Locke. In particular, alternatives are judged equally probable, without there being necessarily any actual experience of their approximately equal frequency of occurrence in the past. And, apart from this, it is evident that judgments based on a somewhat indefinite experience of the past do not easily lend themselves to precise numerical appraisement. Accordingly James Bernoulli,[3] the real founder of the classical school of mathematical probability, while not repudiating the old test of experience, had based many of his conclusions on a quite different criterion—the rule which I have named the Principle of Indifference. The traditional method of the mathematical school essentially depends upon reducing all the possible conclusions to a number of 'equi-probable cases.' And, according to the Principle of

[1] *Treatise of Human Nature*, p. 391 (Green's edition).

[2] *Op. cit.* p. 403.

[3] See especially *Ars Conjectandi*, p. 224. Cf. Laplace, *Théorie analytique*, p. 178.

Indifference, 'cases' are held to be equi-probable when there is no reason for preferring any one to any other, when there is nothing, as with Buridan's ass, to determine the mind in any one of the several possible directions. To take Czuber's example of dice,[1] this principle permits us to assume that each face is equally likely to fall, if there is no reason to suppose any particular irregularity, and it does not require that we should *know* that the construction is regular, or that each face has, as a matter of fact, fallen equally often in the past.

On this Principle, extended by Bernoulli beyond those problems of gaming in which by its tacit assumption Pascal and Huyghens had worked out a few simple exercises, the whole fabric of mathematical probability was soon allowed to rest. The older criterion of experience, never repudiated, was soon subsumed under the new doctrine. First, in virtue of Bernoulli's famous Law of Great Numbers, the fractions representing the probabilities of events were thought to represent also the actual proportion of their occurrences, so that experience, if it were considerable, could be translated into the cyphers of arithmetic. And next, by the aid of the Principle of Indifference, Laplace established his Law of Succession by which the influence of any experience, *however limited*, could be numerically measured, and which purported to prove that, if B has been seen to accompany A twice, it is two to one that B will again accompany A on A's next appearance. No other formula in the alchemy of logic has exerted more astonishing powers. For it has established the existence of God from the premiss of total ignorance; and it has measured with numerical precision the probability that the sun will rise to-morrow.

Yet the new principles did not win acceptance without opposition. D'Alembert,[2] Hume, and Ancillon[3] stand out as the sceptical critics of probability, against the credulity of

[1] *Wahrscheinlichkeitsrechnung*, p. 9.

[2] D'Alembert's scepticism was directed towards the current mathematical theory only, and was not, like Hume's, fundamental and far-reaching. His opposition to the received opinions was, perhaps, more splendid than discriminating.

[3] Ancillon's communication to the Berlin Academy in 1794, entitled *Doutes sur les bases du calcul des probabilités*, is not as well known as it deserves to be. He writes as a follower of Hume, but adds much that is original and interesting. An historian, who also wrote on a variety of philosophical subjects, Ancillon was, at one time, the Prussian Minister of Foreign Affairs.

eighteenth-century philosophers who were ready to swallow without too many questions the conclusions of a science which claimed and seemed to bring an entire new field within the dominion of Reason.[1]

The first effective criticism came from Hume, who was also the first to distinguish the method of Locke and the philosophers from the method of Bernoulli and the mathematicians. "Probability," he says, "or reasoning from conjecture, may be divided into two kinds, viz. that which is founded on *chance* and that which arises from causes."[2] By these two kinds he evidently means the mathematical method of counting the equal chances based on Indifference, and the inductive method based on the experience of uniformity. He argues that 'chance' alone can be the foundation of nothing, and "that there must always be a mixture of causes among the chances, in order to be the foundation of any reasoning."[3] His previous argument against probabilities, which were based on an assumption of cause, is thus extended to the mathematical method also.

But the great prestige of Laplace and the 'verifications' of his principles which his more famous results were supposed to supply had, by the beginning of the nineteenth century, established the science on the Principle of Indifference in an almost unquestioned position. It may be noted, however, that De Morgan, the principal student of the subject in England, seems to have regarded the method of actual experiment and the method of counting cases, which were equally probable on grounds of Indifference, as alternative methods of equal validity.

4. The reaction against the traditional teaching during the past hundred years has not possessed sufficient force to displace

[1] French philosophy of the latter half of the eighteenth century was profoundly affected by the supposed conquests of the Calculus of Probability in all fields of thought. Nothing seemed beyond its powers of prediction, and it almost succeeded in men's minds to the place previously occupied by Revelation. It was under these influences that Condorcet evolved his doctrine of the perfectibility of the human race. The continuity and oneness of modern European thought may be illustrated, if such things amuse the reader, by the reflection that Condorcet derived from Bernoulli, that Godwin was inspired by Condorcet, that Malthus was stimulated by Godwin's folly into stating his famous doctrine, and that from the reading of Malthus on *Population* Darwin received his earliest impulse.

[2] *Treatise of Human Nature*, p. 424 (Green's edition).

[3] *Op. cit.* p. 425.

the established doctrine, and the Principle of Indifference is still very widely accepted in an unqualified form. Criticism has proceeded along two distinct lines; the one, originated by Leslie Ellis, and developed by Dr. Venn, Professor Edgeworth, and Professor Karl Pearson, has been almost entirely confined in its influence to England; the other, of which the beginnings are to be seen in Boole's *Laws of Thought*, has been developed in Germany, where its ablest exponent has been Von Kries. France has remained uninfluenced by either, and faithful, on the whole, to the tradition of Laplace. Even Henri Poincaré, who had his doubts, and described the Principle of Indifference as "very vague and very elastic," regarded it as our only guide in the choice of that convention, "which has always something arbitrary about it," but upon which calculation in probability invariably rests.[1]

5. Before following up in detail these two lines of development, I will summarise again the earlier doctrine with which the leaders of the new schools found themselves confronted.

The earlier philosophers had in mind in dealing with probability the application to the future of the inductions of experience, to the almost complete exclusion of other problems. For the *data* of probability, therefore, they looked only to their own experience and to the recorded experiences of others; their principal refinement was to distinguish these two grounds, and they did not attempt to make a numerical estimate of the chances. The mathematicians, on the other hand, setting out from the simple problems presented by dice and playing cards, and

[1] Poincaré's opinions on Probability are to be found in his *Calcul des Probabilités* and in his *Science et Hypothèse*. Neither of these books appears to me to be in all respects a considered work, but his view is sufficiently novel to be worth a reference. Briefly, he shows that the current mathematical definition is circular, and argues from this that the choice of the particular probabilities, which we are to regard as initially equal before the application of our mathematics, is entirely a matter of 'convention.' Much epigram is, therefore, expended in pointing out that the study of probability is no more than a polite exercise, and he concludes: "Le calcul des probabilités offre une contradiction dans les termes mêmes qui servent à le désigner, et, si je ne craignais de rappeler ici un mot trop souvent répété, je dirais qu'il nous enseigne surtout une chose; c'est de savoir que nous ne savons rien." On the other hand, the greater part of his book is devoted to working out instances of practical application, and he speaks of 'metaphysics' legitimising particular conventions. How this comes about is not explained. He seems to endeavour to save his reputation as a philosopher by the surrender of probability as a valid conception, without at the same time forfeiting his claim as a mathematician to work out probable formulae of practical importance.

requiring for the application of their methods a basis of numerical measurement, dwelt on the negative rather than the positive side of their evidence, and found it easier to measure equal degrees of ignorance than equivalent quantities of experience. This led to the explicit introduction of the Principle of Indifference, or, as it was then termed, the Principle of Non-Sufficient Reason. The great achievement of the eighteenth century was, in the eyes of the early nineteenth, the reconciliation of the two points of view and the measurement of probabilities, which were grounded on experience, by a method whose logical basis was the Principle of Non-Sufficient Reason. This would indeed have been a very astonishing discovery, and would, as its authors declared, have gradually brought almost every phase of human activity within the power of the most refined mathematical analysis.

But it was not long before more sceptical persons began to suspect that this theory proved too much. Its calculations, it is true, were constructed from the *data* of experience, but the more simple and the less complex the experience the better satisfied was the theory. What was required was not a wide experience or detailed information, but a completeness of symmetry in the little information there might be. It seemed to follow from the Laplacian doctrine that the primary qualification for one who would be well informed was an equally balanced ignorance.

6. The obvious reaction from a teaching, which seemed to derive from abstractions results relevant to experience, was into the arms of empiricism; and in the state of philosophy at that time England was the natural home of this reaction. The first protest, of which I am aware, came from Leslie Ellis in 1842.[1] At the conclusion of his *Remarks on an alleged proof of the Method of least squares*,[2] "Mere ignorance," he says, "is no ground for any inference whatever. *Ex nihilo nihil.*" In Venn's *Logic of Chance* Ellis's suggestions are developed into a complete theory:[3] "Experience is our sole guide. If we want to discover what is in reality a series of *things*, not a series of our own conceptions, we must appeal to the things themselves to obtain it, for we cannot find much help elsewhere." Professor Edgeworth[4] was an early disciple of the same school: "The probability," he

[1] *On the Foundations of the Theory of Probabilities.*
[2] Republished in *Miscellaneous Writings.*
[3] *Logic of Chance*, p. 74.
[4] *Metretike*, p. 4.

says, "of head occurring n times if the coin is of the ordinary make is approximately at least $(\frac{1}{2})^n$. This value is rigidly deducible from positive experience, the observations made by gamesters, the experiments recorded by Jevons and De Morgan."

The doctrines of the empirical school will be examined in Chapter VIII., and I postpone my detailed criticism to that chapter. Venn rejects the applications of Bernoulli's theorem, which he describes as "one of the last remaining relics of Realism," as well as the later Laplacian Law of Succession, thus destroying the link between the empirical and the *à priori* methods. But, apart from this, his view that statements of probability are simply a particular class of statements about the actual world of phenomena, would have led him to a closer dependence on actual experience. He holds that the probability of an event's having a certain attribute is simply the fraction expressing the proportion of cases in which, as a matter of actual fact, this attribute is present. Our knowledge, however, of this proportion is often reached inductively, and shares the uncertainty to which all inductions are liable. And, besides, in referring an event to a series we do not postulate that all the members of the series should be identical, but only that they should not be *known* to differ in a relevant manner. Even on this theory, therefore, we are not solely determined by positive knowledge and the direct *data* of experience.

7. The Empirical School in their reaction against the pretentious results, which the Laplacian theory affected to develop out of nothing, have gone too far in the opposite direction. If our experience and our knowledge were complete, we should be beyond the need of the Calculus of Probability. And where our experience is incomplete, we cannot hope to derive from it judgments of probability without the aid either of intuition or of some further *à priori* principle. Experience, as opposed to intuition, cannot possibly afford us a criterion by which to judge whether on given evidence the probabilities of two propositions are or are not equal.

However essential the data of experience may be, they cannot by themselves, it seems, supply us with what we want. Czuber,[1] who prefers what he calls the Principle of Compelling Reason (das Prinzip des zwingenden Grundes), and holds that Probability

[1] *Wahrscheinlichkeitsrechnung*, p. 11.

has an objective and not merely formal interpretation only when it is grounded on definite knowledge, is rightly compelled to admit that we cannot get on altogether without the Principle of Non-Sufficient Reason. On the grounds both of its own intuitive plausibility and of that of some of the conclusions for which it is necessary, we are inevitably led towards this principle as a necessary basis for judgments of probability. In *some* sense, judgments of probability do seem to be based on equally balanced degrees of ignorance.

8. It is from this starting-point that the German logicians have set out. They have perceived that there are few judgments of probability which are altogether independent of some principle resembling that of Non-Sufficient Reason. But they also apprehend, with Boole, that this may be a very arbitrary method of procedure.

It was pointed out in § 18 of Chapter IV. that the cases, in which the Principle of Indifference (or Non-Sufficient Reason) breaks down, have a great deal in common, and that we break up the field of possibility into a number of areas, actually unequal, but indistinguishable on the evidence. Several German logicians, therefore, have endeavoured to determine some rule by which it might be possible to postulate actual equality of area for the fields of the various possibilities.

By far the most complete and closely reasoned solution on these lines is that of Von Kries.[1] He is primarily anxious to discover a proper basis for the numerical measurement of probabilities, and he is thus led to examine with care the grounds of valid judgments of equiprobability. His criticisms of the Principle of Non-Sufficient Reason are searching, and, to meet them, he elaborates a number of qualifying conditions which are, he argues, necessary and sufficient. The value of his book, however, lies, in the opinion of the present writer, in the critical rather than in the constructive parts. The manner in which his qualifying conditions are expressed is often, to an English reader at any rate, somewhat obscure, and he seems sometimes to cover up difficulties, rather than solve them, by the invention of new technical terms. These characteristics render it difficult to expound him adequately in a summary, and the reader must be

[1] *Die Principien der Wahrscheinlichkeitsrechnung. Eine logische Untersuchung.* Freiburg, 1886.

referred to the original for a proper exposition of the Doctrine of *Spielräume*. Briefly, but not very intelligibly perhaps, he may be said to hold that the hypotheses for the probabilities of which we wish to obtain a numerical comparison, must refer to 'fields' (*Spielräume*) which are 'indifferent,' 'comparable' in magnitude, and 'original' (*ursprünglich*). Two fields are 'indifferent' if they are equal before the Principle of Non-Sufficient Reason; they are 'comparable' if it is true that the fields are actually of equal extent; and they are 'original' or ultimate if they are not derived from some other field. The last condition is exceedingly obscure, but it seems to mean that the objects with which we are ultimately dealing must be directly represented by the 'fields' of our hypotheses, and there must not be merely correlation between these objects and the objects of the fields. The qualification of comparability is intended to deal with difficulties such as that connected with the population of different areas of unknown extent; and the qualification of originality with those arising from indirect measurement, as in the case of specific density.

Von Kries's solution is highly suggestive, but it does not seem, so far as I understand it, to supply an unambiguous criterion for all cases. His discussion of the philosophical character of probability is brief and inadequate, and the fundamental error in his treatment of the subject is the *physical*, rather than logical, bias which seems to direct the formulation of his conditions. The condition of *Ursprünglichkeit*, for instance, seems to depend upon physical rather than logical criteria, and is, as a result, much more restricted in its applicability than a condition, which will really meet the difficulties of the case, ought to be. But, although I differ from him in his philosophical conception of probability, the treatment of the Principle of Indifference, which fills the greater part of his book, is, I think, along fruitful lines, and I have been deeply indebted to it in formulating my own conditions in Chapter IV.

Of less closely reasoned and less detailed treatments, which aim at the same kind of result, those of Sigwart and Lotze are worth noticing. Sigwart's[1] position is sufficiently explained by the following extract: "The possibility of a mathematical treatment lies primarily in the fact that in the disjunctive judgment

[1] Sigwart, *Logic* (Eng. edition), vol. ii. p. 220.

the number of terms in the disjunction plays a decisive part. Inasmuch as a limited number of mutually exclusive possibilities is presented, of which one alone is actual, the element of number forms an essential part of our knowledge. . . . Our knowledge must enable us to assume that the particular terms of the disjunction are so far equivalent that they express an equal degree of specialisation of a general concept, or that they cover equal parts of the whole extension of the concept. . . . This equivalence is most intuitable where we are dealing with equal parts of a spatial area, or equal parts of a period of time. . . . But even where this obvious quality is not forthcoming, we may ground our expectations upon a hypothetical equivalence, where we see no reason for considering the extent of one possibility to be greater than that of the others. . . ."

In the beginning of this passage Sigwart seems to be aware of the fundamental difficulty, although exception may be taken to the vagueness of the phrase "equal degree of specialisation of a general concept." But in the last sentence quoted he surrenders the advantages he has gained in the earlier part of his explanation, and, instead of insisting on a knowledge of an equal degree of specialisation, he is satisfied with an absence of any knowledge to the contrary. Hence, in spite of his initial qualifications, he ends unrestrainedly in the arms of Non-Sufficient Reason.[1]

Lotze,[2] in a brief discussion of the subject, throws out some remarks well worth quoting: "We disclaim all knowledge of the circumstances which condition the real issue, so that when we talk of equally possible cases we can only mean *coördinated as equivalent species in the compass of an universal case*; that is to say, if we enumerate the special forms, which the genus can assume, we get a disjunctive judgment of the form: if the condition B is fulfilled, one of the kinds $f_1 f_2 f_3$ . . . of the universal consequent F will occur to the exclusion of the rest. Which of all those different consequents will, in fact, occur, depends in all cases on the special form $b_1 b_2 b_3$ . . . in which that universal condition is fulfilled. . . . A *coördinated* case is a case which answers to one and only one of the mutually exclusive values $b_1 b_2$ . . . of the condition B, and these rival values may occur in

[1] Sigwart's treatment of the subject of probability is curiously inaccurate. Of his four fundamental rules of probability, for instance. three are, as he states them, certainly false.

[2] Lotze, *Logic* (Eng. edition), pp. 364, 365.

reality; it does not answer to a more general form B, of this condition, which can never exist in reality, because it embraces several of the particular values $b_1 b_2$. . . ."

This certainly meets some of the difficulties, and its resemblance to the conditions formulated in Chapter IV. will be evident to the careful reader. But it is not very precise, and not easily applicable to all cases, to those, for instance, of the measurement of continuous quantity. By combining the suggestions of Von Kries, Sigwart, and Lotze, we might, perhaps, patch up a fairly comprehensive rule. We might say, for instance, that if b_1 and b_2 are classes, their members must be finite in number and enumerable or they must compose stretches; that, if they are finite in number, they must be equal in number; and that, if their members compose stretches, the stretches must be equal stretches; and that if b_1 and b_2 are concepts, they must represent concepts of an equal degree of specialisation. But qualifications so worded would raise almost as many difficulties as they solved. How, for instance, are we to know when concepts are of an equal degree of specialisation?

9. That probability is a *relation* has often received incidental recognition from logicians, in spite of the general failure to place proper emphasis on it. The earliest writer, with whom I am acquainted, explicitly to notice this, is Kahle in his *Elementa logicae Probabilium methodo mathematica in usum Scientiarum et Vitae adornata* published at Halle in 1735.[1] Amongst more recent writers casual statements are common to the effect that the probability of a conclusion is relative to the grounds upon which it is based. Take Boole[2] for instance: "It is implied in the definition that probability is always relative to our actual

[1] This work, which seems to have soon fallen into complete neglect and is now extremely rare, is full of interest and original thought. The following quotations will show the fundamental position taken up: "Est cognitio probabilis, si desunt quaedam requisita ad veritatem demonstrativam (p. 15). Propositio probabilis esse potest falsa, et improbabilis esse potest vera; ergo cognitio hodie possibilis, crastina luce mutari potest improbabilem, si accedunt reliqua requisita omnia, in certitudinem (p. 26). . . . Certitudo est terminus relativus: considerare potest ratione representationum in intellectu nostro. . . . Incerta nobis dependent a defectu cognitionis (p. 35). . . . Actionem imprudenter et contra regulas probabilitatis susceptam eventus felix sequi potest. Ergo prudentia actionum ex successu solo non est aestimanda (p. 62). . . . Logica probabilium est scientia dijudicandi gradum certitudinis eorum, quibus desunt requisita ad veritatem demonstrativam (p. 94)."

[2] "On a General Method in the Theory of Probabilities," *Phil. Mag.*, 4th Series, viii., 1854. See also, "On the Application of the Theory of Probabilities

state of information and varies with that state of information." Or Bradley:[1] "Probability tells us what we ought to believe, what we ought to believe *on certain data* . . . Probability is no more 'relative' and 'subjective' than is any other act of logical inference from hypothetical premises. It is relative to the *data* with which it has to deal, and is not relative in any other sense." Or even Laplace, when he is explaining the diversity of human opinions: "Dans les choses qui ne sont que vraisemblables, la différence des données que chaque homme a sur elles, est une des causes principales de la diversité des opinions que l'on voit régner sur les mêmes objets . . . c'est ainsi que le même fait, récité devant une nombreuse assemblée, obtient divers degrés de croyance, suivant l'étendue des connaissances des auditeurs."[2]

10. Here we may leave this account of the various directions in which progress has seemed possible, with the hope that it may assist the reader, who is dissatisfied with the solution proposed in Chapter IV., to determine the line of argument along which he is likeliest to discover the solution of a difficult problem.

---

to the Question of the Combination of Testimonies or Judgments" (*Edin. Phil. Trans.* xxi. p. 600): "Our estimate of the probability of an event varies not absolutely with the circumstances which actually affect its occurrence, but with our knowledge of those circumstances."

[1] *The Principles of Logic*, p. 208.

[2] *Essai philosophique*, p. 7.

CHAPTER VIII

THE FREQUENCY THEORY OF PROBABILITY

1. The theory of probability, outlined in the preceding chapters, has serious difficulties to overcome. There is a theoretical, as well as a practical, difficulty in measuring or comparing degrees of probability, and a further difficulty in determining them *à priori*. We must now examine an alternative theory which is much freer from these troubles, and is widely held at the present time.

2. The theory is in its essence a very old one. Aristotle foreshadowed it when he held that "the probable is that which for the most part happens";[1] and, as we have seen in Chapter VII., an opinion not unlike this was entertained by those philosophers of the seventeenth and eighteenth centuries who approached the problems of probability uninfluenced by the work of mathematicians. But the underlying conception of earlier writers received at the hands of some English logicians during the latter half of the nineteenth century a new and much more complicated form.

The theory in question, which I shall call the Frequency Theory of Probability, first appears[2] as the basis of a proposed logical scheme in a brief essay by Leslie Ellis *On the Foundations of the Theory of Probabilities*, and is somewhat further developed in his *Remarks on the Fundamental Principles of the Theory of*

[1] *Rhet.* i. 2, 1357 a 34.

[2] I give Ellis the priority because his paper, published in 1843, was read on Feb. 14, 1842. The same conception, however, is to be found in Cournot's *Exposition*, also published in 1843: "La théorie des probabilités a pour objet certains rapports numériques qui prendraient des valeurs fixes et complétement déterminées, si l'on pouvait répéter à l'infini les épreuves des mêmes hasards, et qui, pour un nombre fini d'épreuves, oscillent entre des limites d'autant plus resserrées, d'autant plus voisines des valeurs *finales*, que le nombre des épreuves est plus grand."

Probabilities.[1] "If the probability of a given event be correctly determined," he says, "the event will on a long run of trials tend to recur with frequency proportional to their probability. This is generally proved mathematically. It seems to me to be true *à priori*. . . . I have been unable to sever the judgment that one event is more likely to happen than another from the belief that in the long run it will occur more frequently." Ellis explicitly introduces the conception that probability is essentially concerned with a group or series.

Although the priority of invention must be allowed to Leslie Ellis, the theory is commonly associated with the name of Venn. In his *Logic of Chance*[2] it first received elaborate and systematic treatment, and, in spite of his having attracted a number of followers, there has been no other comprehensive attempt to meet the theory's special difficulties or the criticisms directed against it. I shall begin, therefore, by examining it in the form in which Venn has expounded it. Venn's exposition is much coloured by an empirical view of logic, which is not perhaps as necessary to the essential part of his doctrine as he himself implies, and is not shared by all of those who must be classed as in general agreement with him about probability. It will be necessary, therefore, to supplement a criticism of Venn by an account of a more general frequency theory of probability, divested of the empiricism with which he has clothed it.

3. The following quotations from Venn's *Logic of Chance* will show the general drift of his argument: The fundamental conception is that of a series (p. 4). The series is of events which have a certain number of features or attributes in common (p. 10). The characteristic distinctive of probability is this,—the occasional attributes, as distinguished from the permanent, are found on an examination to tend to exist *in a certain definite proportion of the whole number of cases* (p. 11). We require that there should be in nature large classes of objects, throughout all the individual members of which a general resemblance extends. For this

1 These essays were published in the *Transactions* of the Camb. Phil. Soc., the first in 1843 (vol. viii.), and the second in 1854 (vol. ix.). Both were reprinted in *Mathematical and other Writings* (1863), together with three other brief papers on Probability and the Method of Least Squares. All five are full of spirit and originality, and are not now so well known as they deserve to be.

2 The first edition appeared in 1866. Revised editions were issued in 1876 and 1888. References are given to the third edition of 1888.

purpose the existence of natural kinds or groups is necessary (p. 55). The distinctive characteristics of probability prevail principally in the properties of natural kinds, both in the ultimate and in the derivative or accidental properties (p. 63). The same peculiarity prevails again in the force and frequency of most natural agencies (p. 64). There seems reason to believe that it is in such things only, as distinguished from things artificial, that the property in question is to be found (p. 65). How, in any particular case, are we to establish the existence of a probability series? Experience is our sole guide. If we want to discover what is in reality a series of *things*, not a series of our own conceptions, we must appeal to the things themselves to obtain it, for we cannot find much help elsewhere (p. 174). When probability is divorced from direct reference to objects, as it substantially is by not being founded upon experience, it simply resolves itself into the common algebraical doctrine of Permutations and Combinations (p. 87). By assigning an expectation in reference to the individual, we *mean* nothing more than to make a statement about the average of his class (p. 151). When we say of a conclusion within the strict province of probability, that it is not certain, all that we mean is that in some proportion of cases only will such conclusion be right, in the other cases it will be wrong (p. 210).

The essence of this theory can be expressed in a few words. To say, that the probability of an event's having a certain characteristic is $\frac{x}{y}$, is to mean that the event is one of a number of events, a proportion $\frac{x}{y}$ of which have the characteristic in question; and the *fact*, that there *is* such a series of events possessing this frequency in respect of the characteristic, is purely a matter of experience to be determined in the same manner as any other question of fact. That such series do exist happens to be a characteristic of the real world as we know it, and from this the practical importance of the calculation of probabilities is derived.

Such a theory possesses manifest advantages. There is no mystery about it—no new indefinables, no appeals to intuition. Measurement leads to no difficulties; our probabilities or frequencies are ordinary numbers, upon which the arithmetical apparatus can be safely brought to bear. And at the same time it seems to crystallise in a clear, explicit shape the floating opinion of common sense that an event is or is not probable in certain supposed circumstances according as it is or is not usual as a matter of fact and experience.

The two principal tenets, then, of Venn's system are these,—that probability is concerned with series or groups of events, and that all the requisite facts must be determined empirically, a statement in probability merely summing up in a convenient way a group of experiences. Aggregate regularity combined with individual difference happens, he says, to be characteristic of many events in the real world. It will often be the case, therefore, that we can make statements regarding the average of a certain class, or regarding its characteristics in the long run, which we cannot make about any of its individual members without great risk of error. As our knowledge regarding the class as a whole may give us valuable guidance in dealing with an individual instance, we require a convenient way of saying that an individual belongs to a class in which certain characteristics appear on the average with a known frequency; and this the conventional language of probability gives us. The importance of probability depends solely upon the actual existence of such groups or real kinds in the world of experience, and a judgment of probability must necessarily depend for its validity upon our empirical knowledge of them.

4. It is the obvious, as well as the correct, criticism of such a theory, that the identification of probability with statistical frequency is a very grave departure from the established use of words; for it clearly excludes a great number of judgments which are generally believed to deal with probability. Venn himself was well aware of this, and cannot be accused of supposing that all beliefs, which are commonly called probable, are really concerned with statistical frequency. But some of his followers, to judge from their published work, have not always seen, so clearly as he did, that his theory is *not* concerned with the same subject as that with which other writers have dealt under the same title. Venn justifies his procedure by arguing that no other meaning, of which it is possible to take strict logical cognisance, can reasonably be given to the term, and that the other meanings, with which it has been used, have not enough in common to permit their reduction to a single logical scheme. It is useless,

therefore, for a critic of Venn to point out that many supposed judgments of probability are *not* concerned with statistical frequency; for, as I understand the *Logic of Chance*, he admits it; and the critic must show that the sense different from Venn's in which the term probability is often employed *has* an important logical interpretation about which we can generalise. This position I seek to establish. It is, in my opinion, this other sense *alone* which has importance; Venn's theory by itself has few practical applications, and if we allow it to hold the field, we must admit that probability is *not* the guide of life, and that in following it we are not acting according to reason.

5. Part of the plausibility of Venn's theory is derived, I think, from a failure to recognise the narrow limits of its applicability, or to notice his own admissions regarding this. "In every case," he says (p. 124), "in which we extend our inferences by Induction or Analogy, or depend upon the witness of others, or trust to our own memory of the past, or come to a conclusion through conflicting arguments, or even make a long and complicated deduction by mathematics or logic, we have a result of which we can scarcely feel as certain as of the premisses from which it was obtained. In all these cases, then, we are conscious of varying quantities of belief, but are the laws according to which the belief is produced and varied the same? If they cannot be reduced to one harmonious scheme, if, in fact, they can at best be brought to nothing but a number of different schemes, each with its own body of laws and rules, then it is vain to endeavour to force them into one science." All these cases, therefore, in which we are 'not certain,' Venn explicitly excludes from what he chooses to call the science of probability, and he pays no further attention to them. The science of probability is, according to him, *no more* than a method which enables us to express in a convenient form statistical statements of frequency. "The province of probability," he says again on page 160, "is not so extensive as that over which variation of belief might be observed. Probability only considers the case in which this variation is brought about in a certain definite statistical way."[1] He points

[1] Edgeworth uses the term 'probability' widely, as I do; but he makes a distinction corresponding to Venn's by limiting the subject-matter of the *Calculus* of Probabilities. He writes ('Philosophy of Chance,' *Mind*, 1884, p. 223): "The Calculus of Probabilities is concerned with the estimation of degrees of probability; not every species of estimate, but that which is founded

out on p. 194 that for the purposes of probability we must take the statistical frequency from which we start *ready made* and ask no questions about the process or completeness of its manufacture: "It may be obtained by any of the numerous rules furnished by Induction, or it may be inferred deductively, or given by our own observation; its value may be diminished by its depending upon the testimony of witnesses, or its being recalled by our own memory. Its real value may be influenced by these causes or any combinations of them; but all these are preliminary questions with which we have nothing directly to do. We assume our statistical proposition to be true, neglecting the diminution of its value by the processes of attainment."

It must be recognised, therefore, that Venn has deliberately excluded from his survey almost all the cases in which we regard our judgments as 'only probable'; and, whatever the value or consistency of his own scheme, he has left untouched a wide field of study for others.

6. The main grounds, which have induced Venn to regard judgments based on statistical frequency as the only cases of probability which possess logical importance, seem to be two: (i.) that other cases are mainly subjective, and (ii.) that they are incapable of accurate measurement.

With regard to the first it must be admitted that there are many instances in which variation of belief is occasioned by purely psychological causes, and that his argument is valid against those who have defined probability as measuring the degree of subjective belief. But this has not been the usual way of looking at the subject. Probability is the study of the grounds which lead us to entertain a *rational* preference for one belief over another. That there are rational grounds other than statistical frequency, for such preferences, Venn does not deny; he admits in the quotation given above that the '*real value*' of our conclusion is influenced by many other con-

on a particular standard. That standard is the phenomenon of statistical uniformity: the fact that a genus can very frequently be subdivided into species such that the number of individuals in each species bears an approximately constant ratio to the number of individuals in the genus." This use of terms is legitimate, though it is not easy to follow it consistently. But, like Venn's, it leaves aside the most important questions. The Calculus of Probabilities, thus interpreted, is no guide by itself as to which opinion we ought to follow, and is not a measure of the weight we should attach to conflicting arguments.

siderations than that of statistical frequency. Venn's theory, therefore, cannot be fairly propounded by his disciples as *alternative* to such a theory as is propounded here. For my Treatise is concerned with the general theory of arguments from premisses leading to conclusions which are reasonable but not certain; and this is a subject which Venn has, deliberately, not treated in the *Logic of Chance*.

7. Apart from two circumstances, it would scarcely be necessary to say anything further; but in the first place some writers have believed that Venn has propounded a *complete* theory of probability, failing to realise that he is not at all concerned with the sense in which we may say that one induction or analogy, or testimony, or memory, or train of argument is more probable than another; and in the second place he himself has not always kept within the narrow limits, which he has himself laid down as proper to his theory.

For he has not remained content with defining a probability as identical with a statistical frequency, but has often spoken as if his theory told us which alternatives it is *reasonable to prefer*. When he states, for instance, that modality ought to be banished from Logic and relegated to Probability (p. 296), he forgets his own dictum that of premisses, the distinctive characteristic of which is their lack of certainty, Probability takes account of *one class only*, Induction concerning itself with another class, and so forth (p. 321). He forgets also that, when he comes to consider the practical use of statistical frequencies, he has to admit that an event may possess more than one frequency, and that we must decide which of these to prefer on extraneous grounds (p. 213). The device, he says, must be to a great extent arbitrary, and there are no logical grounds of decision; but would he deny that it is often reasonable to found our probability on one statistical frequency rather than on another? And if our grounds are reasonable, are they not in an important sense logical?

Even in those cases, therefore, in which we derive our preference for one alternative over another from a knowledge of statistical frequencies, a statistical frequency by itself is insufficient to determine us. We may call a statistical frequency a probability, if we choose; but the fundamental problem of determining which of several alternatives is logically preferable still awaits solution. We cannot be content with the only counsel Venn can offer, that we should choose a frequency which is derived from a series neither too large nor too small.

The same difficulty, that a probability in Venn's sense is insufficient to determine which alternative is logically preferable, arises in another connection. In most cases the statistical frequency is not given in experience for certain, but is arrived at by a process of *induction*, and inductions, he admits, are not certain. If, in the past, three infants out of every ten have died in their first four years, induction may base on this the doubtful assertion, All infants die in that proportion. But we cannot assert on this ground, as Venn wishes to do, that the probability of the death of an infant in its first four years *is* $\frac{3}{10}$ths. We can say no more than that it is probable (in my sense) that there is such a probability (in his sense). For the purpose of coming to a decision we cannot compare the value of this conclusion with that of others until we know the probability (in my sense) that the statistical frequency really is $\frac{3}{10}$ths. The cases in which we can determine the logical value of a conclusion entirely on grounds of statistical frequency would seem to be extremely few in number.

8. The second main reason which led Venn to develop his theory is to be found in his belief that probabilities which are based on statistical frequencies are alone capable of accurate measurement. The term 'probabilities,' he argues, is properly confined to the case of chances which can be calculated, and all calculable chances can be made to depend upon statistical frequency. In attempting to establish this latter contention he is involved in some paradoxical opinions. "In many cases," he admits, "it is undoubtedly true that we do not resort to direct experience at all. If I want to know what is my chance of holding ten trumps in a game of whist, I do not enquire how often such a thing has occurred before. . . . In practice, *à priori* determination is often easy, whilst *à posteriori* appeal to experience would be not merely tedious but utterly impracticable." But these cases which are usually based on the Principle of Indifference can, he maintains, be justified on statistical grounds. In the case of coin tossing there is a considerable experience of the equally frequent occurrence of heads and tails; the experience gained in this simple case is to be extended to the complex cases by "Induction and Analogy." In one simple case the result to which the Principle of Indifference would lead is that which experience recommends. Therefore in complex cases, where there is no basis of experiment at all, we may assume that Experience, if experience there was, would speak with the same voice as Indifference. This is to assert that, because in one case, where there is no known reason to the contrary, there actually is none, therefore in other cases incapable of verification the absence of known reason to the contrary proves that *actually* there is none.

The attempt to justify the rules of inverse probability on statistical grounds I have failed to understand; and after a careful reading, I am unable to produce an intelligible account of the argument involved in the latter part of chapter vii. of the *Logic of Chance.*[1] I am doubtful whether Venn should not have excluded *à posteriori* arguments in probability from his scheme as well as inductive arguments. The attempt to include them may have been induced by a desire to deal with all cases in which numerical calculation has been commonly thought possible.

9. The argument so far has been solely concerned with the case for the frequency theory developed in the *Logic of Chance.* The criticisms which follow will be directed against a more general form of the same theory which may conceivably have recommended itself to some readers. It is unfortunate that no adherent of the doctrine, with the exception of Venn, has attempted to present the theory of it in detail. Professor Karl Pearson, for instance, probably agrees with Venn in a general way only, and it is very likely that many of the foregoing remarks do not apply to his view of probability; but while I generally disagree with the fundamental premisses upon which his work in probability and statistics seems to rest, I am not clearly aware of the nature of the philosophical theory from which he thinks that he derives them and which makes them appear to him to be satisfactory. A careful exposition of his logical presuppositions would greatly add to the completeness of his work. In the meantime it is only possible to raise general objections to

[1] Let the reader, who is acquainted with this chapter, consider what precise assumption Venn's reasoning requires on p. 187 in the example which seeks to show the efficacy of Lord Lister's antiseptic treatment *à posteriori*. What is the 'inevitable assumption about the bags' when it is translated into the language of this example?

any theory of probability which seeks to found itself upon the conception of statistical frequency.

The generalised frequency theory which I propose to put forward, as perhaps representative of what adherents of this doctrine have in mind, differs from Venn's in several important respects.[1] In the first place, it does not regard probability as being *identical* with statistical frequency, although it holds that all probabilities must be based on statements of frequency, and can be defined in terms of them. It accepts the theory that propositions rather than events should be taken as the subject-matter of probability; and it adopts the comprehensive view of the subject according to which it includes induction and all other cases in which we believe that there are *logical* grounds for preferring one alternative out of a set none of which are certain. Nor does it follow Venn in supposing any special connection to exist between a frequency theory of probability and logical empiricism.

10. A proposition can be a member of many distinct classes of propositions, the classes being merely constituted by the existence of particular resemblances between their members or in some such way. We may know of a given proposition that it is one of a particular class of propositions, and we may also know, precisely or within defined limits, what *proportion* of this class are true, without our being aware whether or not the given proposition is true. Let us, therefore, call the actual proportion of true propositions in a class the truth-frequency[2] of the class, and define the measure of the probability of a proposition relative to a class, of which it is a member, as being equal to the truth-frequency of the class.

The fundamental tenet of a frequency theory of probability is, then, that the probability of a proposition always depends upon referring it to some class whose truth-frequency is known within wide or narrow limits.

Such a theory possesses most of the advantages of Venn's, but escapes his narrowness. There is nothing in it so far which could not be easily expressed with complete precision in the terms of ordinary logic. Nor is it necessarily confined to prob-

[1] In what follows I am much indebted for some suggestions in favour of the frequency theory communicated to me by Dr. Whitehead; but it is not to be supposed that the exposition which follows represents his own opinion.

[2] This is Dr. Whitehead's phrase.

abilities which are numerical. In some cases we may know the exact number which expresses the truth-frequency of our class; but a less precise knowledge is not without value, and we may say that one probability is greater than another, without knowing how much greater, and that it is large or small or negligible, if we have knowledge of corresponding accuracy about the truth-frequencies of the classes to which the probabilities refer. The magnitudes of some pairs of probabilities we shall be able to compare numerically, others in respect of more and less only, and others not at all. A great deal, therefore, of what has been said in Chapter III. would apply equally to the present theory, with this difference that the probabilities would, as a matter of fact, have numerical values *in all cases*, and the less complete comparisons would only hold the field in cases where the real probabilities were partially unknown. On the frequency theory, therefore, there is an important sense in which probabilities can be unknown, and the relative vagueness of the probabilities employed in ordinary reasoning is explained as belonging not to the probabilities themselves but only to our knowledge of them. For the probabilities are relative, not to our knowledge, but to some objective class, possessing a perfectly definite truth-frequency, to which we have chosen to refer them.

The frequency theory expounded in this manner cannot easily avoid mention of the relativity of probabilities which is implicit here, as it is in Venn's. Whether or not the probability of a proposition is relative to given *data*, it is clearly relative to the particular class or series to which we choose to refer it. A given proposition has a great variety of different probabilities corresponding to each of the various distinct classes of which it is a member; and before an intelligible meaning can be given to a statement that the probability of a proposition is so-and-so, the class must be specified to which the proposition is being referred. Most adherents of the frequency theory would probably go further, and agree that the class of reference must be determined in any particular case by the *data* at our disposal. Here, then, is another point on which it is not necessary for the frequency theory to diverge from the theory of this Treatise. It should, I think, be generally agreed by every school of thought that the probability of a conclusion is in an important sense *relative to given premisses*. On this issue and also on the point that our knowledge of many probabilities is not numerically definite, there might well be for the future an end of disagreement, and disputation might be reserved for the philosophical interpretation of these settled facts, which it is unreasonable to deny, however we may explain them.

11. I now proceed to those contentions upon which my fundamental criticism of the frequency theory is founded. The first of these relates to the method by which the class of reference is to be determined. The magnitude of a probability is always to be measured by the truth-frequency of some class; and this class, it is allowed, must be determined by reference to the premisses, on which the probability of the conclusion is to be determined. But, as a given proposition belongs to innumerable different classes, how are we to know which class the premisses indicate as appropriate? What substitute has the frequency theory to offer for judgments of relevance and indifference? And without something of this kind, what principle is there for uniquely determining the class, the truth-frequency of which is to measure the probability of the argument? Indeed the difficulties of showing how given premisses determine the class of reference, by means of rules expressed in terms of previous ideas, and without the introduction of any notion, which is new and peculiar to probability, appear to me insurmountable.

Whilst no general criterion of choice seems to exist, where of two alternative classes neither includes the other, it might be thought that where one does include the other, the obvious course would be to take the narrowest and most specialised class. This procedure was examined and rejected by Venn: though the objection to it is due, not, as he supposed, to the lack of sufficient statistics in such cases upon which to found a generalisation, but to the inclusion in the class-concept of marks characteristic of the proposition in question, but nevertheless *not relevant* to the matter in hand. If the process of narrowing the class were to be carried to its furthest point, we should generally be left with a class whose *only* member is the proposition in question, for we generally know something about it which is true of no other proposition. We cannot, therefore, define the class of reference as being the class of propositions of which everything is true which is *known* to be true of the proposition whose probability we seek to determine. And, indeed, in those examples for which the frequency theory possesses the greatest *prima facie* plausibility, the class of reference is selected by taking account of *some only* of the known characteristics of the *quaesitum*, those characteristics, namely, which are *relevant* in the circumstances. In those cases in which one can admit that the probability can be measured by reference to a known truth-frequency, the class of reference is formed of propositions about which our *relevant* knowledge is the same as about the proposition under consideration. In these special cases we get the same result from the frequency theory as from the Principle of Indifference. But this does not serve to rehabilitate the frequency theory as a *general* explanation of probability, and goes rather to show that the theory of this Treatise is the generalised theory, comprehending within it such applications of the idea of statistical truth-frequency as have validity.

'Relevance' is an important term in probability, of which the meaning is readily intelligible. I have given my own definition of it already. But I do not know how it is to be explained in terms of the frequency theory. Whether supporters of this theory have fully appreciated the difficulty I much doubt. It is a fundamental issue involving the essence of the *peculiarity* of probability, which prevents its being explained away in terms of statistical frequency or anything else.

12. Yet perhaps a modified view of the frequency theory could be evolved which would avoid this difficulty, and I proceed, therefore, to some further criticisms. It might be agreed that a novel element must be admitted at this point, and that relevancy must be determined in some such manner as has been explained in earlier chapters. With this admission, it might be argued, the theory would still stand, divested, it is true, of some of its original simplicity, but nevertheless a substantial theory differing in important respects, although not quite so fundamentally as before, from alternative schemes.

The next important objection, then, is concerned with the manner in which the principal theorems of probability are to be established on a theory of frequency. This will involve an anticipation in some part of later arguments; and the reader may be well advised to return to the following paragraph after he has finished Part II.

13. Let us begin by a consideration of the 'Addition Theorem.'

If a/h denotes the probability of a on hypothesis h, this theorem may be written $(a+b)/h = a/h + b/h - ab/h$, and may be read 'On hypothesis h the probability of "a or b" is equal to the probability of a + the probability of b − the probability of "both a and b."' This theorem, interpreted in some way or other, is universally assumed; and we must, therefore, inquire what proof of it the frequency theory can afford. A little symbolism will assist the argument: Let α_f represent the truth-frequency of any class α, and let a_α/h stand for 'the probability of a on hypothesis h, α being the class of reference determined by this hypothesis.'[1] We then have $a_\alpha/h = \alpha_f$, and we require to prove a proposition, for values of γ and δ not yet determined, which will be of the form:

$$(a+b)_\delta/h = a_\alpha/h + b_\beta/h - ab_\gamma/h.$$

Now if δ' is the class of propositions $(a+b)$ such that a is an α and b a β, it is easily shown by the ordinary arithmetic of classes that $\delta'_f = \alpha_f + \beta_f - \alpha\beta_f$ where $\alpha\beta$ is the class of propositions which are members of both α and β. In the case, therefore, where $\delta = \delta'$ and $\gamma = \alpha\beta$, an addition theorem of the required kind has been established.

But it does not follow by any reasonable rule that, if h determines α and β as the appropriate classes of reference for a and b, h must necessarily determine δ' and $\alpha\beta$ as the appropriate classes of reference for $(a+b)$ and ab; it may, for instance, be the case that h, while it renders α and β determinate, yields no information whatever regarding $\alpha\beta$, and points to some quite different class μ as the suitable class of reference for ab. On the frequency theory, therefore, we cannot maintain that the addition theorem is true in general, but only in those special cases where it happens that $\delta = \delta'$ and $\gamma = \alpha\beta$.

The following is a good example: We are given that the proportion of black-haired men in the population is $\frac{p_1}{q}$ and the proportion of colour-blind men $\frac{p_2}{q}$, and there is no *known* connection between black-hair and colour-blindness: what is the probability that a man, about whom nothing special

[1] The question, previously at issue, as to *how* the class of reference is determined by the hypothesis, is now ignored.

is known, is [1] *either* black-haired *or* colour-blind? If we represent the hypotheses by h and the alternatives by a and b, it would usually be held that, colour-blindness and black hair being *independent for knowledge* [2] relative to the given data, $ab/h = \frac{p_1 p_2}{q^2}$, and that, therefore, by the addition theorem, $(a+b)/h = \frac{p_1}{q} + \frac{p_2}{q} - \frac{p_1 p_2}{q^2}$. But, on the frequency theory, this result might be invalid; for $a\beta_f = \frac{p_1 p_2}{q^2}$, *only* if this is the *actual* proportion in fact of persons who are both colour-blind and black-haired, and that this is the actual proportion cannot possibly be inferred from the *independence for knowledge* of the characters in question.[3]

Precisely the same difficulty arises in connection with the multiplication theorem $ab/h = a/bh . b/h$.[4] In the frequency notation, which is proposed above, the corresponding theorem will be of the form $ab_\delta/h = a_\gamma/bh . b_\beta/h$. For this equation to be satisfied it is easily seen that δ must be the class of propositions xy such that x is a member of a and y of β, and γ the class of propositions xb such that x is a member of a; and, as in the case of the addition theorem, we have no guarantee that these classes γ and δ will be those which the hypotheses bh and h will respectively determine as the appropriate classes of reference for a and ab.

In the case of the theorem of inverse probability [5]

$$\frac{b/ah}{c/ah} = \frac{a/bh}{a/ch} \cdot \frac{b/h}{c/h}$$

the same difficulty again arises, with an additional one when practical applications are considered. For the relative probabilities of our *à priori* hypotheses, b and c, will scarcely ever be capable of determination by means of known frequencies, and in the most legitimate instances of the inverse principle's operation

[1] In the course of the present discussion the disjunctive $a+b$ is never interpreted so as to *exclude* the conjunctive ab.

[2] For a discussion of this term see Chapter XVI. § 2.

[3] Venn argues (*Logic of Chance*, pp. 173, 174) that there is an inductive ground for making this inference. The question of extending the fundamental theorems of a frequency theory of probability by means of induction is discussed in § 14 below.

[4] *Vide* Chapter XII. § 6, and Chapter XIV. § 4.

[5] *Vide* Chapter XIV. § 5.

we depend either upon an inductive argument or upon the Principle of Indifference. It is hard to think of an example in which the frequency conditions are even approximately satisfied.

Thus an important class of case, in which arguments in probability, generally accepted as satisfactory, do not satisfy the frequency conditions given above, are those in which the notion is introduced of two propositions being, on certain *data*, independent for knowledge. The meaning and definition of this expression is discussed more fully in Part II.; but I do not see what interpretation the frequency theory can put upon it. Yet if the conception of 'independence for knowledge' is discarded, we shall be brought to a standstill in the vast majority of problems, which are ordinarily considered to be problems in probability, simply from the lack of sufficiently detailed data. Thus the frequency theory is not adequate to explain the processes of reasoning which it sets out to explain. If the theory restricts its operation, as would seem necessary, to those cases in which we *know* precisely how far the true members of a and β overlap, the vast majority of arguments in which probability has been employed must be rejected.

14. An appeal to some further principle is, therefore, required before the ordinary apparatus of probable inference can be established on considerations of statistical frequency; and it may have occurred to some readers that assistance may be obtained from the principles of induction. Here also it will be necessary to anticipate a subsequent discussion. If the argument of Part III. is correct, nothing is more fatal than Induction to the theory now under criticism. For, so far from Induction's lending support to the fundamental rules of probability, it is itself dependent on them. In any case, it is generally agreed that an inductive conclusion is only probable, and that its probability increases with the number of instances upon which it is founded. According to the frequency theory, this belief is only justified if the majority of inductive conclusions actually are true, and it will be false, even on our existing data, that *any* of them are even probable, if the acknowledged possibility that a majority are false is an actuality. Yet what possible reason can the frequency theory offer, which does not beg the question, for supposing that a majority *are true*? And failing this, what ground have we for believing the inductive process to be reasonable? Yet we invariably assume that with our existing knowledge it is logically reasonable to attach some weight to the inductive method, even if future experience shows that *not one* of its conclusions is verified in fact. The frequency theory, therefore, in its present form at any rate, entirely fails to explain or justify the most important source of the most usual arguments in the field of probable inference.

15. The failure of the frequency theory to explain or justify arguments from induction or analogy suggests some remarks of a more general kind. While it is undoubtedly the case that many valuable judgments in probability are partly based on a knowledge of statistical frequencies, and that many more can be held, with some plausibility, to be indirectly derived from them, there remains a great mass of probable argument which it would be paradoxical to justify in the same manner. It is not sufficient, therefore, even if it is possible, to show that the theory can be developed in a self-consistent manner; it must also be shown how the body of probable argument, upon which the greater part of our generally accepted knowledge seems to rest, can be explained in terms of it; for it is certain that much of it does not appear to be derived from premisses of statistical frequency.

Take, for instance, the intricate network of arguments upon which the conclusions of *The Origin of Species* are founded: how impossible it would be to transform them into a shape in which they would be seen to rest upon statistical frequency! Many individual arguments, of course, are explicitly founded upon such considerations; but this only serves to differentiate them more clearly from those which are not. Darwin's own account of the nature of the argument may be quoted: "The belief in Natural Selection must at present be grounded entirely on general considerations: (1) on its being a *vera causa*, from the struggle for existence and the certain geological fact that species do somehow change; (2) from the analogy of change under domestication by man's selection; (3) and chiefly from this view connecting under an intelligible point of view a host of facts. When we descend to details . . . we cannot prove that a single species has changed; nor can we prove that the supposed changes are beneficial, which is the groundwork of the theory; nor can we explain why some species have changed and others have not."[1] Not only in the main argument, but in many of the subsidiary discussions,[2] an elaborate combination of induction and analogy is superimposed upon a narrow and limited knowledge of statistical frequency. And this is equally the case in almost all everyday arguments of any degree of complexity. The class of judgments, which a theory of statistical frequency can comprehend, is too narrow to justify its claim to present a complete theory of probability.

16. Before concluding this chapter, we should not overlook the element of truth which the frequency theory embodies and which provides its plausibility. In the first place, it gives a true account, so long as it does not argue that probability and frequency are *identical*, of a large number of the most *precise* arguments in probability, and of those to which mathematical treatment is easily applicable. It is this characteristic which has recommended it to statisticians, and explains the large measure of its acceptance in England at the present time; for the popularity in this country of an opinion, which has, so far as I know, no thorough supporters abroad, may reasonably be attributed to the chance which has led most of the English writers, who have paid much attention to probability in recent years, to approach the subject from the statistical side.

In the second place, the statement that the probability of an event is measured by its actual frequency of occurrence 'in the long run' has a very close connection with a valid conclusion which can be derived, *in certain cases*, from Bernoulli's theorem. This theorem and its connection with the theory of frequency will be the subject of Chapter XXIX.

17. The absence of a recent exposition of the logical basis of the frequency theory by any of its adherents has been a great disadvantage to me in criticising it. It is possible that some of the opinions, which I have examined at length, are now held by no one; nor am I absolutely certain, at the present stage of the inquiry, that a partial rehabilitation of the theory may not be possible. But I am sure that the objections which I have raised cannot be met without a great complication of the theory, and without robbing it of the simplicity which is its greatest

[1] Letter to G. Bentham, *Life and Letters*, vol. iii. p. 25.

[2] *E.g.* in the discussion on the relative effect of disuse and selection in reducing unnecessary organs to a rudimentary condition.

preliminary recommendation. Until the theory has been given new foundations, its logical basis is not so secure as to permit controversial applications of it in practice. A good deal of modern statistical work may be based, I think, upon an inconsistent logical scheme, which, avowedly founded upon a theory of frequency, introduces principles which this theory has no power to justify.

CHAPTER IX

THE CONSTRUCTIVE THEORY OF PART I. SUMMARISED

1. That part of our knowledge which we obtain directly, supplies the premisses of that part which we obtain by argument. From these premisses we seek to justify some degree of rational belief about all sorts of conclusions. We do this by perceiving certain logical relations between the premisses and the conclusions. The kind of rational belief which we *infer* in this manner is termed *probable* (or in the limit *certain*), and the logical relations, by the perception of which it is obtained, we term *relations of probability.*

The probability of a conclusion a derived from premisses h we write a/h ; and this symbol is of fundamental importance.

2. The object of the Theory or Logic of Probability is to systematise such processes of inference. In particular it aims at elucidating rules by means of which the probabilities of different arguments can be compared. It is of great practical importance to determine which of two conclusions is on the evidence the more probable.

The most important of these rules is the Principle of Indifference. According to this Principle we must rely upon direct judgment for discriminating between the relevant and the irrelevant parts of the evidence. We can only discard those parts of the evidence which are irrelevant by *seeing* that they have no logical bearing on the conclusion. The irrelevant evidence being thus discarded, the Principle lays it down that if the evidence for either conclusion is the same (*i.e.* symmetrical), then their probabilities also are the same (*i.e.* equal).

If, on the other hand, there is additional evidence (*i.e.* in addition to the symmetrical evidence) for one of the conclusions, and this evidence is *favourably relevant,* then that conclusion is the more probable. Certain rules have been given by which to judge whether or not evidence is favourably relevant. And by combinations of these judgments of preference with the judgments of indifference warranted by the Principle of Indifference more complicated comparisons are possible.

3. There are, however, many cases in which these rules furnish no means of comparison; and in which it is certain that it is not actually within our power to make the comparison. It has been argued that in these cases the probabilities are, in fact, *not comparable.* As in the example of similarity, where there are different orders of increasing and diminishing similarity, but where it is not possible to say of every pair of objects which of them is on the whole the more like a third object, so there are different orders of probability, and probabilities, which are not of the same order, cannot be compared.

4. It is sometimes of practical importance, when, for example, we wish to evaluate a chance or to determine the amount of our expectation, to say not only that one probability is greater than another, but by how much it is greater. We wish, that is to say, to have a numerical measure of degrees of probability.

This is only occasionally possible. A rule can be given for numerical measurement when the conclusion is one of a number of equiprobable, exclusive, and exhaustive alternatives, but not otherwise.

5. In Part II. I proceed to a symbolic treatment of the subject, and to the greater systematisation, by symbolic methods on the basis of certain axioms, of the rules of probable argument.

In Parts III., IV., and V. the nature of certain very important types of probable argument of a complex kind will be treated in detail; in Part III. the methods of Induction and Analogy, in Part IV. certain semi-philosophical problems, and in Part V. the logical foundations of the methods of inference now commonly known as *statistical.*

PART II

FUNDAMENTAL THEOREMS

CHAPTER X

INTRODUCTORY

1. In Part I. we have been occupied with the epistemology of our subject, that is to say, with what we know about the characteristics and the justification of probable Knowledge. In Part II. I pass to its Formal Logic. I am not certain of how much positive value this Part will prove to the reader. My object in it is to show that, starting from the philosophical ideas of Part I., we can deduce by rigorous methods out of simple and precise definitions the usually accepted results, such as the theorems of the addition and multiplication of probabilities and of inverse probability. The reader will readily perceive that this Part would never have been written except under the influence of Mr. Russell's *Principia Mathematica*. But I am sensible that it may suffer from the over-elaboration and artificiality of this method without the justification which its grandeur of scale affords to that great work. In common, however, with other examples of formal method, this attempt has had the negative advantage of compelling the author to make his ideas precise and of discovering fallacies and mistakes. It is a part of the spade-work which a conscientious author has to undertake; though the process of doing it may be of greater value to him than the results can be to the reader, who is concerned to know, as a safeguard of the reliability of the rest of the construction, that the thing can be done, rather than to examine the architectural plans in detail. In the development of my own thought, the following chapters have been of great importance. For it was through trying to prove the fundamental theorems of the subject on the hypothesis that Probability was a *relation* that I first worked my way into the subject; and the rest of this Treatise has arisen out of attempts to solve the successive questions to which the ambition to treat Probability as a branch of Formal Logic first gave rise.

A further occasion of diffidence and apology in introducing this Part of my Treatise arises out of the extent of my debt to Mr. W. E. Johnson. I worked out the first scheme in complete independence of his work and ignorant of the fact that he had thought, more profoundly than I had, along the same lines; I have also given the exposition its final shape with my own hands. But there was an intermediate stage, at which I submitted what I had done for his criticism, and received the benefit not only of criticism but of his own constructive exercises. The result is that in its final form it is difficult to indicate the exact extent of my indebtedness to him. When the following pages were first in proof, there seemed little likelihood of the appearance of any work on Probability from his own pen, and I do not now proceed to publication with so good a conscience, when he is announcing the approaching completion of a work on Logic which will include "Problematic Inference."

I propose to give here a brief summary of the five chapters following, without attempting to be rigorous or precise. I shall then be free to write technically in Chapters XI.-XV., inviting the reader, who is not specially interested in the details of this sort of technique, to pass them by.

2. Probability is concerned with *arguments*, that is to say, with the "bearing" of one set of propositions upon another set. If we are to deal formally with a generalised treatment of this subject, we must be prepared to consider relations of probability between *any* pair of sets of propositions, and not only between sets which are actually the subject of knowledge. But we soon find that some limitation must be put on the character of sets of propositions which we can consider as the hypothetical subject of an argument, namely, that they must be *possible* subjects of knowledge. We cannot, that is to say, conveniently apply our theorems to premisses which are self-contradictory and formally inconsistent with themselves.

For the purpose of this limitation we have to make a distinction between a set of propositions which is merely false in fact and a set which is formally inconsistent with itself.[1] This leads

[1] Spinoza had in mind, I think, the distinction between Truth and Probability in his treatment of Necessity, Contingence, and Possibility. *Res enim omnes ex data Dei natura necessario sequutae sunt, et ex necessitate naturae Dei determinatae sunt ad certo modo existendum et operandum* (*Ethices* i. 33). That is to say, everything is, without qualification, true or false. *At res*

us to the conception of a *group* of propositions, which is defined as a set of propositions such that—(i.) if a logical principle belongs to it, all propositions which are instances of that logical principle also belong to it; (ii.) if the proposition p and the proposition 'not-p or q' both belong to it, then the proposition q also belongs to it; (iii.) if any proposition p belongs to it, then the contradictory of p is *excluded* from it. If the group defined by one part of a set of propositions excludes a proposition which belongs to a group defined by another part of the set, then the set taken as a whole is *inconsistent with itself* and is incapable of forming the premiss of an argument.

The conception of a group leads on to a precise definition of one proposition *requiring* another (which in the realm of assertion corresponds to *relevance* in the realm of probability), and of logical priority as being an order of propositions arising out of their relation to those special groups, or *real groups*, which are in fact the subject of knowledge. Logical priority has no absolute signification, but is relative to a specific body of knowledge, or, as it has been termed in the traditional logic, to the *Universe of Reference.*

It also enables us to reach a definition of *inference* distinct from *implication*, as defined by Mr. Russell. This is a matter of very great importance. Readers who are acquainted with the work of Mr. Russell and his followers will probably have noticed that the contrast between his work and that of the traditional logic is by no means wholly due to the greater precision and more mathematical character of his technique. There is a difference also in the design. His object is to discover what assumptions are required in order that the formal propositions generally accepted by mathematicians and logicians may be obtainable

---

aliqua nulla alia de causa contingens dicitur, nisi respectu defectus nostrae cognitionis (*Ethices* i. 33, scholium). That is to say, Contingence, or, as I term it, Probability, solely arises out of the limitations of our knowledge. Contingence in this wide sense, which includes every proposition which, in relation to our knowledge, is only probable (this term covering all intermediate degrees of probability), may be further divided into Contingence in the strict sense, which corresponds to an *à priori* or formal probability exceeding zero, and Possibility; that is to say, into formal possibility and empirical possibility. *Res singulares voco contingentes, quatenus, dum ad earum solam essentiam attendimus, nihil invenimus, quod earum existentiam necessario ponat, vel quod ipsam necessario secludat. Easdem res singulares voco possibiles, quatenus, dum ad causas, ex quibus produci debent, attendimus, nescimus, an ipsae determinatae sint ad easdem producendum* (*Ethices* iv. Def 3, 4).

as the result of successive steps or substitutions of a few very simple types, and to lay bare by this means any inconsistencies which may exist in received results. But beyond the fact that the conclusions to which he seeks to lead up are those of common sense, and that the uniform type of argument, upon the validity of which each step of his system depends, is of a specially obvious kind, he is not concerned with analysing the methods of valid reasoning which we actually employ. He concludes with familiar results, but he reaches them from premisses, which have never occurred to us before, and by an argument so elaborate that our minds have difficulty in following it. As a method of setting forth the system of formal truth, which shall possess beauty, inter-dependence, and completeness, his is vastly superior to any which has preceded it. But it gives rise to questions about the relation in which ordinary reasoning stands to this ordered system, and, in particular, as to the precise connection between the process of inference, in which the older logicians were principally interested but which he ignores, and the relation of implication on which his scheme depends.

'p implies q' is, according to his definition, exactly equivalent to the disjunction 'q is true or p is false.' If q is true, 'p implies q' holds for all values of p; and similarly if p is false, the implication holds for all values of q. This is not what we mean when we say that q can be inferred or follows from p. For whatever the exact meaning of inference may be, it certainly does not hold between *all* pairs of true propositions, and is not of such a character that *every* proposition follows from a false one. It is not true that 'A male now rules over England' follows or can be inferred from 'A male now rules over France'; or 'A female now rules over England' from 'A female now rules over France'; whereas, on Mr. Russell's definition, the corresponding implications hold simply in virtue of the facts that 'A male now rules over England' is true and 'A female now rules over France' is false.

The distinction between the Relatival Logic of Inference and Probability, and Mr. Russell's Universal Logic of Implication, seems to be that the former is concerned with the relations of propositions in general to a particular limited *group*. Inference and Probability depend for their importance upon the fact that in actual reasoning the limitation of our knowledge presents us with a particular set of propositions, to which we must relate any other proposition about which we seek knowledge. The course of an argument and the results of reasoning depend, not simply on what is true, but on the particular body of knowledge from which we have set out. Ultimately, indeed, Mr. Russell cannot avoid concerning himself with groups. For his aim is to discover the smallest set of propositions which specify our formal knowledge, and then to show that they do in fact specify it. In this enterprise, being human, he must confine himself to that part of formal truth which we know, and the question, how far his axioms comprehend *all* formal truth, must remain insoluble. But his object, nevertheless, is to establish a train of implications between formal truths ; and the character and the justification of rational argument as such is not his subject.

3. Passing on from these preliminary reflections, our first task is to establish the axioms and definitions which are to make operative our symbolical processes. These processes are almost entirely a development of the idea of representing a probability by the symbol a/h, where h is the premiss of an argument and a its conclusion. It might have been a notation more in accordance with our fundamental ideas, to have employed the symbol a/h to designate the *argument* from h to a, and to have represented the probability of the argument, or rather the degree of rational belief about a which the argument authorises, by the symbol $P(a/h)$. This would correspond to the symbol $V(a/h)$ which has been employed in Chapter VI. for the evidential value of the argument as distinct from its probability. But in a section where we are only concerned with probabilities, the use of $P(a/h)$ would have been unnecessarily cumbrous, and it is, therefore, convenient to drop the prefix P and to denote the probability itself by a/h.

The discovery of a convenient symbol, like that of an essential word, has often proved of more than verbal importance. Clear thinking on the subject of Probability is not possible without a symbol which takes an explicit account of the premiss of the argument as well as of its conclusion ; and endless confusion has arisen through discussions about the probability of a conclusion without reference to the argument as a whole. I claim, therefore, the introduction of the symbol a/h as an essential step towards any progress in the subject.

4. Inasmuch as relations of Probability cannot be assumed to possess the properties of numbers, the terms *addition* and *multiplication* of probabilities have to be given appropriate meanings by definition. It is convenient to employ these familiar expressions, rather than to invent new ones, because the properties which arise out of our definitions of addition and multiplication in Probability are analogous to those of addition and multiplication in Arithmetic. But the process of establishing these properties is a little complicated and occupies the greater part of Chapter XII.

The most important of the definitions of Chapter XII. are the following (the numbers referring to the numbers of Chapter XII.):

II. The Definition of *Certainty*: $a/h=1$.

III. The Definition of *Impossibility*: $a/h=0$.

VI. The Definition of *Inconsistency*: ah is inconsistent if $a/h=0$.

VII. The Definition of a *Group*: the class of propositions a such that $a/h=1$ is the group h.

VIII. The Definition of *Equivalence*: if $b/ah=1$ and $a/bh=1$ $(a \equiv b)/h=1$.

IX. The Definition of *Addition*: $ab/h+a\bar{b}/h$ [1] $=a/h$.

X. The Definition of *Multiplication*: $ab/h=a/bh \, . \, b/h= b/ah \, . \, a/h$. The symbolical development of the subject largely proceeds out of these definitions of Addition and Multiplication. It is to be observed that they give a meaning, not to the addition and multiplication of *any* pairs of probabilities, but only to pairs which satisfy a certain form. The definition of Multiplication may be read: 'the probability of both a and b given h is equal to the probability of a given bh, multiplied by the probability of b given h.'

XI. The Definition of *Independence*: if $a_1/a_2h=a_1/h$ and $a_2/a_1h=a_2/h$, a_1/h and a_2/h are independent.

XII. The Definition of *Irrelevance*: if $a_1/a_2h=a_1/h$, a_2 is irrelevant to a_1/h.

5. In Chapter XIII. these definitions, supplemented by a few axioms, are employed to demonstrate the fundamental theorems of *Certain* or *Necessary Inference*. The interest of this chiefly lies in the fact that these theorems include those which the

[1] $\bar{b}$ stands for the contradictory of b.

traditional Logic has termed the *Laws of Thought*, as for example the Law of Contradiction and the Law of Excluded Middle. These are here exhibited as a part of the generalised theory of Inference or Rational Argument, which includes probable Inference as well as certain Inference. The object of this chapter is to show that the ordinarily accepted rules of Inference can in fact be deduced from the definitions and axioms of Chapter XII.

6. In Chapter XIV. I proceed to the fundamental Theorems of Probable Inference, of which the following are the most interesting:

Addition Theorem: $(a+b)/h = a/h + b/h - ab/h$, which reduces to $(a+b)/h = a/h + b/h$, where a and b are mutually exclusive; and, if $p_1 p_2 \ldots p_n$ form, relative to h, a set of exclusive and exhaustive alternatives, $a/h = \sum_1^n p_r a/h$.

Theorem of Irrelevance: If $a/h_1h_2 = a/h_1$, then $a/h_1\bar{h}_2 = a/h_1$; *i.e.* if a proposition is irrelevant, its contradictory also is irrelevant.

Theorem of Independence: If $a_2/a_1h = a_2/h$, $a_1/a_2h = a_1/h$; *i.e.* if a_1 is irrelevant to a_2/h, it follows that a_2 is irrelevant to a_1/h and that a_1/h and a_2/h are independent.

Multiplication Theorem: If a_1/h and a_2/h are independent, $a_1a_2/h = a_1/h \cdot a_2/h$.

Theorem of Inverse Probability: $\frac{a_1/bh}{a_2/bh} = \frac{b/a_1h}{b/a_2h} \cdot \frac{a_1/h}{a_2/h}$. Further, if $a_1/h = p_1$, $a_2/h = p_2$, $b/a_1h = q_1$, $b/a_2h = q_2$, and $a_1/bh + a_2/bh = 1$, then $a_1/bh = \frac{p_1q_1}{p_1q_1 + p_2q_2}$; and if $a_1/h = a_2/h$, $a_1/bh = \frac{q_1}{q_1+q_2}$, which is equivalent to the statement that the probability of a_1 when we know b is equal to $\frac{q_1}{q_1+q_2}$, where q_1 is the probability of b when we know a_1 and q_2 its probability when we know a_2. This theorem enunciated with varying degrees of inaccuracy appears in all Treatises on Probability, but is not generally proved.

Chapter XIV. concludes with some elaborate theorems on the combination of premisses based on a technical symbolic device, known as the *Cumulative Formula*, which is the work of Mr. W. E. Johnson.

7. In Chapter XV. I bring the non-numerical theory of probability developed in the preceding chapters into connection with the usual numerical conception of it, and demonstrate how

and in what class of cases a meaning can be given to a numerical measure of a relation of probability. This leads on to what may be termed numerical approximation, that is to say, the relating of probabilities, which are not themselves numerical, to probabilities, which are numerical, by means of *greater* and *less*, by which in some cases numerical limits may be ascribed to probabilities which are not capable of numerical measures.

CHAPTER XI

THE THEORY OF GROUPS, WITH SPECIAL REFERENCE TO LOGICAL CONSISTENCE, INFERENCE, AND LOGICAL PRIORITY

1. THE Theory of Probability deals with the relation between two sets of propositions, such that, if the first set is known to be true, the second can be known with the appropriate degree of probability by argument from the first.[1] The relation, however, also exists when the first set is not known to be true and is hypothetical.

In a symbolical treatment of the subject it is important that we should be free to consider *hypothetical* premisses, and to take account of relations of probability as existing between *any* pair of sets of propositions, whether or not the premiss is actually part of knowledge. But in acting thus we must be careful to avoid two possible sources of error.

2. The first is that which is liable to arise wherever *variables* are concerned. This was mentioned in passing in § 18 of Chapter IV. We must remember that whenever we substitute for a variable some particular value of it, this may so affect the relevant evidence as to modify the probability. This danger is always present except where, as in the first half of Chapter XIII., the conclusions respecting the variable are *certain*.

3. The second difficulty is of a different character. Our premisses may be hypothetical and not actually the subject of knowledge. But must they not be *possible* subjects of knowledge? How are we to deal with hypothetical premisses which are self-contradictory or formally inconsistent with themselves, and which cannot be the subject of rational belief of any degree?

[1] Or more strictly, "perception of which, together with knowledge of the first set, justifies an appropriate degree of rational belief about the second."

Whether or not a relation of probability can be held to exist between a conclusion and a self-inconsistent premiss, it will be convenient to exclude such relations from our scheme, so as to avoid having to provide for anomalies which can have no interest in an account of the actual processes of valid reasoning. Where a premiss is inconsistent with itself it cannot be required.

4. Let us term the collection of propositions, which are logically involved in the premisses in the sense that they follow from them, or, in other words, stand to them in the relation of certainty,[1] the *group* specified by the premisses. That is to say, we define a group as *containing* all the propositions logically involved in any of the premisses or in any conjunction of them; and as *excluding* all the propositions the contradictories of which are logically involved in any of the premisses or in any conjunction of them.[2] To say, therefore, that a proposition follows from a premiss, is the same thing as to say that it belongs to the group which the premiss specifies.

The idea of a 'group' will then enable us to define 'logical consistency.' If any part of the premisses specifies a group containing a proposition, the contradictory of which is contained in a group specified by some other part, the premisses are *logically inconsistent*; otherwise they are logically consistent. In short, premisses are inconsistent if a proposition 'follows from' one part of them, and its contradictory from another part.

5. We have still, however, to make precise what we mean in this definition by one proposition *following from* or being *logically involved in* the truth of another. We seem to intend by these expressions some kind of transition by means of a *logical principle.* A logical principle cannot be better defined, I think, than in terms of what in Mr. Russell's *Logic of Implication* is termed a *formal implication.* 'p implies q' is a formal implication if 'not-p or q' is formally true; and a proposition is formally true, if it is a value of a propositional function, in which all the constituents other

[1] 'a can be inferred from b,' 'a follows from b,' 'a is certain in relation to b,' 'a is logically involved in b,' I regard as equivalent expressions, the precise meaning of which will be defined in succeeding paragraphs. 'a is implied by b,' I use in a different sense, namely, in Mr. Russell's sense, as the equivalent of 'b or not-a.'

[2] For the conception of a *group*, and for many other notions and definitions in the course of this chapter—those, for example, of a real group and of logical priority—I am largely indebted to Mr. W. E. Johnson. The origination of the theory of groups is due to him.

than the arguments are logical constants, and of which all the values are true.

We might define a *group* in such a way that all logical principles belonged to every group. In this case *all* formally true propositions would belong to every group. This definition is logically precise and would lead to a coherent theory. But it possesses the defect of not closely corresponding to the methods of reasoning we actually employ, because all logical principles are not in fact known to us. And even in the case of those which we do know, there seems to be a logical order (to which on the above definition we cannot give a sense) amongst propositions, which are about logical constants and are formally true, just as there is amongst propositions which are not formally true. Thus, if we were to assume the premisses in every argument to include all formally true propositions, the sphere of probable argument would be limited to what (in contradistinction to formally true propositions) we may term empirical propositions.

6. For this reason, therefore, I prefer a narrower definition—which shall correspond more exactly to what we seem to mean when we say that one proposition follows from another. Let us define a *group* of propositions as a set of propositions such that:

(i.) if the proposition 'p is formally true' belongs to the group, all propositions which are instances of the same formal propositional function also belong to it;

(ii.) if the proposition p and the proposition 'p implies q' both belong to it, then the proposition q also belongs to it;

(iii.) if any proposition p belongs to it, then the contradictory of p is excluded from it.

According to this definition all processes of certain inference are wholly composed of steps each of which is of one of two simple types (and if we like we might perhaps regard the first as comprehending the other). I do not feel certain that these conditions may not be narrower than what we mean when we say that one proposition follows from another. But it is not necessary for the purpose of defining a group, to dogmatise as to whether any other additional methods of inference are, or are not, open to us. If we define a group as the propositions logically involved in the premisses in the above sense, and prescribe that the premisses of an argument in probability must specify a group not *less* extensive than this, we are placing the *minimum* amount of restriction upon the form of our premisses. If, sometimes or as a rule, our premisses in fact include some more powerful principle of argument, so much the better.

In the formal rules of probability which follow, it will be postulated that the set of propositions, which form the premiss of any argument, must not be inconsistent. The premiss must, that is to say, specify a 'group' in the sense that no part of the premiss must exclude a proposition which follows from another part. But for this purpose we do not need to dogmatise as to what the *criterion* is of inference or certainty.

7. It will be convenient at this point to define a term which expresses the relation converse to that which exists between a set of propositions and the group which they specify. The propositions $p_1 p_2 \ldots p_n$ are said to be *fundamental* to the group h if (i.) they themselves belong to the group (which involves their being consistent with one another); (ii.) if between them they completely specify the group; and (iii.) if none of them belong to the group specified by the rest (for if p_r belongs to the group specified by the rest, this term is redundant).

When the fundamental set is *uniquely* determined, a group h' is a sub-group to the group h, if the set fundamental to h' is included in the set fundamental to h.

Logically there can be more than one distinct set of propositions fundamental to a given group; and some extra-logical test must be applied before the fundamental set is determined uniquely. On the other hand, a group is completely determined when the constituent propositions of the fundamental set are given. Further, any consistent set of propositions evidently specifies some group, although such a set may contain propositions *additional* to those which are fundamental to the group it specifies. It is clear also that only one group can be specified by a given set of consistent propositions. The members of a group are, we may say, *rationally bound up* with the set of propositions fundamental to it.

8. If Mr. Bertrand Russell is right, the whole of pure mathematics and of formal logic follows, in the sense defined above, from a small number of primitive propositions. The group, therefore, which is specified by these primitive propositions, includes the most remote deductions not only amongst those known to mathematicians, but amongst those which time

and skill have not yet served to solve. If we define certainty in a logical and not a psychological sense, it seems necessary, if our premisses include the essential axioms, to regard as certain all propositions which follow from these, whether or not they are known to us. Yet it seems as if there must be some logical sense in which unproved mathematical theorems—some of those, for instance, which deal with the theory of numbers—can be likely or unlikely, and in which a proposition of this kind, which has been suggested to us by analogy or supported by induction, can possess an intermediate degree of probability.

There can be no doubt, I think, that the logical relation of certainty does exist in these cases in which lack of skill or insight prevents our apprehending it, in spite of the fact that sufficient premisses, including sufficient logical principles, are known to us. In these cases we must say, what we are not permitted to say when the indeterminacy arises from lack of premisses, that the probability is *unknown*. There is still a sense, however, in which in such a case the knowledge we actually possess can be, in a logical sense, only probable. While the relation of certainty exists between the fundamental axioms and every mathematical hypothesis (or its contradictory), there are other data in relation to which these hypotheses possess intermediate degrees of probability. If we are unable through lack of skill to discover the relation of probability which an hypothesis does in fact bear towards one set of data, this set is practically useless, and we must fix our attention on some other set in relation to which the probability is not unknown. When Newton held that the binomial theorem possessed for empirical reasons sufficient probability to warrant a further investigation of it, it was not in relation to the axioms of mathematics, whether he knew them or not, that the probability existed, but in relation to his empirical evidence combined, perhaps, with *some* of the axioms. There is, in short, an exception to the rule that we must always consider the probability of any conclusion in relation to the whole of the data in our possession. When the relation of the conclusion to the whole of our evidence cannot be known, then we must be guided by its relation to some part of the evidence. When, therefore, in later chapters I speak of a formal proposition as possessing an intermediate degree of probability, this will always be in relation to evidence from which the proposition does not logically follow in the sense defined in § 6.

9. It follows from the preceding definitions that a proposition is *certain* in relation to a given premiss, or, in other words, *follows from* this premiss if it is included in the group which that premiss specifies. It is *impossible* if it is excluded from the group—if, that is to say, its contradictory follows from the premiss. We often say, somewhat loosely, that two propositions are contradictory to one another, when they are inconsistent in the sense that, relative to our evidence, they cannot belong to the same group. On the other hand, a proposition, which is not itself included in the group specified by the premiss and whose contradictory is not included either, has in relation to the premiss an intermediate degree of probability.

If a follows from h and is, therefore, included in the group specified by h, this is denoted by $a/h=1$. The relation of certainty, that is to say, is denoted by the symbol of unity. The reason why this notation is useful and has been adopted by common consent will appear when the meaning of the *product* of a pair of relations of probability has been explained. If we represent the relation of certainty by γ and any other probability by a, the product $a \cdot \gamma = a$. Similarly, if a is excluded from the group specified by h and is impossible in relation to it, this is denoted by $a/h=0$. The use of the symbol zero to denote impossibility arises out of the fact that, if ω denotes impossibility and a any other relation of probability, then, in the senses of multiplication and addition to be defined later, the product $a \cdot \omega = \omega$, and the sum $a + \omega = a$. Lastly, if a is not included in the group specified by h, this is written $a/h \neq 1$ or $a/h < 1$; and if it is not excluded, this is written $a/h \neq 0$ or $a/h > 0$.

10. The theory of groups now enables us to give an account, with the aid of some further conceptions, of logical priority and of the true nature of inference. The groups, to which we refer the arguments by which we actually reason, are not arbitrarily chosen. They are determined by those propositions of which we have direct knowledge. Our group of reference is specified by those direct judgments in which we personally *rationally certify* the truth of some propositions and the falsity of others. So long as it is undetermined, or not determined uniquely, which propositions are fundamental, it is not possible to discover

a necessary order amongst propositions or to show in what way a true proposition 'follows from' one true premiss rather than another. But when we have determined what propositions are fundamental, by selecting those which we know directly to be true, or in some other way, then a meaning can be attached to priority and to the distinction between inference and implication. When the propositions which we know directly are given, there is a logical order amongst those other propositions which we know indirectly and by argument.

11. It will be useful to distinguish between those groups which are hypothetical and those of which the fundamental set is known to be true. We will term the former *hypothetical groups*, and the latter *real groups*. To the real group, which contains *all* the propositions which are known to be true, we may assign the old logical term *Universe of Reference.* While knowledge is here taken as the criterion of a real group, what follows will be equally valid whatever criterion is taken, so long as the fundamental set is in some manner or other determined uniquely.

If it is impossible for us to know a proposition p except by inference from a knowledge of q, so that we cannot know p to be true unless we already know q, this may be expressed by saying that 'p requires q.' More precisely *requirement* is defined as follows:

p does not require q if there is some real group to which p belongs and q does not belong, *i.e.* if there is a real group h such that $p/h=1$, $q/h \neq 1$; hence

p requires q if there is *no* real group to which p belongs and q does not belong.

p does not require q within the group h, if the group h, to which p belongs, contains a subgroup[1] h' to which p belongs and q does not belong; *i.e.* if there is a group h' such that $h'/h=1$, $p/h'=1$, $q/h' \neq 1$. This reduces to the proposition next but one above if h is the Universe of Reference. In § 13 these definitions will be generalised to cover intermediate degrees of probability.

12. Inference and logical priority can be defined in terms of requirement and real groups. It is convenient to distinguish two types of inference corresponding to hypothetical and real

[1] Subgroups have only been defined, it must be noticed (see § 7 above) when the fundamental set of the group has been, in some way, uniquely determined.

groups—*i.e.* to cases where the argument is only hypothetical, and cases where the conclusion can be asserted :

Hypothetical Inference.—'If p, q,' which may also be read 'q is hypothetically inferrible from p,' means that there is a real group h such that $q/ph=1$, and $q/h \neq 1$. In order that this may be the case, ph must specify a group; *i.e.* $p/h \neq 0$, or in other words p must not be excluded from h. Hypothetical inference is also equivalent to: 'p implies q,' and 'p implies q' does not require 'q.' In other words, q is hypothetically inferrible from p, if we know that q is true or p is false and if we can know this without first knowing either that q is true or that p is false.

Assertoric Inference.—'$p \therefore q$,' which may be read 'p therefore q' or 'q may be asserted by inference from p,' means that 'if p, q' is true, and in addition 'p' belongs to a real group; *i.e.* there are proper groups h and h' such that $p/h=1$, $q/ph'=1$, $q/h' \neq 1$, and $p/h' \neq 0$.

p is prior to q when p does not require q, and q requires p, when, that is to say, we can know p without knowing q, but not q unless we first know p.

p is prior to q within the group h when p does not require q within the group, and q does require p within the group.

It follows from this and from the preceding definitions that, if a proposition is fundamental in the sense that we can only know it directly, there is no proposition prior to it; and, more generally, that, if a proposition is fundamental to a given group, there is no proposition prior to it within the group.

13. We can now apply the conception of requirement to intermediate degrees of probability. The notation adopted is, it will be remembered, as follows:

$p/h=a$ means that the proposition p has the probable relation of degree a to the proposition h; while it is postulated that h is self-consistent and therefore specifies a group.

$p/h=1$ means that p follows from h and is, therefore, included in the group specified by h.

$p/h=0$ means that p is excluded from the group specified by h.

If h specifies the Universe of Reference, *i.e.* if its group comprehends the whole of our knowledge, p/h is called *the absolute probability of p*, or (for short) *the probability of p*; and if $p/h=1$ and h specifies any real group, p is said to be *absolutely certain* or (for short) *certain*. Thus p is 'certain' if it is a member of a real group, and a 'certain' proposition is one which we know to be true. Similarly if $p/h=0$ under the same conditions, p is *absolutely impossible*, or (for short) *impossible*. Thus an 'impossible' proposition is one which we know to be false.

The definition of requirement, when it is generalised so as to take account of intermediate degrees of probability, becomes, it will be seen, equivalent to that of relevance:

The probability of p does not require q within the group h, if there is a subgroup h' such that, for every subgroup h'' which includes h' and is included in h (*i.e.* $h'/h''=1, h''/h=1$), $p/h''=p/h'$, and $q/h' \neq q/h$.

When p is included in the group h, this definition reduces to the definition of requirement given in § **11**.

14. The importance of the theory of groups arises as soon as we admit that there are *some* propositions which we take for granted without argument, and that all arguments, whether demonstrative or probable, consist in the relating of other conclusions to these as premisses.

The particular propositions, which are in fact fundamental to the Universe of Reference, vary from time to time and from person to person. Our theory must also be applicable to hypothetical Universes. Although a particular Universe of Reference may be defined by considerations which are partly psychological, when once the Universe is given, our theory of the relation in which other propositions stand towards it is entirely logical.

The formal development of the theory of argument from imposed and limited premisses, which is attempted in the following chapters, resembles in its general method other parts of formal logic. We seek to establish implications between our primitive axioms and the derivative propositions, without specific reference to what particular propositions are fundamental in our actual Universe of Reference.

It will be seen more clearly in the following chapters that the laws of inference are the laws of probability, and that the former is a particular case of the latter. The relation of a proposition to a group depends upon the relevance to it of the group, and a group is relevant in so far as it contains a necessary or sufficient condition of the proposition, or a necessary or sufficient condition of a necessary or sufficient condition, and so on; a condition being necessary if every hypothetical group, which includes the proposition together with the Universe of Reference, includes the condition, and sufficient if every hypothetical group, which includes the condition together with the Universe of Reference, includes the proposition.

CHAPTER XII

THE DEFINITIONS AND AXIOMS OF INFERENCE AND PROBABILITY

1. It is not necessary for the validity of what follows to decide in what manner the set of propositions is determined, which is fundamental to our Universe of Reference, or to make definite assumptions as to what propositions are included in the group which is specified by the *data.* When we are investigating an empirical problem, it will be natural to include the whole of our logical apparatus, the whole body, that is to say, of formal truths which are known to us, together with that part of our empirical knowledge which is relevant. But in the following formal developments, which are designed to display the logical rules of probability, we need only assume that our data always include those logical rules, of which the steps of our proofs are instances, together with the axioms relating to probability which we shall enunciate.

The object of this and the chapters immediately following is to show that all the usually assumed conclusions in the fundamental logic of inference and probability follow rigorously from a few axioms, in accordance with the fundamental conceptions expounded in Part I. This body of axioms and theorems corresponds, I think, to what logicians have termed the *Laws of Thought*, when they have meant by this something narrower than the whole system of formal truth. But it goes beyond what has been usual, in dealing at the same time with the laws of probable, as well as of necessary, inference.

2. This and the following chapters of Part II. are largely independent of many of the more controversial issues raised in the preceding chapters. They do not prejudge the question as

to whether or not all probabilities are theoretically measurable; and they are not dependent on our theories as to the part played by direct judgment in establishing relations of probability or inference between particular propositions. Their premisses are all *hypothetical*. *Given* the existence of certain relations of probability, others are inferred. Of the conclusions of Chapter III., of the criteria of equiprobability and of inequality discussed in Chapters IV. and V., and of the criteria of inference discussed in §§ 5, 6 of Chapter XI., they are, I think, wholly independent. They deal with a different part of the subject, not so closely connected with epistemology.

3. In this chapter I confine myself to Definitions and Axioms.

Propositions will be denoted by small letters, and relations by capital letters. In accordance with common usage, a disjunctive combination of propositions is represented by the sign of addition, and a conjunctive combination by simple juxtaposition (or, where it is necessary for clearness, by the sign of multiplication): *e.g.* 'a or b or c' is written '$a+b+c$,' and 'a and b and c' is written 'abc.' '$a+b$' is not so interpreted as to exclude 'a and b.' The contradictory of a is written $\bar{a}$.

4. *Preliminary Definitions:*

I. If there exists a relation of probability P between the proposition a and the premiss h

$a/h = \mathrm{P}$ Def.

II. If P is the relation of certainty [1]

$\mathrm{P} = 1$ Def.

III. If P is the relation of impossibility [1]

$\mathrm{P} = 0$ Def.

IV. If P is a relation of probability, but not the relation of certainty $\mathrm{P} < 1$. Def.

V. If P is a relation of probability, but not the relation of impossibility $\mathrm{P} > 0$. Def.

VI. If $a/h = 0$, the conjunction ah is *inconsistent*. Def.

VII. The class of propositions a such that $a/h = 1$ is *the group specified by h* or (for short) *the group h*. Def.

VIII. If $b/ah = 1$ and $a/bh = 1$, $(a \equiv b)/h = 1$. Def.

This may be regarded as the definition of *Equivalence*. Thus we see that equivalence is relative to a premiss h. a is equivalent to b, given h, if b follows from ah, and a from bh.

[1] These symbols were first employed by Leibnitz. See p. 155 below.

5. *Preliminary Axioms :*

We shall assume that there is included in every premiss with which we are concerned the formal implications which allow us to assert the following axioms :

(i.) Provided that a and h are propositions or conjunctions of propositions or disjunctions of propositions, and that h is not an inconsistent conjunction, there exists one and only one relation of probability P between a as conclusion and h as premiss. Thus any conclusion a bears to any consistent premiss h one and only one relation of probability.

(ii.) If $(a \equiv b)/h = 1$, and x is a proposition, $x/ah = x/bh$. This is the Axiom of Equivalence.

(iii.)
$$(\overline{a+b} \equiv \bar{a}\bar{b})/h = 1$$
$$(aa \equiv a)/h = 1$$
$$(\bar{\bar{a}} \equiv a)/h = 1$$
$$(ab + \bar{a}b \equiv b)/h = 1.$$

If $a/h = 1$, $ah \equiv h$. That is to say, if a is included in the group specified by h, h and ah are equivalent.

6. *Addition and Multiplication.*—If we were to assume that probabilities are numbers or ratios, these operations could be given their usual arithmetical signification. In adding or multiplying probabilities we should be simply adding or multiplying numbers. But in the absence of such an assumption, it is necessary to give a meaning by definition to these processes. I shall define the addition and multiplication of relations of probabilities only for certain types of such relations. But it will be shown later that the limitation thus placed on our operations is not of practical importance.

We define the *sum* of the probable relations ab/h and $a\bar{b}/h$ as being the probable relation a/h ; and the *product* of the probable relations a/bh and b/h as being the probable relation ab/h. That is to say :

IX. $$ab/h + a\bar{b}/h = a/h. \qquad \text{Def.}$$

X. $$ab/h = a/bh \, . \, b/h = b/ah \, . \, a/h. \qquad \text{Def.}$$

Before we proceed to the axioms which will make these symbols operative, the definitions may be restated in more familiar language. IX. may be read : "The sum of the probabilities of 'both a and b' and of 'a but not b,' relative to the same hypothesis, is equal to the probability of 'a' relative to this hypothesis." X. may be read: "The probability of 'both a and b,' assuming h, is equal to the product of the probability of b, assuming h, and the probability of a, assuming both b and h." Or in the current terminology[1] we should have: "The probability that both of two events will occur is equal to the probability of the first multiplied by the probability of the second, assuming the occurrence of the first." It is, in fact, the ordinary rule for the multiplication of the probabilities of events which are not 'independent.' It has, however, a much more central position in the development of the theory than has been usually recognised.

Subtraction and division are, of course, defined as the inverse operations of addition and multiplication:

XI. If $PQ = R$, $P = \frac{R}{Q}$. Def.

XII. If $P + Q = R$, $P = R - Q$. Def.

Thus we have to introduce as definitions what would be axioms if the meaning of addition and multiplication were already defined. In this latter case we should have been able to apply the ordinary processes of addition and multiplication without any further axioms. As it is, we need axioms in order to make these symbols, to which we have given our own meaning, operative. When certain properties are associated, it is often more or less arbitrary which we take as defining properties and which we associate with these by means of axioms. In this case I have found it more convenient, for the purposes of formal development, to reverse the arrangement which would come most natural to commonsense, full of preconceptions as to the meaning of addition and multiplication. I define these processes, for the theory of probability, by reference to a comparatively unfamiliar property, and associate the more familiar properties with this one by means of axioms. These axioms are as follows:

(iv.) If P, Q, R are relations of probability such that the products PQ, PR and the sums $P + Q$, $P + R$ exist, then:

(iv. a) If PQ exists, QP exists, and $PQ = QP$. If $P + Q$ exists, $Q + P$ exists and $P + Q = Q + P$.

(iv. b) $PQ < P$ unless $Q = 1$ or $P = 0$; $P + Q > P$ unless $Q = 0$.
$PQ = P$ if $Q = 1$ or $P = 0$; $P + Q = P$ if $Q = 0$.

(iv. c) If $PQ \gtreqless PR$, then $Q \gtreqless R$ unless $P = 0$. If $P + Q \gtreqless P + R$, then $Q \gtreqless R$ and conversely.

[1] *E.g.* Bertrand, *Calcul des probabilités*, p. 26.

A meaning has not been given, it is important to notice, to the signs of addition and multiplication between probabilities *in all cases*. According to the definitions we have given, $P+Q$ and PQ have not an interpretation whenever P and Q are relations of probability, but in certain conditions only. Furthermore, if $P+Q=R$ and $Q=S+T$, it does not follow that $P+S+T=R$, since no meaning has been assigned to such an expression as $P+S+T$. The equation must be written $P+(S+T)=R$, and we cannot infer from the foregoing axioms that $(P+S)+T=R$. The following axioms allow us to make this and other inferences in cases in which the sum $P+S$ exists, *i.e.* when $P+S=A$ and A is a relation of probability.

(v.) $[\pm P\pm Q]+[\pm R\pm S]=[\pm P\pm R]-[\mp Q\mp S]=[\pm P\pm R]+[\pm Q\pm S]=[\pm P\pm Q]-[\mp R\mp S]$

in every case in which the probabilities $[\pm P\pm Q]$, $[\pm R\pm S]$, $[\pm P\pm R]$, etc., exist, *i.e.*, in which these sums satisfy the conditions necessary in order that a meaning may be given to them in the terms of our definition.

(vi.) $P(R\pm S)=PR\pm PS$, if the sum $R\pm S$ and the products PR and PS exist as probabilities.

7. From these axioms it is possible to derive a number of propositions respecting the addition and multiplication of probabilities. They enable us to prove, for instance, that if $P+Q=R+S$ then $P-R=S-Q$, provided that the differences $P-R$ and $S-Q$ exist; and that $(P+Q)(R+S)=(P+Q)R+(P+Q)S=[PR+QR]+[PS+QS]=[PR+QS]+[QR+PS]$, provided that the sums and products in question exist. In general any rearrangement which would be legitimate in an equation between arithmetic quantities is also legitimate in an equation between probabilities, provided that our initial equation and the equation which finally results from our symbolic operations can both be expressed in a form which contains only products and sums which have an interpretation as probabilities in accordance with the definitions. If, therefore, this condition is observed, we need not complicate our operations by the insertion of brackets at every stage, and no result can be obtained as a result of leaving them out, if it is of the form prescribed above, which could not be obtained if they had been rigorously inserted throughout. We can only be interested in our final results when they deal with actually existent and intelligible probabilities—for our object is,

always, to compare one probability with another—and we are not incommoded, therefore, in our symbolic operations by the circumstance that sums and products do not exist between every pair of probabilities.

8. *Independence :*

XIII. If $a_1/a_2h = a_1/h$ and $a_2/a_1h = a_2/h$, the probabilities a_1/h and a_2/h are *independent*. Def.

Thus the probabilities of two arguments having the same premisses are independent, if the addition to the premisses of the conclusion of either leaves them unaffected.

Irrelevance : [1]

XIV. If $a_1/a_2h = a_1/h$, a_2 is *irrelevant on the whole*, or, for short, *irrelevant* to a_1/h. Def.

[1] This is repeated for convenience of reference from Chapter IV. § 14. It is only necessary here to take account of *irrelevance on the whole*, not of the more precise sense.

CHAPTER XIII

THE FUNDAMENTAL THEOREMS OF NECESSARY INFERENCE

1. In this chapter we shall be mainly concerned with deducing the existence of relations of certainty or impossibility, given other relations of certainty or impossibility,—with the rules, that is to say, of *Certain* or, as De Morgan termed it, of *Necessary* Inference. But it will be convenient to include here a few theorems dealing with intermediate degrees of probability. Except in one or two important cases I shall not trouble to translate these theorems from the symbolism in which they are expressed, since their interpretation presents no difficulty.

2. (1) $a/h + \bar{a}/h = 1$.

For $\quad ab/h + \bar{a}b/h = b/h \quad$ by IX.,

$\quad a/bh \,.\, b/h + \bar{a}/bh \,.\, b/h = b/h \quad$ by X.

Put $\quad b/h = 1$, then $a/bh + \bar{a}/bh = 1 \quad$ by (iv. b),

since $\quad b/h = 1,\ bh \equiv h \quad$ by (iii.).

Thus $\quad a/h + \bar{a}/h = 1 \quad$ by (ii.).

(1.1) If $a/h = 1$, $\bar{a}/h = 0$,

$\quad a/h + \bar{a}/h = 1 \quad$ by (1),

$\quad \therefore\ a/h + \bar{a}/h = a/h = a/h + 0 \quad$ by (iv. b),

$\quad \therefore\ \bar{a}/h = 0 \quad$ by (iv. c).

(1.2) Similarly, if $\bar{a}/h = 1$, $a/h = 0$.

(1.3) If $a/h = 0$, $\bar{a}/h = 1$,

$\quad a/h + \bar{a}/h = 1 \quad$ by (1),

$\quad \therefore\ 0 + \bar{a}/h = 0 + 1 \quad$ by (iv. b),

$\quad \therefore\ \bar{a}/h = 1 \quad$ by (iv. c).

(1.4) Similarly, if $\bar{a}/h = 0$, $a/h = 1$.

(2) $\quad a/h < 1$ or $a/h = 1 \quad$ by IV.

(3) $\quad a/h > 0$ or $a/h = 0 \quad$ by V.,

i.e. there are no negative probabilities.

(4) $ab/h < b/h$ or $ab/h = b/h$ by X. and (iv. *b*).

(5) If P and Q are relations of probability and $P+Q=0$, then $P=0$ and $Q=0$.

$$P+Q>P \text{ unless } Q=0 \quad \text{by (iv. } b\text{)},$$

and

$$P>0 \text{ unless } P=0 \quad \text{by V.}$$

$$\therefore P+Q>0 \text{ unless } Q=0.$$

Hence, if $P+Q=0$, $Q=0$ and similarly $P=0$.

(6) If $PQ=0$, $P=0$ or $Q=0$,

$$Q>0 \text{ unless } Q=0 \quad \text{by V.}$$

Hence $PQ>P\,.\,0$ unless $Q=0$ or $P=0$ by (iv. *c*),

i.e. $PQ>0$ unless $Q=0$ or $P=0$ by (iv. *b*).

Whence, if $PQ=0$, the result follows.

(7) If $PQ=1$, $P=1$ and $Q=1$,

$$PQ<P \text{ unless } P=0 \text{ or } Q=1 \quad \text{by (iv. } b\text{)},$$

$$PQ=P \text{ if } P=0 \text{ or } Q=1 \quad \text{by (iv. } b\text{)},$$

and

$$P<1 \text{ unless } P=1 \quad \text{by IV.},$$

$$\therefore PQ<1 \text{ unless } P=1.$$

Hence $P=1$; similarly $Q=1$.

(8) If $a/h=0$, $ab/h=0$ and $a/bh=0$ if bh is not inconsistent.

For $ab/h=b/ah\,.\,a/h=a/bh\,.\,b/h$ by X.,

and since $a/h=0$, $b/ah\,.\,a/h=0$ by (iv. *b*),

$$\therefore ab/h=0 \text{ and } a/bh\,.\,b/h=0,$$

$$\therefore \text{unless } b/h=0,\ a/bh=0 \quad \text{by (5)},$$

whence the result by VI.

Thus, if a conclusion is impossible, we may add to the conclusion or add consistently to the premisses without affecting the argument.

(9) If $a/h=1$, $a/bh=1$ if bh is not inconsistent.

Since $a/h=1$, $\bar{a}/h=0$ by (1.1),

$\therefore \bar{a}/bh=0$ by (8) if bh is not inconsistent,

whence $a/bh=1$ by (1.4).

Thus we may add to premisses, which make a conclusion certain, any other premisses not inconsistent with them, without affecting the result.

(10) If $a/h=1$, $ab/h=b/ah=b/h$,

$$ab/h=b/ah\,.\,a/h=a/bh\,.\,b/h \quad \text{by X.}$$

Since $a/h=1$, $a/bh=1$ by (9) unless $b/h=0$,

$$\therefore b/ah\,.\,a/h=b/ah \text{ and } a/bh\,.\,b/h=b/h \quad \text{by (iv. } b\text{)},$$

whence the result, unless $b/h=0$.

If $b/h=0$, the result follows from (8).

(11) If $ab/h=1$, $a/h=1$.

For $ab/h=b/ah \,.\, a/h$ by X.,

$\therefore\ a/h=1$ by (7).

(12) If $(a\equiv b)/h=1$, $a/h=b/h$,

$b/ah \,.\, a/h=a/bh \,.\, b/h$ by X.

and $b/ah=1$, $a/bh=1$ by VIII.,

$\therefore\ a/h=b/h$ by (iv. b).

(12.1) If $(a\equiv b)/h=1$ and hx is not inconsistent,

$$a/hx=b/hx.$$

$a/hx \,.\, x/h=x/ah \,.\, a/h$,

and $b/hx \,.\, x/h=x/bh \,.\, b/h$ by X.,

$x/ah=x/bh$ by (ii.),

and $a/h=b/h$ by (12),

$\therefore\ a/hx=b/hx$ unless $x/h=0$.

This is the *principle of equivalence*. In virtue of it and of axiom (ii.), if $(a\equiv b)/h=1$, we can substitute a for b and *vice versa*, wherever they occur in a probability whose premisses include h.

(13) $a/a=1$, unless a is inconsistent.

For $a/a=aa/a=a/aa \,.\, a/a$ by (iii.), (12), and X.,

whence $a/aa=1$ by (ii.), unless $a/a=0$,

i.e. $a/a=1$, unless a is inconsistent by (iii.), (12), and VI.

(13.1) $\bar{a}/a=0$, unless a is inconsistent. This follows from (13) and (1.1).

(13.2) $a/\bar{a}=0$, unless $\bar{a}$ is inconsistent. This follows from (iii.) by writing $\bar{a}$ for a in (13.1).

(14) If $a/b=0$ and a is not inconsistent, $b/a=0$.

Let f be the group of assumptions, common to a and b, which we have supposed to be included in every real group;

then $a/b=a/bf$ and $b/a=b/af$ by (iii.) and (12),

and $a/bf \,.\, b/f=b/af \,.\, a/f$ by X.

Since $a/bf=0$ by hypothesis,

and $a/f\neq 0$, since a is not inconsistent,

$\therefore\ b/af=0$,

whence $b/a=0$.

Thus, if a is impossible given b, then b is impossible given a.

(15) If $h_1/h_2=0$, $h_1h_2/h=0$,

$h_1h_2/h=h_1/h_2h \,.\, h_2/h$ by II.,

and since $h_1/h_2=0$, $h_1/h_2h=0$ by (8), unless $h/h_2=0$, whence the result by (iv. b), unless $h/h_2=0$.

If $$h/h_2=0,\ h_2/h=0 \qquad \text{by (14)},$$
since we assume that h is not inconsistent, and hence
$$h_1h_2/h=0 \qquad \text{by (8)}.$$
Thus, if h_1 is impossible given h_2, h_1h_2 is always impossible and is excluded from every group.

(15.1) If $h_1h_2/h=0$ and h_2h is not inconsistent, $h_1/h_2h=0$. This, which is the converse of (15), follows from X. and (6).

(16) If $h_1/h_2=1$, $(h_1+\bar{h}_2)/h=1$,
$$\bar{h}_1/h_2=0 \qquad \text{by (1)},$$
$$\therefore\ \bar{h}_1h_2/h=0 \qquad \text{by (15)},$$
$$\therefore\ \overline{\bar{h}_1h_2}/h=1 \qquad \text{by (1.3)},$$
$$\therefore\ (h_1+\bar{h}_2)/h=1 \qquad \text{by (12) and (iii.)}.$$

(16.1) We may write (16):

If $h_1/h_2=1$, $(h_2 \supset h_1)/h=1$, where '$\supset$' symbolises 'implies.' Thus if h_1 follows from h_2, then it is always certain that h_2 implies h_1.

(16.2) If $(h_1+\bar{h}_2)/h=1$ and h_2h is not inconsistent, $h_1/h_2h=1$.
$$\bar{h}_1h_2/h=0, \text{ as in (16)},$$
$$\therefore\ \bar{h}_1/h_2h=0 \text{ by (15.1), since } h_2h \text{ is not inconsistent},$$
$$\therefore\ h_1/h_2h=1 \qquad \text{by (1.4)}.$$
This is the converse of (14).

(16.3) We may write (16.2):

If $(h_2 \supset h_1)/h=1$ and h_2h is not inconsistent, $h_1/h_2h=1$. Thus, if we define a 'group' as a set of propositions, which follow from and are certain relatively to the proposition which specifies them, this proposition proves that, if $h_2 \supset h_1$ and h_2 belong to a group h_2h, then h_1 also belongs to this group.

(17) If $(h_1 \supset: a\equiv b)/h=1$ and h_1h is not inconsistent, $a/h_1h=b/h_1h$. This follows from (16.3) and (12).

(18) $a/a=1$ or $\bar{a}/\bar{a}=1$.
$$a/a=1, \text{ unless } a \text{ is inconsistent}, \qquad \text{by (13)}.$$
If a is inconsistent, $a/h=0$, where h is not inconsistent, and therefore
$$\bar{a}/h=1 \qquad \text{by (1.3)}.$$
Thus unless a is inconsistent, $\bar{a}$ is not inconsistent, and therefore
$$\bar{a}/\bar{a}=1 \qquad \text{by (13)}.$$

(19) $a\bar{a}/h=0$,
$$\bar{a}/\bar{a}=1 \text{ or } a/a=1 \qquad \text{by (18)};$$
$$\therefore\ a/\bar{a}=0 \text{ or } \bar{a}/a=0 \qquad \text{by (1.1) and (1.2)}.$$
In either case $a\bar{a}/h=0$ by (15).

Thus it is impossible that both a and its contradictory should be true. This is the Law of Contradiction.

(20) $(a+\bar{a})/h=1$.

Since $(a\bar{a}\equiv\overline{a+\bar{a}})/h=1$ by (iii.),

$\overline{a+\bar{a}}/h=0$ by (19) and (12),

$\therefore\ (a+\bar{a})/h=1$ by (1.3).

Thus it is certain that either a or its contradictory is true. This is the Law of Excluded Middle.

(21) If $a/h_1=1$ and $a/h_2=0$, $h_1h_2/h=0$.

For $a/h_1h_2 \,.\, h_1/h_2=h_1/ah_2 \,.\, a/h_2$,

and $\bar{a}/h_1h_2 \,.\, h_2/h_1=h_2/\bar{a}h_1 \,.\, \bar{a}/h_1$ by X.,

$\therefore\ a/h_1h_2 \,.\, h_1/h_2=0$ and $\bar{a}/h_1h_2 \,.\, h_2/h_1=0$,

since, by hypothesis and (1), $\bar{a}/h_1=0$ and $a/h_2=0$,

$\therefore\ a/h_1h_2=0$ or $h_1/h_2=0$,

and $a/h_1h_2=1$ or $h_2/h_1=0$,

$\therefore\ h_1/h_2=0$ or $h_2/h_1=0$.

In either case $h_1h_2/h=0$ by (15).

Thus, if a proposition is certain relatively to one set of premisses, and impossible relatively to another set, the two sets are incompatible.

(22) If $a/h_1=0$ and $h_1/h=1$, $a/h=0$,

$ah_1/h=0$ by (15), $\therefore\ h_1/ah \,.\, a/h=0$,

$h_1/ah=1$ by (9), unless $a/h=0$.

$\therefore$ in any case $a/h=0$.

(23) If $b/a=0$ and $b/\bar{a}=0$, $b/h=0$.

$ab/h=0$ and $\bar{a}b/h=0$ by (15),

$\therefore\ a/bh=0$ or $b/h=0$,

and $\bar{a}/bh=0$ or $b/h=0$ by II. and (iv.),

whence $b/h=0$ by (1.4).

CHAPTER XIV

THE FUNDAMENTAL THEOREMS OF PROBABLE INFERENCE

1. I SHALL give proofs in this chapter of most of the fundamental theorems of Probability, with very little comment. The bearing of some of them will be discussed more fully in Chapter XVI.

2. *The Addition Theorems:*

(24) $(a+b)/h = a/h + b/h - ab/h$.

In IX. write $(a+b)$ for a, and $\bar{a}b$ for b.

Then $\quad (a+b)\bar{a}b/h + (a+b)\overline{\bar{a}b}/h = (a+b)/h$,

whence $\quad \bar{a}b/h + (a+b)(a+\bar{b})/h = (a+b)/h \qquad$ by (iii.),

$\bar{a}/bh \,.\, b/h + a/h = (a+b)/h \qquad$ by (iii.) and IX.

That is to say, $(a+b)/h = a/h + (1 - a/bh) \,.\, b/h$,

$= a/h + b/h - ab/h$.

In accordance with the principles of Chapter XII. § 6, this should be written, strictly, in the form $a/h + (b/h - ab/h)$, or in the form $b/h + (a/h - ab/h)$. The argument is valid, since the probability $(b/h - ab/h)$ is equal to $\bar{a}b/h$, as appears from the preceding proof, and, therefore, exists. This important theorem gives the probability of 'a or b' relative to a given hypothesis in terms of the probabilities of 'a,' 'b,' and 'a and b' relative to the same hypothesis.

(24.1) If $ab/h = 0$, *i.e.* if a and b are exclusive alternatives relative to the hypothesis, then

$$(a+b)/h = a/h + b/h.$$

This is the ordinary rule for the addition of the probabilities of exclusive alternatives.

(24.2) $ab/h + \bar{a}b/h = b/h$,

since $\quad ab + \bar{a}b \equiv b \qquad$ by (iii.),

and $\quad a\bar{a}b/h = 0 \qquad$ by (19) and (8).

(24.3) $(a+b)/h = a/h + b\bar{a}/h$. This follows from (24) and (24.2).

(24.4) $(a+b+c)/h=(a+b)/h+c/h-(ac+bc)/h$
$=a/h+b/h+c/h-ab/h-bc/h-ca/h+abc/h.$

(24.5) And in general
$(p_1+p_2+\ldots+p_n)/h=\Sigma p_r/h-\Sigma p_s p_t/h+\Sigma p_r p_s p_t/h\ldots$
$+(-1)^{n-1}p_1p_2\ldots p_n/h.$

(24.6) If $p_s p_t/h=0$ for all pairs of values of s and t, it follows by repeated application of X. that

$$(p_1+p_2+\ldots+p_n)/h=\sum_1^n p_r/h.$$

(24.7) If $p_s p_t/h=0$, etc., and $(p_1+p_2+\ldots+p_n)/h=1$, *i.e.* if $p_1p_2\ldots p_n$ form, relatively to h, a set of exclusive and exhaustive alternatives, then

$$\sum_1^n p_r/h=1.$$

(25) If $p_1p_2\ldots p_n$ form, relative to h, a set of exclusive and exhaustive alternatives,

$$a/h=\sum_1^n p_r a/h.$$

Since $(p_1+p_2+\ldots+p_n)/h=1$ by hypothesis,
$\therefore\ (p_1+p_2+\ldots+p_n)/ah=1$ by (9) if ah is not inconsistent;
and since $p_s p_t/h=0$ by hypothesis,
$\therefore\ p_s p_t/ah=0$ by (9), if ah is not inconsistent.

Hence $\sum_1^n p_r/ah=(p_1+p_2+\ldots+p_n)/ah$ by (24.6)
$=1.$

Also $p_r a/h=p_r/ah\ .\ a/h.$

Summing $\sum_1^n p_r a/h=a/h\ .\ \sum_1^n p_r/ah,$

$$\therefore\ a/h=\sum_1^n p_r a/h, \text{ if } ah \text{ is not inconsistent.}$$

If ah is inconsistent, *i.e.* if $a/h=0$ (for h is by hypothesis consistent), the result follows at once by (8).

(25.1) If $p_r a/h=X_r$, the above may be written

$$p_r/ah=\frac{X_r}{\sum_1^n X_r}.$$

(26) $a/h = (a + \bar{h})/h$.

For $(a+\bar{h})/h = a/h + \bar{h}/h - a\bar{h}/h$ by (24),

$= a/h$ by (13.1) and (8).

(26.1) This may be written

$$a/h = (h \supset a)/h.$$

(27) If $(a+b)/h = 0$, $a/h = 0$.

$a/h + [b/h - ab/h] = 0$, by (24) and hypothesis

$\therefore\ a/h = 0$ by (v.).

(27.1) If $a/h = 0$ and $b/h = 0$, $(a+b)/h = 0$. This follows from (24).

(28) If $a/h = 1$, $(a+\bar{b})/h = 1$,

$(a+\bar{b})/h = a/h + \bar{b}\bar{a}/h$ by (24.3),

whence $(a+\bar{b})/h = a/h = 1$ by (1.1) and (8), together with the hypothesis. That is to say, a certain proposition is implied by every proposition.

(28.1) If $a/h = 0$, $(\bar{a}+b)/h = 1$ by substituting $\bar{a}$ for a and b for $\bar{b}$ in (28). That is to say, a certainly false proposition implies every proposition.

(29) If $a/(h_1+h_2) = 1$, $a/h_1 = 1$,

$\bar{a}/(h_1+h_2) = 0$,

and $\therefore\ \bar{a}(h_1+h_2)/h_1 = 0$ by (15).

Hence $\bar{a}h_1/h_1 = 0$ by (27),

whence the result.

(29.1) If $a/h_1 = 1$ and $a/h_2 = 1$, $a/(h_1+h_2) = 1$.

As in (20) $\bar{a}h_1/(h_1+h_2) = 0$ and $\bar{a}h_2/(h_1+h_2) = 0$.

Hence $\bar{a}(h_1+h_2)/(h_1+h_2) = 0$ by (27.1),

whence the result.

(29.2) If $a/(h_1+h_2) = 0$, $a/h_1 = 0$. This follows from (29).

(29.3) If $a/h_1 = 0$ and $a/h_2 = 0$, $a/(h_1+h_2) = 0$. This follows from (29.1).

3. *Irrelevance and Independence:*

(30) If $a/h_1h_2 = a/h_1$, then $a/h_1\bar{h}_2 = a/h_1$, if $h_1\bar{h}_2$ is not inconsistent.

$$\begin{aligned} a/h_1 &= ah_2/h_1 + a\bar{h}_2/h_1 \quad \text{by (24.2)},\\ &= a/h_1h_2 \cdot h_2/h_1 + a/h_1\bar{h}_2 \cdot \bar{h}_2/h_1,\\ &= a/h_1 \cdot h_2/h_1 + a/h_1\bar{h}_2 \cdot \bar{h}_2/h_1,\\ \therefore\ a/h_1 \cdot \bar{h}_2/h_1 &= a/h_1\bar{h}_2 \cdot \bar{h}_2/h_1, \end{aligned}$$

whence $a/h_1 = a/h_1\bar{h}_2$, unless $\bar{h}_2/h_1 = 0$, *i.e.* if $h_1\bar{h}_2$ is not inconsistent.

Thus, if a proposition is irrelevant to an argument, then the contradictory of the proposition is also irrelevant.

(31) If $a_2/a_1h = a_2/h$ and a_2h is not inconsistent, $a_1/a_2h = a_1/h$.

This follows by (iv. c), since $a_2/a_1h . a_1/h = a_1/a_2h . a_2/h$ by X. If, that is to say, a_1 is irrelevant to the argument a_2/h (see XIV.), and a_2 is not inconsistent with h: then a_2 is irrelevant to the argument a_1/h; and a_1/h and a_2/h are independent (see XIII.).

4. *Theorems of Relevance:*

(32) If $a/hh_1 > a/h$, $h_1/ah > h_1/h$.

ah is consistent since, otherwise, $a/hh_1 = a/h = 0$.

Therefore $\qquad a/h . h_1/ah = a/hh_1 . h_1/h \qquad$ by X.,

$\qquad > a/h . h_1/h \qquad$ by hypothesis;

so that $\qquad h_1/ah > h_1/h$.

Thus if h_1 is favourably relevant to the argument a/h, a is favourably relevant to the argument h_1/h.

This constitutes a formal demonstration of the generally accepted principle that if a hypothesis helps to explain a phenomenon, the fact of the phenomenon supports the reality of the hypothesis.

In the following theorems p will be said to be more favourable to a/h, than q is to b/h, if $\frac{a/ph}{a/h} > \frac{b/qh}{b/h}$, *i.e.* if, in the language of § 8 below, the coefficient of influence of p on a/h is greater than the coefficient of influence of q on b/h.

(33) If x is favourable to a/h, and h_1 is not less favourable to a/hx than x is to a/hh_1, then h_1 is favourable to a/h.

For $a/hh_1 = a/h . \frac{a/hx}{a/h} . \frac{a/hh_1x}{a/hx} . \frac{a/hh_1}{a/hh_1x}$; and by hypothesis the second term on the right is greater than unity and the product of the third and fourth terms is greater than or equal to unity.

(33.1) *A fortiori*, if x is favourable to a/h and not favourable to a/hh_1, and if h_1 is not unfavourable to a/hx, then h_1 is favourable to a/h.

(34) If x is favourable to a/h, and h_1 is not less favourable to x/ha than x is to h_1/ha, then h_1 is favourable to a/h.

This follows by the same reasoning as (33), since by an application of the Multiplication Theorem

$$\frac{a/hh_1x}{a/hx} \cdot \frac{a/hh_1}{a/hh_1x} = \frac{x/hh_1a}{x/ha} \cdot \frac{h_1/ha}{h_1/hax}.$$

(35) If x is favourable to a/h, but not more favourable to it than h_1x is, and not less favourable to it than to a/hh_1, then h_1 is favourable to a/h.

For $$a/hh_1 = a/h \cdot \left\{\frac{a/h}{a/hx} \cdot \frac{a/hh_1x}{a/h}\right\} \cdot \left\{\frac{a/hx}{a/h} \cdot \frac{a/hh_1}{a/hh_1x}\right\}.$$

This result is a little more substantial than the two preceding. By judging the influence of x and h_1x on the arguments a/h and a/hh_1, we can infer the influence of h_1 by itself on the argument a/h.

5. *The Multiplication Theorems:*

(36) If a_1/h and a_2/h are independent, $a_1a_2/h = a_1/h \cdot a_2/h$.

For $$a_1a_2/h = a_1/a_2h \cdot a_2/h = a_2/a_1h \cdot a_1/h \qquad \text{by X.},$$

and since a_1/h and a_2/h are independent,

$$a_1/a_2h = a_1/h \text{ and } a_2/a_1h = a_2/h \qquad \text{by XIII.}$$

Therefore $$a_1a_2/h = a_1/h \cdot a_2/h.$$

Hence, when a_1/h and a_2/h are independent, we can arrive at the probability of a_1 and a_2 jointly on the same hypothesis by simple multiplication of the probabilities a_1/h and a_2/h taken separately.

(37) If $p_1/h = p_2/p_1h = p_3/p_1p_2h = \ldots$,

$$p_1p_2p_3 \ldots p_n/h = \{p_1/h\}^n.$$

For $p_1p_2p_3 \ldots /h = p_1/h \cdot p_2/p_1h \cdot p_3/p_1p_2h \ldots$ by repeated applications of X.

6. *The Inverse Principle:*

(38) $\dfrac{a_1/bh}{a_2/bh} = \dfrac{b/a_1h}{b/a_2h} \cdot \dfrac{a_1/h}{a_2/h}$, provided bh, a_1h, and a_2h are each consistent.

For $$a_1/bh \cdot b/h = b/a_1h \cdot a_1/h,$$

and $$a_2/bh \cdot b/h = b/a_2h \cdot a_2/h \qquad \text{by X.},$$

whence the result follows, since $b/h \neq 0$, unless bh is inconsistent.

(38.1) If $a_1/h = p_1$, $a_2/h = p_2$, $b/a_1h = q_1$, $b/a_2h = q_2$, and $a_1/bh + a_2/bh = 1$, then it easily follows that

$$a_1/bh = \frac{p_1q_1}{p_1q_1 + p_2q_2},$$

and $$a_2/bh = \frac{p_2q_2}{p_1q_1 + p_2q_2}.$$

(38.2) If $a_1/h = a_2/h$ the above reduces to

$$a_1/bh = \frac{q_1}{q_1 + q_2},$$

and $$a_2/bh = \frac{q_2}{q_1 + q_2},$$

since $a_1/h \neq 0$, unless a_1h is inconsistent.

The proposition is easily extended to the cases in which the number of a's is greater than two.

It will be worth while to translate this theorem into familiar language. Let b represent the occurrence of an event B, a_1 and a_2 the hypotheses of the existence of two possible causes A_1 and A_2 of B, and h the general data of the problem. Then p_1 and p_2 are the *à priori* probabilities of the existence of A_1 and A_2 respectively, when it is not known whether or not the event B has occurred; q_1 and q_2 the probabilities that each of the causes A_1 and A_2, if it exists, will be followed by the event B. Then $\frac{p_1q_1}{p_1q_1 + p_2q_2}$ and $\frac{p_2q_2}{p_1q_1 + p_2q_2}$ are the probabilities of the existence of A_1 and A_2 respectively *after* the event, *i.e.* when, in addition to our other data, we know that the event B has occurred. The initial condition, that bh must not be inconsistent, simply ensures that the problem is a possible one, *i.e.* that the occurrence of the event B is on the initial data at least possible.

The reason why this theorem has generally been known as the Inverse Principle of Probability is obvious. The causal problems to which the Calculus of Probability has been applied are naturally divided into two classes—the direct in which, given the cause, we deduce the effect; the indirect or inverse in which, given the effect, we investigate the cause. The Inverse Principle has been usually employed to deal with the latter class of problem.

7. *Theorems on the Combination of Premisses*:

The Multiplication Theorems given above deal with the combination of *conclusions*; given a/h_1 and a/h_2 we considered the relation of a_1a_2/h to these probabilities. In this paragraph the corresponding problem of the combination of *premisses* will be treated; given a/h_1 and a/h_2 we shall consider the relation of a/h_1h_2 to these probabilities.

$$(39)\quad a/h_1h_2h=\frac{ah_1h_2/h}{h_1h_2/h}=\frac{ah_1h_2/h}{ah_1h_2/h+\bar{a}h_1h_2/h}\quad \text{by X. and (24.2)}$$

$$=\frac{u}{u+v},$$

where u is the *à priori* probability of the conclusion a and both hypotheses h_1 and h_2 jointly, and v is the *à priori* probability of the contradictory of the conclusion and both hypotheses h_1 and h_2 jointly.

$$(40)\quad a/h_1h_2=\frac{ah_1/h_2}{ah_1/h_2+\bar{a}h_1/h_2}=\frac{h_1/ah_2\cdot q}{h_1/ah_2\cdot q+h_1/\bar{a}h_2\cdot(1-q)},$$

$$=\frac{h_2/ah_1\cdot p}{h_2/ah_1\cdot p+h_2/\bar{a}h_1\cdot(1-p)},$$

where $p=a/h_1$ and $q=a/h_2$.

$$(40.1)\quad \text{If}\quad p=\tfrac{1}{2},\quad a/h_1h_2=\frac{h_2/ah_1}{h_2/ah_1+h_2/\bar{a}h_1},$$

and increases with $\dfrac{h_2/ah_1}{h_2/\bar{a}h_1}$.

These results are not very valuable and show the need of an original method of reduction. This is supplied by Mr. W. E. Johnson's *Cumulative Formula*, which is at present unpublished but which I have his permission to print below.[1]

8. It is first of all necessary to introduce a new symbol. Let us write

XV. $$a/bh=\{a^hb\}a/h \qquad \text{Def.}$$

We may call $\{a^hb\}$ *the coefficient of influence* of b upon a on hypothesis h.

XVI. $$\{a^hb\}\cdot\{ab^hc\}=\{a^hb^hc\} \qquad \text{Def.}$$

and similarly $$\{a^hb\}\cdot\{ab^hcd^he\}=\{a^hb^hcd^he\}.$$

These coefficients thus belong by definition to a general class of operators, which we may call *separative factors*.

(41) $$ab/h=\{a^hb\}\,.\,a/h\,.\,b/h,$$

since $$ab/h=a/bh\,.\,b/h.$$

[1] The substance of propositions (41) to (49) below is derived in its entirety from his notes,—the exposition only is mine.

Thus we may also call $\{a^h b\}$ *the coefficient of dependence* between a and b on hypothesis h.

(41·1) $abc/h = \{a^h b^h c\} \,.\, a/h \,.\, b/h \,.\, c/h.$

For $abc/h = \{ab^h c\} ab/h \,.\, c/h$ by (41),

$= \{ab^h c\} \,.\, \{a^h b\} \,.\, a/h \,.\, b/h \,.\, c/h$ by (41).

(41.2) And in general

$$abcd \ldots /h = \{a^h b^h c^h d^h \ldots\} \,.\, a/h \,.\, b/h \,.\, c/h \,.\, d/h \ldots$$

(42) $\{a^h b\} = \{b^h a\},$

since $a/bh \,.\, b/h = b/ah \,.\, a/h.$

(42.1) $\{a^h b^h c\} = \{a^h c^h b\},$

since $a/h \,.\, b/h \,.\, c/h = a/h \,.\, c/h \,.\, b/h.$

(42.2) And in general we have *a commutative rule*, by which the order of the terms may be always commuted—

e.g. $\{a^h bc^h \, def^h g\} = \{bc^h a^h g^h \, def\}$

$\{a^h bc^h \, def^h g\} = \{a^h cb^h \, fed^h g\}.$

(43) As a multiplier the separative factor operates so as to separate the terms that may be associated (or joined) in the multiplicand.

Thus $\{ab^h cd^h e\} \,.\, \{a^h b\} = \{a^h b^h cd^h e\},$

for $abcde/h = \{ab^h cd^h e\} \,.\, ab/h \,.\, cd/h \,.\, e/h$

$= \{ab^h cd^h e\} \,.\, \{a^h b\} \,.\, a/h \,.\, b/h \,.\, cd/h \,.\, e/h,$

and also $abcde/h = \{a^h b^h cd^h e\} \,.\, a/h \,.\, b/h \,.\, cd/h \,.\, e/h.$

Similarly (for example)

$$\{abc^h d^h ef\} \,.\, \{ab^h c\} \,.\, \{a^h b\} = \{a^h b^h c^h d^h ef\}.$$

(44) $\{a^h b\} \,.\, \{ab\} = \{a^h b\}.$

For $ab/h = \{ab\} ab/h.$

By a symbolic convention, therefore, we may put $\{ab\} = 1$.

(44.1) If $\{a^h b\} = 1$, it follows that a/h and b/h are independent arguments; and conversely.

(45) Rule of Repetition $\{aa^h b\} = \{a^h b\}.$

For $aab/h = ab/h$ by (vi.) and (12).

(46) *The Cumulative Formula:*

$x/ah : x'/ah : x''/ah : \ldots$

$= x/h \,.\, a/xh : x'/h \,.\, a/x'h : x''/h \,.\, a/x''h : \ldots$ by (38).

Take $n+1$ propositions $a, b, c \ldots$ Then by repetition

$x/ah \,.\, x/bh \,.\, x/ch \ldots : x'/a \,.\, x'b/ \,.\, x'/c \ldots : x''/a \,.\, x''/b \,.\, x''/c \ldots : \ldots$

$= (x/h)^{n+1} a/xh \,.\, b/xh \ldots : (x'/h)^{n+1} a/x'h \,.\, b/x'h \ldots$

$: (x''/h)^{n+1} a/x''h \,.\, b/x''h \ldots$

which may be written

$$\overset{n+1}{\Pi} x/ah : \overset{n+1}{\Pi} x'/ah : \overset{n+1}{\Pi} x''/ah : \ldots$$
$$= (x/h)^{n+1} \overset{n+1}{\Pi} a/xh : (x'/h)^{n+1} \overset{n+1}{\Pi} a/x'h : \ldots$$

Now

$$x/habc \ldots : x'/habc \ldots : x''/habc \ldots$$
$$= x/h \,.\, (abc \ldots)/xh : x'/h \,.\, (abc \ldots)/x'h : \ldots \quad \text{by (38),}$$

and

$$abc \ldots /xh = \{a^{xh}b^{xh}c \ldots\} \overset{n+1}{\Pi} a/xh \qquad \text{by (41.2),}$$

$$\therefore\ (x/h)^n \,.\, x/habc \ldots : (x'/h)^n \,.\, x'/habc \ldots : (x''/h)^n \,.\, x''/habc \ldots : \ldots$$
$$= \{a^{xh}b^{xh}c \ldots\} x/ah \,.\, x/bh \,.\, x/ch \ldots : \{a^{x'h}b^{x'h}c \ldots\} x'/ah \,.\, x'/bh \,.\, x'/ch \ldots : \ldots$$

which may be written

$$(x/h)^n \,.\, x/habc \ldots \propto \{a^{xh}b^{xh}c \ldots\} \,.\, x/ah \,.\, x/bh \,.\, x/ch \ldots$$

where variations of x are involved.

The cumulative formula is to be applied when, having accumulated the evidence a, b, c . . ., we desire to know the comparative probabilities of the various possible inferences x, x' . . . which may be drawn, and already know determinately the force of each of the items a, b, c . . . separately as evidence for x, x'. . . .

Besides the factors x/ah, x/bh, etc., we require to know two other sets of values, viz.: (1) x/h, etc., *i.e.* the *à priori* probabilities of x, etc., and (2) $\{a^{xh}b^{xh}c \ldots\}$, etc., *i.e.* the coefficients of dependence between a, b, and c . . . on hypotheses xh, etc. It may be remarked that the values $\{a^{xh}b^{xh}c \ldots\}$, $\{a^{x'h}b^{x'h}c \ldots\}$ . . . are not in any way related, even when $x' \equiv \bar{x}$.

What corresponds to the cumulative formula has been employed, sometimes, by mathematicians in a simplified form which is, except under special conditions, incorrect. First, it has been tacitly assumed that $\{a^{xh}b^{xh}c \ldots\}$, $\{a^{x'h}b^{x'h}c \ldots\}$ . . . are all unity: so that

$$(x/h)^n x/habc \ldots \propto x/ah \,.\, x/bh \,.\, x/ch \ldots$$

Secondly, the factor $(x/h)^n$ has been omitted, so that

$$x/habc \ldots \propto x/ah \,.\, x/bh \,.\, x/ch \ldots$$

It is this second incorrect statement of the formula which leads to the fallacious rule for the combination of the testimonies of independent witnesses ordinarily given in the text-books.[1]

(46.1). If $abc \ldots /xh = \{a^{xh}b^{xh}c \ldots\}\, a/xh \,.\, b/xh \,.\, c/xh \ldots$

then $\quad x/habc \ldots \propto \{a^{xh}b^{xh}c \ldots\}\, x/ah \,.\, x/bh \,.\, x/ch \ldots$

[1] See p 180 below.

This result is exceedingly interesting. Mr. Johnson is the first to arrive at the simple relation, expressed above, between the direct and the inverse formulæ: viz. that the *same* coefficient is required for correcting the simple formulæ of multiplication in both cases. As he remarks, however, while the direct formula gives the required probability directly by multiplication, the inverse formula gives only the *comparative* probability.

(46.2) If $x, x', x'' \ldots$ are exclusive and exhaustive alternatives,

$$x/habc\ldots = \frac{(x/h)^{-n} \cdot \{a^{xh}b^{xh}c\ldots\}\overset{n-1}{\Pi} x/ah}{\Sigma[(x'/h)^{-n} \cdot \{a^{x'h}b^{x'h}c\ldots\}\overset{n-1}{\Pi} x'/ah]},$$

since $\qquad x/habc\ldots \propto (x/h)^{-n}\{a^{xh}b^{xh}c\ldots\}\overset{n-1}{\Pi} x/ah,$

and $\qquad \Sigma x'/habc\ldots = 1 \qquad$ by (24.7).

(47). $$\frac{x/habc\ldots}{x/h} = \frac{a/h \cdot b/h \cdot c/h\ldots}{abc\ldots/h} \cdot \frac{abc\ldots/xh}{a/xh \cdot b/xh \cdot c/xh\ldots} \cdot \left[\frac{x/ah}{x/h} \cdot \frac{x/bh}{x/h}\ldots\right].$$

For $\qquad abc\ldots x/h = x/h \cdot abc\ldots/xh,$

$$\therefore \frac{abc\ldots x/h}{abc\ldots/h \cdot x/h} = \frac{abc\ldots/xh}{abc\ldots/h} = \frac{a/h \cdot b/h \cdot c/h\ldots}{abc\ldots/h} \cdot \frac{abc\ldots/xh}{a/xh \cdot b/xh \cdot c/xh\ldots} \cdot \left[\frac{a/xh}{a/h} \cdot \frac{b/xh}{b/h}\ldots\right],$$

whence the result, since $\frac{a/xh}{a/h} = \frac{x/ah}{x/h}$, etc.

(47.1) The above formula may be written in the condensed form

$$\{abc\ldots{}^h x\} = \frac{1}{\{a^h b^h c^h\ldots\}} \cdot \{a^{xh}b^{xh}c^{xh}\ldots\} \cdot [\{a^h x\} \cdot \{b^h x\} \cdot \{c^h x\}\ldots].$$

(48). $$\frac{\{x/h\}^n x/habc\ldots}{\{\bar{x}/h\}^n \bar{x}/habc\ldots} = \frac{\{a^{xh}b^{xh}c^{xh}\ldots\}}{\{a^{\bar{x}h}b^{\bar{x}h}c^{\bar{x}h}\ldots\}} \cdot \frac{x/ah \cdot x/bh \cdot x/ch\ldots}{\bar{x}/ah \cdot \bar{x}/bh \cdot \bar{x}/ch\ldots}.$$

This follows at once from (46.2), since x and $\bar{x}$ are exclusive and exhaustive alternatives. (It is assumed that xh, $\bar{x}h$, and ah, etc., are not inconsistent.)

This formula gives $x/habc\ldots$ in terms of x/ah, x/bh, etc., together with the *three* values x/h, $\{a^{xh}b^{xh}c^{xh}\ldots\}$, and $\{a^{\bar{x}h}b^{\bar{x}h}c^{\bar{x}h}\ldots\}$.

$$(48.1)\qquad \frac{x/habcd\ldots}{\bar{x}/habcd\ldots}:\frac{x/hbcd\ldots}{\bar{x}/hbcd\ldots}=\frac{\{a^{xh}bcd\ldots\}\cdot x/ah}{\{a^{\bar{x}h}bcd\ldots\}\cdot\bar{x}/ah}:\frac{x/h}{\bar{x}/h}.$$

This gives the effect on the odds (prob. x : prob. $\bar{x}$) of the extra knowledge a.

(49) When several data co-operate as evidence in favour of a proposition, they continually strengthen their own mutual probabilities, on the assumption that when the proposition is known to be true or to be false the data jointly are not counterdependent.

I.e. if $\{a^{xh}b^{xh}c\ldots\}$ and $\{a^{\bar{x}h}b^{\bar{x}h}c\ldots\}$ are not less than unity, and $x/kh>x/h$ where k is any of the data $a, b, c\ldots$, then $\{a^hb^hc^hd\ldots\}$ beginning with unity, continually increases, as the number of its terms is increased.

$$abc\ldots/h=xabc\ldots/h+\bar{x}abc\ldots/h \qquad \text{by (24.2).}$$
$$=x/h\,.\,abc\ldots/xh+\bar{x}/h\,.\,abc\ldots/\bar{x}h.$$
$$\geq x/h\,.\,\Pi a/xh\,.\,b/xh\ldots+\bar{x}/h\Pi a/\bar{x}h\,.\,b/\bar{x}h\ldots$$

(since $\{a^{xh}b^{xh}c\ldots\}$ and $\{a^{\bar{x}h}b^{\bar{x}h}c\ldots\}$ are not less than unity),

$$\geq x/h\,.\,\Pi\left[\frac{ax/h}{x/h}\cdot\frac{bx/h}{x/h}\cdots\right]+\bar{x}/h\,.\,\Pi\left[\frac{a\bar{x}/h}{\bar{x}/h}\cdot\frac{b\bar{x}/h}{\bar{x}/h}\cdots\right],$$

$$\therefore\ \frac{abc\ldots/h}{\Pi[a/h\,.\,b/h\ldots]}\geq x/h\,.\,\Pi\left[\frac{x/ah}{x/h}\cdot\frac{x/bx}{x/h}\cdots\right]+\bar{x}/h\,.\,\Pi\left[\frac{\bar{x}/ah}{\bar{x}/h}\cdot\frac{\bar{x}/bh}{\bar{x}/h}\cdots\right].$$

We can show that each additional piece of evidence $a, b, c\ldots$ increases the value of this expression. For let $x/h\,.\,\mathrm{G}+\bar{x}/h\,.\,\mathrm{G}'$ be its value when all the evidence up to k exclusive is taken, so that

$$x/kh\,.\,\mathrm{G}+\bar{x}/kh\,.\,\mathrm{G}'$$

is its value when k is taken. Now $\mathrm{G}>\mathrm{G}'$ since $x/ah>x/h$, etc., and $\bar{x}/ah<\bar{x}/h$, etc., by the hypothesis that the evidence favours x; and for the same reason $x/kh-x/h$, which is equal to $\bar{x}/h-\bar{x}/kh$, is positive.

$$\therefore\ \mathrm{G}\,(x/kh-x/h)>\mathrm{G}'(\bar{x}/h-\bar{x}/kh),$$

i.e. $$x/kh\,.\,\mathrm{G}+\bar{x}/kh\,.\,\mathrm{G}'>x/h\,.\,\mathrm{G}+\bar{x}/h\,.\,\mathrm{G}',$$

whence the result.

(49.1) The above proposition can be generalised for the case of exclusive alternatives x, x', x'' ... (in place of x, $\bar{x}$).

For $\{a^h b^h c^h \ldots\}$

$$= x/h \,.\, \{a^{xh} b^{xh} c \ldots\}\{a^h x\}\ \{b^h x\}\{c^h x\} \ldots$$
$$+ x'/h \,.\, \{a^{x'h} b^{x'h} c \ldots\}\{a^h x'\}\ \{b^h x'\}\{c^h x'\} \ldots$$
$$+ x''/h \,.\, \{a^{x''h} b^{x''h} c \ldots\}\{a^h x''\}\ \{b^h x''\}\{c^h x''\} \ldots + \ldots ;$$

from which it follows that, if $\{a^{xh} b^{xh} c \ldots\}$etc. $\not< 1$, and if $\{a^h x\} - 1$, $\{b^h x - 1\}$, $\{c^h x - 1\}$, etc., have the same sign, then $\{a^h b^h c \ldots\}$ is increasing (with the number of letters) from unity.

Mr. Johnson describes this result as a generalisation of the corrected "middle term fallacy" (see Chap. V. § 4).

APPENDIX

ON SYMBOLIC TREATMENTS OF PROBABILITY

THE use of the symbol 0 for impossibility and 1 for certainty was first introduced by Leibnitz in a very early pamphlet, entitled *Specimen certitudinis seu demonstrationum in jure, exhibitum in doctrina conditionum*, published in 1665 (*vide* Couturat, *Logique de Leibnitz*, p. 553). Leibnitz represented intermediate degrees of probability by the sign ½, meaning, however, by this symbol a *variable* between 0 and 1.

Several modern writers have made some attempt at a symbolic treatment of Probability. But with the exception of Boole, whose methods I have discussed in detail in Chapters XV., XVI., and XVII., no one has worked out anything very elaborate.

Mr. McColl published a number of brief notes on Probability of considerable interest—see especially his *Symbolic Logic, Sixth Paper on the Calculus of Equivalent Statements*, and *On the Growth and Use of a Symbolical Language*. The conception of probability as a relation between propositions underlies his symbolism, as it does mine.[1] The probability of a, relative to the *à priori* premiss h, he writes $\frac{a}{\epsilon}$; and the probability, given b in addition to the *à priori* premiss, he writes $\frac{a}{b}$. Thus $\frac{a}{\epsilon} = a/h$, and $\frac{a}{b} = a/bh$. The difference $\frac{a}{b} - \frac{a}{\epsilon}$, *i.e.* the change in the probability of a brought about by the addition of b to the evidence, he calls 'the dependence of the statement a upon the state-

[1] I did not come across these notes until my own method was considerably developed. Mr. McColl has been the first to use the fundamental symbol of Probability.

ment b,' and denotes it by $\delta\frac{a}{b}$. Thus $\delta\frac{a}{b}=0$, where, in my terminology, b is *irrelevant* to a on evidence h. The multiplication and addition formulæ he gives as follows: $\frac{ab}{\epsilon}=\frac{a}{\epsilon}\cdot\frac{b}{a}=\frac{b}{\epsilon}\cdot\frac{a}{b}$.

$$\frac{a+b}{\epsilon}=\frac{a}{\epsilon}+\frac{b}{\epsilon}-\frac{ab}{\epsilon}.$$

Also $$\delta\frac{a}{b}=\frac{A}{B}\delta\frac{b}{a}, \text{ where } A=\frac{a}{\epsilon}.$$

It is surprising how little use he succeeds in making of these good results. He arrives, however, at the inverse formula in the shape—

$$\frac{c_r}{v}=\frac{\frac{c_r}{\epsilon}\frac{v}{c_r}}{\sum_{r=1}^{r=n}\frac{c_r}{\epsilon}\cdot\frac{v}{c_r}},$$

where $c_1 \ldots c_n$ are a series of mutually exclusive causes of the event v and include all possible causes of it; reaching it as a generalisation of the proposition

$$\frac{a}{b}=\frac{\frac{a}{\epsilon}\cdot\frac{b}{a}}{\frac{a}{\epsilon}\cdot\frac{b}{a}+\frac{\bar{a}}{\epsilon}\cdot\frac{b}{\bar{a}}}.$$

In a paper entitled "Operations in Relative Number with Applications to the Theory of Probabilities,"[1] Mr. B. I. Gilman attempted a symbolic treatment based on a frequency theory similar to Venn's, but made more precise and more consistent with itself: "Probability has to do, not with individual events, but with classes of events; and not with one class, but with a pair of classes,—the one containing, the other contained. The latter being the one with which we are principally concerned, we speak, by an ellipsis, of its probability without mentioning the containing class; but in reality probability is a ratio, and to define it we must have both correlates given." But Mr. Gilman's symbolic treatment leads to very little. More recently R. Laemmel, in his *Untersuchungen über die Ermittlung von Wahrscheinlichkeiten*, made a beginning on somewhat similar lines; but in his case also the symbolic treatment leads to no substantial results.

Apart from the writers mentioned above, there are a few who have incidentally made use of a probability symbol. It will be sufficient to cite Czuber.[2] He denotes the probability of an event

[1] Published in the volume of Johns Hopkins *Studies in Logic*.

[2] *Wahrscheinlichkeitsrechnung*, vol. i. pp. 43-48.

E by W(E), and the probability of the event E given the occurrence of an event F by $W_F(E)$. He uses this symbol to give $W_F(E) = W_{\overline{F}}(E)$ as the criterion of the independence of the events E and F ($\overline{F}$ denoting the non-occurrence of F); $W_F(E)=1$, as the expression of the fact that E is a necessary consequence of F; and one or two other similar results.

Finally there is in the *Bulletin of the Physico-mathematical Society of Kazan* for 1887 a memoir in Russian by Platon S. Porctzki entitled "A Solution of the General Problem of the Theory of Probability by Means of Mathematical Logic." I have seen it stated that Schröder intended to publish ultimately a symbolic treatment of Probability. Whether he had prepared any manuscript on the subject before his death I do not know.

CHAPTER XV

NUMERICAL MEASUREMENT AND APPROXIMATION OF PROBABILITIES

1. The possibility of numerical measurement, mentioned at the close of Chapter III., arises out of the Addition Theorem (24.1). In introducing the definitions and the axiom, which are required in order to make the convention of numerical measurement operative, we may appear, as in the case of the original definitions of Addition and Multiplication, to be arguing in an artificial way. This appearance is due, here as in Chapter XII., to our having given the names of addition and multiplication to certain processes of compounding probabilities *in advance of* postulating that the processes in question have the properties commonly associated with these names. As common sense is hasty to impute the properties as soon as it hears the names, it may overlook the necessity of formally introducing them.

2. The definitions and the axiom which are needed in order to give a meaning to numerical measurement are the following:—

XVII. $a/h + \{a/h + [a/h + (a/h + \ldots r \text{ terms})]\} = r \,.\, a/h.$ Def.

XVIII. If $r \,.\, a/h = b/f$, then $a/h = \frac{1}{r} \,.\, b/f.$ Def.

XIX. If $b/f = q \,.\, c/g$, then $\frac{1}{r} \,.\, b/f = \frac{q}{r} c/g.$ Def.

Thus if $b/h = a/h + a/h + \ldots$ to r terms, then the probability b/h is said to be r times the probability a/h; hence if $ab/h = 0$ and $a/h = b/h$, the probability $(a+b)/h$ is *twice* the probability a/h. If a and b are exhaustive as well as exclusive alternatives relatively to h, so that $(a+b)/h = 1$, since we take the relation of certainty as our unit, then $a/h = b/h = \frac{1}{2}$.

We also need the following axiom postulating the existence of relations of probability corresponding to all proper fractions:

(vii.) If q and r are any finite integers and $q<r$, there exists a relation of probability which can be expressed, by means of the convention of the foregoing definitions, as $\frac{q}{r}$.

3. From these axioms and definitions combined with those of Chapter XII., it is easy to show (certainty being represented by unity and impossibility by zero) that we can manipulate according to the ordinary laws of arithmetic the "numbers" which by means of a special convention we have thus introduced to represent probabilities. Of the kind of proofs necessary for the complete demonstration of this the following is given as an example:

(50) If $a/f=\frac{1}{m}$ and $b/h=\frac{1}{n}$, $a/f+b/h=\frac{m+n}{mn}$.

Let the probability $\frac{1}{mn}=\mathrm{P}$, which exists by (vii.),

then $$n\,.\,\mathrm{P}=\frac{1}{m}=a/f \qquad \text{by (XIX.)},$$

and $$m\,.\,\mathrm{P}=\frac{1}{n}=b/h,$$

$$\begin{aligned}\therefore\ a/f+b/h &= n\,.\,\mathrm{P}+m\,.\,\mathrm{P}, \text{ if this probability exists,}\\ &=\mathrm{P}+\mathrm{P}\ldots \text{to } n \text{ terms}+\mathrm{P}+\mathrm{P}\ldots \text{to } m \text{ terms},\\ &=\mathrm{P}+\mathrm{P}\ldots \text{to } m+n \text{ terms},\\ &=(m+n)\mathrm{P}=\frac{m+n}{mn} \qquad \text{by (XIX.)}.\end{aligned}$$

This probability exists in virtue of (vii.).

4. Many probabilities—in fact all those which are equal to the probability of some other argument which has the same premiss and of which the conclusion is incompatible with that of the original argument—are numerically measurable in the sense that there is *some* other probability with which they are comparable in the manner described above. But they are not numerically measurable in the most usual sense, unless the probability with which they are thus comparable is the relation of certainty. The conditions under which a probability a/h is numerically measurable and equal to $\frac{q}{r}$ are easily seen. It

is necessary that there should exist probabilities $a_1/h_1, a_2/h_2 \ldots$, $a_q/h_q \ldots a_r/h_r$, such that

$$a_1/h_1 = a_2/h_2 = \ldots = a_q/h_q = \ldots = a_r/h_r,$$

$$a/h = \sum_1^q a_s/h_s, \text{ and } \sum_1^r a_s/h_s = 1.$$

If $a/h = \frac{q_1}{r_1}$ and $b/h = \frac{q_2}{r_2}$, it follows from (32) that $ab/h = \frac{q_1 q_2}{r_1 r_2}$ only if a/h and b/h are independent arguments. Unless, therefore, we are dealing with independent arguments, we cannot apply detailed mathematical reasoning even when the individual probabilities are numerically measurable. The greater part of mathematical probability, therefore, is concerned with arguments which are *both* independent *and* numerically measurable.

5. It is evident that the cases in which exact numerical measurement is possible are a very limited class, generally dependent on evidence which warrants a judgment of equiprobability by an application of the Principle of Indifference. The fuller the evidence upon which we rely, the less likely is it to be perfectly symmetrical in its bearing on the various alternatives, and the more likely is it to contain some piece of relevant information favouring one of them. In actual reasoning, therefore, perfectly equal probabilities, and hence exact numerical measures, will occur comparatively seldom.

The sphere of inexact numerical comparison is not, however, quite so limited. Many probabilities, which are incapable of numerical measurement, can be placed nevertheless *between* numerical limits. And by taking particular non-numerical probabilities as standards a great number of comparisons or approximate measurements become possible. If we can place a probability in an order of magnitude with some standard probability, we can obtain its approximate measure by comparison.

This method is frequently adopted in common discourse. When we ask how probable something is, we often put our question in the form—Is it more or less probable than so and so ?—where 'so and so' is some comparable and better known probability. We may thus obtain information in cases where it would be impossible to ascribe any *number* to the probability in question. Darwin was giving a numerical limit to a non-numerical probability when he said of a conversation with Lyell that he thought it no more likely that he should be right in nearly all points than that he should toss up a penny and get heads twenty times running.[1] Similar cases and others also, where the probability which is taken as the standard of comparison is itself non-numerical and not, as in Darwin's instance, a numerical one, will readily occur to the reader.

A specially important case of approximate comparison is that of 'practical certainty.' This differs from logical certainty since its contradictory is not impossible, but we are in practice completely satisfied with any probability which approaches such a limit. The phrase has naturally not been used with complete precision; but in its most useful sense it is essentially non-numerical—we cannot measure practical certainty in terms of logical certainty. We can only explain how great practical certainty is by giving instances. We may say, for instance, that it is measured by the probability of the sun's rising to-morrow. The type which we shall be most likely to take will be that of a well-verified induction.

6. Most of such comparisons must be based on the principles of Chapter V. It is possible, however, to develop a systematic method of approximation which may be occasionally useful. The theorems given below are chiefly suggested by some work of Boole's. His theorems were introduced for a different purpose, and he does not seem to have realised this interesting application of them; but analytically his problem is identical with that of approximation.[2] This method of approximation is also substantially the same analytically as that dealt with by Mr. Yule under the heading of *Consistence*.[3]

[1] *Life and Letters*, vol. ii. p. 240.

[2] In Boole's *Calculus* we are apt to be left with an equation of the second or of an even higher degree from which to derive the probability of the conclusion; and Boole introduced these methods in order to determine which of the several roots of his equation should be taken as giving the true solution of the problem in probability. In each case he shows that that root must be chosen which lies between certain limits, and that only one root satisfies this condition. The general theory to be applied in such cases is expounded by him in Chapter XIX. of *The Laws of Thought*, which is entitled "On Statistical Conditions." But the solution given in that chapter is awkward and unsatisfactory, and he subsequently published a much better method in the *Philosophical Magazine* for 1854 (4th series, vol. viii.) under the title "On the Conditions by which the Solutions of Questions in the Theory of Probabilities are limited."

[3] *Theory of Statistics*, chap. ii.

(51) xy/h always lies between[1] x/h and $x/h+y/h-1$ and between y/h and $x/h+y/h-1$.

For
$$xy/h=x/h-x\bar{y}/h \qquad \text{by (24.2)},$$
$$=x/h-\bar{y}/h\,.\,x/\bar{y}h \qquad \text{by X.}$$
Now $x/\bar{y}h$ lies between 0 and 1 by (2) and (3),
$$\therefore\ xy/h \text{ lies between } x/h \text{ and } x/h-\bar{y}/h,$$
$$i.e. \text{ between } x/h \text{ and } x/h+y/h-1.$$
As $xy/h \not< 0$, the above limits may be replaced by x/h and 0, if $x/h+y/h-1<0$.

We thus have limits for xy/h, close enough sometimes to be useful, which are available whether or not x/h and y/h are *independent* arguments. For instance, if y/h is nearly certain, $xy/h=x/h$ nearly, quite independently of whether or not x and y are independent. This is obvious; but it is useful to have a simple and general formula for all such cases.

(52) $x_1x_2\ldots x_{n+1}/h$ is always greater than $\sum_1^{n+1} x_r/h-n$.

For by (51) $x_1x_2\ldots x_{n+1}/h>x_1x_2\ldots x_n/h+x_{n+1}/h-1$
$$>x_1x_2\ldots x_{n-1}/h+x_n/h+x_{n+1}/h-2,$$
and so on.

(53) $xy/h+\bar{x}\bar{y}/h$ is always less than $x/h-y/h+1$, and less than $y/h-x/h+1$.

For as in (51) $xy/h=x/h-x\bar{y}/h$
and $\bar{x}\bar{y}/h=\bar{y}/h-x\bar{y}/h$,
$$\therefore\ xy/h+\bar{x}\bar{y}/h=x/h-y/h+1-2x\bar{y}/h,$$
whence the required result.

(54) $xy/h-\bar{x}\bar{y}/h=x/h+y/h-1$.

This proposition, which follows immediately from the above, is really out of place here. But its close connection with conclusions (51) and (53) is obvious. It is slightly unexpected, perhaps, that the difference of the probabilities that both of two events will occur and that neither of them will, is independent of whether or not the events themselves are independent.

7. It is not worth while to work out more of these results here. Some less systematic approximations of the same kind are given in the course of the solutions in Chapter XVII.

In seeking to compare the degree of one probability with that of another we may desire to get rid of one of the terms, on account

[1] In this and the following theorems the term 'between' includes the limits.

of its not being comparable with any of our standard probabilities. Thus our object in general is to eliminate a given symbol of quantity from a set of equations or inequations. If, for instance, we are to obtain numerical limits within which our probability must lie, we must eliminate from the result those probabilities which are non-numerical. This is the general problem for solution.

(55) A general method of solving these problems when we can throw our equations into a linear shape so far as all symbols of probability are concerned, is best shown in the following example :—

Suppose we have

$$\lambda+\nu=a \qquad \text{(i.)}$$
$$\lambda+\sigma=b \qquad \text{(ii.)}$$
$$\lambda+\nu+\sigma=c \qquad \text{(iii.)}$$
$$\lambda+\mu+\nu+\rho=d \qquad \text{(iv.)}$$
$$\lambda+\mu+\sigma+\tau=e \qquad \text{(v.)}$$
$$\lambda+\mu+\nu+\rho+\sigma+\tau+\upsilon=1 \qquad \text{(vi.)}$$

where λ, μ, ν, ρ, σ, τ, υ represent probabilities which are to be eliminated, and limits are to be found for c in terms of the standard probabilities a, b, d, e, and 1.

λ, μ, etc., must all lie between 0 and 1.

From (i.) and (iii.) $\sigma=c-a$; from (ii.) and (iii.) $\nu=c-b$.

From (i.), (ii.), and (iii.) $\lambda=a+b-c$,

whence $\quad c-a\geqslant 0, \ c-b\geqslant 0, \ a+b-c\geqslant 0,$

substituting for σ, ν, λ in (iv.), (v.), and (vi.)

$$\mu+\rho=d-a, \ \mu+\tau=e-b, \ \mu+\rho+\tau+\upsilon=1-c,$$

whence $\rho=d-a-\mu, \ \tau=e-b-\mu, \ \upsilon=1-c-d+a-e+b+\mu,$

$$\therefore \ d-a-\mu\geqslant 0, \ e-b-\mu\geqslant 0, \ 1-c-d+a-e+b+\mu\geqslant 0.$$

We have still to eliminate μ. $\quad \mu\leqslant d-a, \ \mu\leqslant e-b,$

$$\mu\geqslant c+d+e-a-b-1,$$

$$\therefore \ d-a\geqslant c+d+e-a-b-1 \text{ and } e-b\geqslant c+d+e-a-b-1.$$

Hence we have :

Upper limits of c:—$b+1-e, a+1-d, a+b$ (whichever is least);

Lower limits of c :—a, b (whichever is greatest).

This example, which is only slightly modified from one given by Boole, represents the actual conditions of a well-known problem in probability.

CHAPTER XVI

OBSERVATIONS ON THE THEOREMS OF CHAPTER XIV. AND THEIR DEVELOPMENTS, INCLUDING TESTIMONY

1. In Definition XIII. of Chapter XII. a meaning was given to the statement that a_1/h and a_2/h are independent arguments. In Theorem (33) of Chapter XIV. it was shown that, if a_1/h and a_2/h are independent, $a_1a_2/h = a_1/h \cdot a_2/h$. Thus where on given evidence there is independence between a_1 and a_2, the probability on this evidence of a_1a_2 jointly is the product of the probabilities of a_1 and a_2 separately. It is difficult to apply mathematical reasoning to the Calculus of Probabilities unless this condition is fulfilled; and the fulfilment of the condition has often been assumed too lightly. A good many of the most misleading fallacies in the theory of Probability have been due to a use of the Multiplication Theorem in its simplified form in cases where this is illegitimate.

2. These fallacies have been partly due to the absence of a clear understanding as to what is meant by *Independence*. Students of Probability have thought of the independence of events, rather than of the independence of arguments or propositions. The one phraseology is, perhaps, as legitimate as the other; but when we speak of the dependence of events, we are led to believe that the question is one of direct causal dependence, two events being dependent if the occurrence of one is a part cause or a possible part cause of the occurrence of the other. In this sense the result of tossing a coin is dependent on the existence of bias in the coin or in the method of tossing it, but it is independent of the actual results of other tosses; immunity from smallpox is dependent on vaccination, but is independent of statistical returns relating to immunity; while the testimonies of two witnesses about the same occurrence are independent, so long as there is no collusion between them.

This sense, which it is not easy to define quite precisely, is at any rate not the sense with which we are concerned when we deal with independent probabilities. We are concerned, not with direct causation of the kind described above, but with 'dependence for knowledge,' with the question whether the *knowledge* of one fact or event affords any rational ground for expecting the existence of the other. The dependence for knowledge of two events usually arises, no doubt, out of causal connection, or what we term such, of *some* kind. But two events are not independent for knowledge merely because there is an absence of direct causal connection between them; nor, on the other hand, are they necessarily dependent because there is in fact a causal train which brings them into an indirect connection. The question is whether there is any *known* probable connection, direct or indirect. A knowledge of the results of other tossings of a coin may be hardly less relevant than a knowledge of the bias of the coin; for a knowledge of these results may be a ground for a probable knowledge of the bias. There is a similar connection between the statistics of immunity from smallpox and the causal relations between vaccination and smallpox. The truthful testimonies of two witnesses about the same occurrence have a common cause, namely the occurrence, however independent (in the legal sense of the absence of collusion) the witnesses may be. For the purposes of probability two facts are only independent if the existence of one is no *indication* of anything which might be a part cause of the other.

3. While dependence and independence may be thus connected with the conception of causality, it is not convenient to found our definition of independence upon this connection. A partial or possible cause involves ideas which are still obscure, and I have preferred to define independence by reference to the conception of relevance, which has been already discussed. Whether there really are material external causal laws, how far causal connection is distinct from logical connection, and other such questions, are profoundly associated with the ultimate problems of logic and probability and with many of the topics, especially those of Part III., of this treatise. But I have nothing useful to say about them. Nearly everything with which I deal can be expressed in terms of logical relevance. And the relations between logical relevance and material cause must be left doubtful.

4. It will be useful to give a few examples out of writers who, as I conceive, have been led into mistakes through misapprehending the significance of Independence.

Cournot,[1] in his work on Probability, which after a long period of neglect has come into high favour with a modern school of thought in France, distinguishes between 'subjective probability' based on ignorance and 'objective probability' based on the calculation of 'objective possibilities,' an 'objective possibility' being a chance event brought about by the combination or convergence of phenomena belonging to *independent* series. The existence of objectively chance events depends on his doctrine that, as there are series of phenomena causally dependent, so there are others between the causal developments of which there is independence. These objective possibilities of Cournot's, whether they be real or fantastic, can have, however, small importance for the theory of probability. For it is not known to us what series of phenomena are thus independent. If we had to wait until we knew phenomena to be independent in this sense before we could use the simplified multiplication theorem, most mathematical applications of probability would remain hypothetical.

5. Cournot's 'objective probability,' depending wholly on objective fact, bears some resemblances to the conception in the minds of those who adopt the frequency theory of probability. The proper definition of independence on this theory has been given most clearly by Mr. Yule[2] as follows:

"Two attributes A and B are usually defined to be independent, within any given field of observation or 'universe,' when the chance of finding them together is the product of the chances of finding either of them separately. The physical meaning of the definition seems rather clearer in a different form of statement, viz. if we define A and B to be independent *when the proportion of A's amongst the B's of the given universe is the same as in that universe at large.* If, for instance, the question were put, 'What is the test for independence of smallpox attack and vaccination?' the natural reply would be, 'The percentage of vaccinated amongst the attacked should be the same as in the general population.' . . ."

[1] For some account of Cournot, see Chapter XXIV. § 3.

[2] "Notes on the Theory of Association of Attributes in Statistics," *Biometrika*, vol. ii. p. 125.

This definition is consistent with the rest of the theory to which it belongs, but is, at the same time, open to the general objections to it.[1] Mr. Yule admits that A and B may be independent in the world at large but not in the world of C's. The question therefore arises as to what world given evidence specifies, and whether any step forward is possible when, as is generally the case, we do not know for certain what the proportions in a given world actually are. As in the case of Cournot's independent series, it is in general impossible that we should know whether A and B are or are not independent in this sense. The logical independence for knowledge which justifies our reasoning in a certain way must be something different from either of these objective forms of independence.

6. I come now to Boole's treatment of this subject. The central error in his system of probability arises out of his giving two inconsistent definitions of 'independence.'[2] He first wins the reader's acquiescence by giving a perfectly correct definition: "Two events are said to be independent when the probability of the happening of either of them is unaffected by our *expectation* of the occurrence or failure of the other."[3] But a moment later he interprets the term in quite a different sense; for, according to Boole's second definition, we must regard the events as independent unless we are told either that they *must* concur or that they *cannot* concur. That is to say, they are independent unless we know for certain that there is, in fact, an invariable connection between them. "The simple events, x, y, z, will be said to be *conditioned* when they are not free to occur in every possible combination; in other words, when some compound event depending upon them is precluded from occurring.

[1] See Chapter VIII.

[2] Boole's mistake was pointed out, accurately though somewhat obscurely, by H. Wilbraham in his review "On the Theory of Chances developed in Professor Boole's *Laws of Thought*" (*Phil. Mag.* 4th series, vol. vii., 1854). Boole failed to understand the point of Wilbraham's criticism, and replied hotly, challenging him to impugn any individual results ("Reply to some Observations published by Mr. Wilbraham," *Phil. Mag.* 4th series, vol. viii., 1854). He returned to the same question in a paper entitled "On a General Method in the Theory of Probabilities," *Phil. Mag.* 4th series, vol. viii., 1854, where he endeavours to support his theory by an appeal to the Principle of Indifference. McColl, in his "Sixth Paper on Calculus of Equivalent Statements," saw that Boole's fallacy turned on his definition of Independence; but I do not think he understood, at least he does not explain, where precisely Boole's mistake lay.

[3] *Laws of Thought*, p. 255. The italics in this quotation are mine.

. . . Simple unconditioned events are by definition independent."[1] In fact as long as xz is *possible*, x and z are independent. This is plainly inconsistent with Boole's first definition, with which he makes no attempt to reconcile it. The consequences of his employing the term independence in a double sense are far-reaching. For he uses a method of reduction which is only valid when the arguments to which it is applied are independent in the first sense, and assumes that it is valid if they are independent in the second sense. While his theorems are true if all the propositions or events involved are independent in the first sense, they are not true, as he supposes them to be, if the events are independent only in the second sense. In some cases this mistake involves him in results so paradoxical that they might have led him to detect his fundamental error.[2] Boole was almost certainly led into this error through supposing that the *data* of a problem can be of the form, "Prob. $x=p$," *i.e.* that it is sufficient to state that the probability of a proposition is such and such, without stating to what premisses this probability is referred.[3]

It is interesting that De Morgan should have given, incidentally, a definition of independence almost identical with Boole's second definition: "Two events are independent if the latter might have existed without the former, or the

[1] *Op. cit.* p. 258.

[2] There is an excellent instance of this, *Laws of Thought*, p. 286. Boole discusses the problem: Given the probability p of the disjunction 'either Y is true, or X and Y are false,' required the probability of the conditional proposition, 'If X is true, Y is true.' The two propositions are formally equivalent; but Boole, through the error pointed out above, arrives at the result $\frac{cp}{1-p+cp}$, where c is the probability of 'If either Y is true, or X and Y false, X is true.' His explanation of the paradox amounts to an assertion that, so long as two propositions, which are formally equivalent when true, are only probable, they are not necessarily equivalent.

[3] In studying and criticising Boole's work on Probability, it is very important to take into account the various articles which he contributed to the *Philosophical Magazine* during 1854, in which the methods of *The Laws of Thought* are considerably improved and modified. His last and most considered contribution to Probability is his paper "On the application of the Theory of Probabilities to the question of the combination of testimonies or judgments," to be found in the *Edin. Phil. Trans.* vol. xxi., 1857. This memoir contains a simplification and general summary of the method originally proposed in *The Laws of Thought*, and should be regarded as superseding the exposition of that book. In spite of the error already alluded to, which vitiates many of his conclusions, the memoir is as full as are his other writings of genius and originality.

former without the latter, for anything that we know to the contrary."[1]

7. In many other cases errors have arisen, not through a misapprehension of the meaning of independence, but merely through careless assumptions of it, or through enunciating the Theorem of Multiplication without its qualifying condition. Mathematicians have been too eager to assume the legitimacy of those complicated processes of multiplying probabilities, for which the greater part of the mathematics of probability is engaged in supplying simplifications and approximate solutions. Even De Morgan was careless enough in one of his writings[2] to enunciate the Multiplication Theorem in the following form: "The probability of the happening of two, three, or more events is the product of the probabilities of their happening separately (p. 398). . . . Knowing the probability of a compound event, and that of one of its components, we find the probability of the other by dividing the first by the second. This is a mathematical result of the last too obvious to require further proof (p. 401)."

An excellent and classic instance of the danger of wrongful assumptions of independence is given by the problem of determining the probability of throwing heads twice in two consecutive tosses of a coin. The plain man generally assumes without hesitation that the chance is $(\frac{1}{2})^2$. For the *à priori* chance of heads at the first toss is $\frac{1}{2}$, and we might naturally suppose that the two events are independent,—since the mere fact of heads having appeared once can have no influence on the next toss. But this is not the case unless we know for certain that the coin is free from bias. If we do not know whether there is bias, or which way the bias lies, then it is reasonable to put the probability somewhat higher than $(\frac{1}{2})^2$. The *fact* of heads having appeared at the first toss is not the cause of heads appearing at the second also, but the *knowledge*, that the coin has fallen heads already, affects our forecast of its falling thus in the future, since heads in the past may have been due to a cause which will favour heads in the future. The possibility of bias in a coin, it may be noticed,

[1] "Essay on Probabilities" in the *Cabinet Encyclopaedia*, p. 26. De Morgan is not very consistent with himself in his various distinct treatises on this subject, and other definitions may be found elsewhere. Boole's second definition of Independence is also adopted by Macfarlane, *Algebra of Logic*, p. 21.

[2] "Theory of Probabilities" in the *Encyclopaedia Metropolitana*.

always favours 'runs'; this possibility increases the probability both of 'runs' of heads and of 'runs' of tails.

This point is discussed at some length in Chapter XXIX. and further examples will be given there. In this chapter, therefore, I will do more than refer to an investigation by Laplace and to one real and one supposed fallacy of Independence of a type with which we shall not be concerned in Chapter XXIX.

8. Laplace, in so far as he took account at all of the considerations explained in § 7, discussed them under the heading of *Des inégalités inconnues qui peuvent exister entre les chances que l'on suppose égales.*[1] In the case, that is to say, of the coin with unknown bias, he held that the true probability of heads even at the first toss differed from $\frac{1}{2}$ by an amount unknown. But this is not the correct way of looking at the matter. In the supposed circumstances the *initial* chances for heads and tails respectively at the first toss really are equal. What is not true is that the initial probability of 'heads twice' is equal to the probability of 'heads once' squared.

Let us write 'heads at first toss' $=h_1$; 'heads at second toss' $=h_2$. Then $h_1/h=h_2/h=\frac{1}{2}$, and $h_1h_2/h=h_2/h_1h \,.\, h_1/h$. Hence $h_1h_2/h=\{h_1/h\}^2$ only if $h_2/h_1h=h_2/h$, *i.e.* if the knowledge that heads has fallen at the first toss does not affect in the least the probability of its falling at the second. In general, it is true that h_2/h_1h will not differ greatly from h_2/h (for relative to most hypotheses heads at the first toss will *not much* influence our expectation of heads at the second), and $\frac{1}{4}$ will, therefore, give a good approximation to the required probability. Laplace suggests an ingenious method by which the divergence may be diminished. If we throw two coins and define 'heads' at any toss as the face thrown by the second coin, he discusses the probability of 'heads twice running' with the first coin. The solution of this problem involves, of course, particular assumptions, but they are of a kind more likely to be realised in practice than the complete absence of bias. As Laplace does not state them, and as his proof is incomplete, it may be worth while to give a proof in detail.

Let h_1, t_1, h_2, t_2 denote heads and tails respectively with the first and second coins respectively at the first toss, and h_1', t_1', h_2', t_2' the corresponding events at the second toss, then

[1] *Essai philosophique*, p. 49. See also "Mémoire sur les Probabilités," *Mém. de l'Acad.* p. 228, and cp. D'Alembert, "Sur le calcul des probabilités," *Opuscules mathématiques* (1780), vol. vii.

the probability (with the above convention) of 'heads twice running,' *i.e.* agreement between the two coins twice running, is

$$(h_2h_2' + t_2t_2')(h_1h_1' + t_1t_1')/h = (h_2h_2' + t_2t_2')/(h_1h_1' + t_1t_1', h) \,.\, (h_1h_1' + t_1t_1')/h.$$

Since $h_2h_2'/(h_1h_1' + t_1t_1', h) = t_2t_2'/(h_1h_1' + t_1t_1', h)$ by the Principle of Indifference, and $h_2h_2't_2t_2'/h = 0$.

$\therefore\ (h_2h_2' + t_2t_2')/(h_1h_1' + t_1t_1', h) = 2.h_2h_2'/(h_1h_1' + t_1t_1', h)$ by (24.1).

Similarly $\qquad (h_1h_1' + t_1t_1')/h = 2h_1h_1'/h.$

We may assume that $h_1/h_1'h = h_1/h$, *i.e.* that heads with one coin is irrelevant to the probability of heads with the other: and $h_1/h = h_1'/h = \frac{1}{2}$ by the Principle of Indifference, so that

$$(h_1h_1' + t_1t_1')/h = 2(\tfrac{1}{2})^2 = \tfrac{1}{2}.$$

$$\begin{aligned}\therefore\ (h_2h_2' + t_2t_2')(h_1h_1' + t_1t_1')/h &= 2h_2h_2'/(h_1h_1' + t_1t_1', h)\,.\,\tfrac{1}{2}\\ &= h_2h_2'/(h_1h_1' + t_1t_1', h)\\ &= \tfrac{1}{2}h_2/(h_2', h_1h_1' + t_1t_1', h),\end{aligned}$$

since, $(h_1h_1' + t_1t_1')$ being irrelevant to h'_2/h, $h'_2/(h_1h_1' + t_1t_1', h) = h_2'/h = \frac{1}{2}$.

Now $h_2/(h_2', h_1h_1' + t_1t_1', h)$ is greater than $\frac{1}{2}$, since the fact of the coins having agreed once may be *some* reason for supposing they will agree again. But it is less than h_2/h_1h: for we may assume that $h_2/(h_2', h_1h_1' + t_1t_1', h)$ is less than $h_2/(h_2', h_1h_1', h)$, and also that $h_2/(h_2', h_1h_1', h) = h_2/h_1h$, *i.e.* that heads twice running with one coin does not increase the probability of heads twice running with a different coin. Laplace's method of tossing, therefore, yields with these assumptions, more or less legitimate according to the content of h, a probability nearer to $\frac{1}{4}$ than is h_1h_2/h. If $h_2/(h_2', h_1h_1' + t_1t_1', h) = \frac{1}{2}$, then the probability is exactly $\frac{1}{4}$.

9. Two other examples will complete this rather discursive commentary. It has been supposed that by the Principle of Indifference the probability of the existence of iron upon Sirius is $\frac{1}{2}$, and that similarly the probability of the existence there of any other element is also $\frac{1}{2}$. The probability, therefore, that not one of the 68 terrestrial elements will be found on Sirius is $(\frac{1}{2})^{68}$, and that at least one will be found there is $1 - (\frac{1}{2})^{68}$ or approximately certain. This argument, or a similar one, has been seriously advanced. It would seem to prove also, amongst

many other things, that at least one college exactly resembling some college at either Oxford or Cambridge will almost certainly be found on Sirius. The fallacy is partly due, as has been pointed out by Von Kries and others, to an illegitimate use of the Principle of Indifference. The probability of iron on Sirius is *not* $\frac{1}{2}$. But the result is also due to the fallacy of false independence. It is assumed that the known existence of 67 terrestrial elements on Sirius would not increase the probability of the sixty-eighth's being found there also, and that their known absence would not decrease the sixty-eighth's probability.[1]

10. The other example is that of Maxwell's classic mistake in the theory of gases.[2] According to this theory molecules of gas move with great velocity in every direction. Both the directions and velocities are unknown, but the probability that a molecule has a given velocity is a function of that velocity and is independent of the direction. The maximum velocity and the mean velocity vary with the temperature. Maxwell seeks to determine, on these conditions alone, the probability that a molecule has a given velocity. His argument is as follows:

If $\phi(x)$ represents the probability that the component of velocity parallel to the axis of X is x, the probability that the velocity has components x, y, z parallel to the three axes is $\phi(x)\phi(y)\phi(z)$. Thus if $F(v)$ represents the probability of a total velocity v, we have $\phi(x)\phi(y)\phi(z) = F(v)$, where $v^2 = x^2 + y^2 + z^2$. It is not difficult to deduce from this (assuming that the

[1] See Von Kries, *Die Principien der Wahrscheinlichkeitsrechnung*, p. 10. Stumpf (*Über den Begriff der mathem. Wahrscheinlichkeit*, pp. 71-74) argues that the fallacy results from not taking into account the fact that there might be as many metals as atomic weights, and that therefore the chance of iron is $\frac{1}{z}$, where z is the number of possible atomic weights. A. Nitsche (*Vierteljsch. f. wissensch. Philos.*, 1892) thinks that the real alternatives are 0, or only 1, or only 2 . . . or 68 terrestrial elements on Sirius, and that these are equally probable, the chance of each being $\frac{1}{69}$.

[2] I take the statement of this from Bertrand's *Calcul des probabilités*, p. 30. Let me here quote a precocious passage on Probability regarded as a branch of Logic, from a letter written by Maxwell in his nineteenth year (1850), before he came up to Cambridge: "They say that Understanding ought to work by the rules of right reason. These rules are, or ought to be, contained in Logic; but the actual science of logic is conversant at present only with things either certain, impossible, or entirely doubtful, none of which (fortunately) we have to reason on. Therefore the true logic for this world is the calculus of Probabilities, which takes account of the magnitude of the probability which is, or ought to be, in a reasonable man's mind" (*Life*, page 143).

functions are analytical) that $\phi(x)$ must be of the form $Ge^{-k^2x^2}$.

It is generally agreed at the present time that this result is erroneous. But the nature of the error is, I think, quite different from what it is commonly supposed to be.

Bertrand,[1] Poincaré,[2] and Von Kries,[3] all cite this argument of Maxwell's as an illustration of the fallacy of Independence; and argue that $\phi(x)$, $\phi(y)$, and $\phi(z)$ cannot, as he assumes, represent independent probabilities, if, as he also assumes, the probability of a velocity is a function of that velocity. But it is not in this way that the error in the result really arises. If we do not know *what* function of the velocity the probability of that velocity is, a knowledge of the velocity parallel to the axes of x and y tells us nothing about the velocity parallel to the axis of z. Maxwell was, I think, quite right to hold that a mere assumption that the probability of a velocity is *some* function of that velocity, does *not* interfere with the mutual independence of statements as to the velocity parallel to each of the three axes. Let us denote the proposition, 'the velocity parallel to the axis of X is x' by $X(x)$, the corresponding propositions relative to the axes of Y and Z by $Y(y)$ and $Z(z)$, and the proposition 'the total velocity is v' by $V(v)$; and let h represent our *à priori* data. Then if $X(x)/h = \phi(x)$ it is a justifiable inference from the Principle of Indifference that $Y(y)/h = \phi(y)$ and $Z(z)/h = \phi(z)$. Maxwell infers from this that $X(x)Y(y)Z(z)/h = \phi(x)\phi(y)\phi(z)$. That is to say, he assumes that $Y(y)/X(x) \,.\, h = Y(y)/h$ and that $Z(z)/Y(y) \,.\, X(x) \,.\, h = Z(z)/h$. I do not agree with the authorities cited above that this is illegitimate. So long as we do not know what function of the total velocity the probability of that velocity is, a knowledge of the velocities parallel to the axes of x and y has no bearing on the probability of a given velocity parallel to the axis of z. But Maxwell goes on to infer that $X(x)Y(y)Z(z)/h = V(v)/h$, where $v^2 = x^2 + y^2 + z^2$. It is here, and in a very elementary way, that the error creeps in. The propositions $X(x)Y(y)Z(z)$ and $V(v)$ are *not* equivalent. The latter follows from the former, but the former does not follow from the latter. There is more than one set of values x, y, z,

[1] *Calcul des probabilités*, p. 30.
[2] *Calcul des probabilités* (2nd ed.), pp. 41–44.
[3] *Wahrscheinlichkeitsrechnung*, p. 199.

which will yield the same value v. Thus the probability $V(v)/h$ is much greater than the probability $X(x)Y(y)Z(z)/h$. As we do not know the direction of the total velocity v, there are many ways, not inconsistent with our *data*, of resolving it into components parallel to the axes. Indeed I think it is a legitimate extension of the preceding argument to put $V(v)/h = \phi(v)$; for there is no reason for thinking differently about the direction V from what we think about the direction X.

A difficulty analogous to this occurs in discussing the problem of the dispersion of bullets over a target—a subject round which, on account of a curiosity which it seems to have raised in the minds of many students of probability, a literature has grown up of a bulk disproportionate to its importance.

11. I now pass to the Principle of Inverse Probability, a theorem of great importance in the history of the subject. With various arguments which have been based upon it I shall deal in Chapter XXX. But it will be convenient to discuss here the history of the Principle itself and of attempts at proving it.

It first makes its appearance somewhat late in the history of the subject. Not until 1763, when Bayes's theorem was communicated to the Royal Society,[1] was a rule for the determination of inverse probabilities explicitly enunciated. It is true that solutions to inductive problems requiring an implicit and more or less fallacious use of the inverse principle had already been propounded, notably by Daniel Bernoulli in his investigations into the statistical evidence in favour of inoculation.[2] But the appearance of Bayes's *Memoir* marks the beginning of a new stage of development. It was followed in 1767 by a contribution from Michell[3] to the *Philosophical Transactions* on the distribu-

[1] Published in the *Phil. Trans.* vol. liii., 1763, pp. 376-398. This Memoir was communicated by Price after Bayes's death; there was a second Memoir in the following year (vol. liv. pp. 298-310), to which Price himself made some contributions. See Todhunter's *History*, pp. 299 *et seq.* Thomas Bayes was a dissenting minister of Tunbridge Wells, who was a Fellow of the Royal Society from 1741 until his death in 1761. A German edition of his contributions to Probability has been edited by Timerding.

[2] "Essai d'une nouvelle analyse de la mortalité causée par la petite vérole, et des avantages de l'inoculation pour la prévenir," *Hist. de l'Acad.*, Paris, 1760 (published 1766). Bernoulli argued that the recorded results of inoculation rendered it a probable cause of immunity. This is an inverse argument, though Bayes's theorem is not used in the course of it. See also D. Bernoulli's *Memoir on the Inclinations of the Planetary Orbits.*

[3] Michell's argument owes more, perhaps, to Daniel Bernoulli than to Bayes.

tion of the stars, to which further reference will be made in Chapter XXV. And in 1774 the rule was clearly, though not quite accurately, enunciated by Laplace in his "Mémoire sur la probabilité des causes par les évènemens" (*Mémoires présentés à l'Académie des Sciences*, vol. vi., 1774). He states the principle as follows (p. 623):

"Si un évènement peut être produit par un nombre n de causes différentes, les probabilités de l'existence de ces causes prises de l'évènement sont entre elles comme les probabilités de l'évènement prises de ces causes; et la probabilité de l'existence de chacune d'elles est égale à la probabilité de l'évènement prise de cette cause, divisée par la somme de toutes les probabilités de l'évènement prises de chacune de ces causes."

He speaks as if he intended to prove this principle, but he only give explanations and instances without proof. The principle is not strictly true in the form in which he enunciates it, as will be seen on reference to theorems (38) of Chapter XIV.; and the omission of the necessary qualification has led to a number of fallacious arguments, some of which will be considered in Chapter XXX.

12. The value and originality of Bayes's *Memoir* are considerable, and Laplace's method probably owes much more to it than is generally recognised or than was acknowledged by Laplace. The principle, often called by Bayes's name, does not appear in his *Memoir* in the shape given it by Laplace and usually adopted since; but Bayes's enunciation is strictly correct and his method of arriving at it shows its true logical connection with more fundamental principles, whereas Laplace's enunciation gives it the appearance of a *new* principle specially introduced for the solution of causal problems. The following passage[1] gives, in my opinion, a right method of approaching the problem: "If there be two subsequent events, the probability of the second $\frac{b}{N}$ and the probability of both together $\frac{P}{N}$, and, it being first discovered that the second event has happened, from hence I guess that the first event has also happened, the probability I am in the right is $\frac{P}{b}$." If the occurrence of the first event

[1] Quoted by Todhunter, *op. cit.* p. 296. Todhunter underrates the importance of this passage, which he finds unoriginal, yet obscure.

is denoted by a and of the second by b, this corresponds to $ab/h = a/bh \, . \, b/h$ and therefore $a/bh = \frac{ab/h}{b/h}$; for $ab/h = \frac{\text{P}}{\text{N}}$, $b/h = \frac{\text{b}}{\text{N}}$, $a/bh = \frac{\text{P}}{b}$. The direct and indeed fundamental dependence of the inverse principle on the rule for compound probabilities was not appreciated by Laplace.

13. A number of proofs of the theorem have been attempted since Laplace's time, but most of them are not very satisfactory, and are generally couched in such a form that they do no more than recommend the plausibility of their thesis. Mr. McColl[1] gave a symbolic proof, closely resembling theorem (38) when differences of symbolism are allowed for; and a very similar proof has also been given by A. A. Markoff.[2] I am not acquainted with any other rigorous discussion of it.

Von Kries[3] presents the most interesting and careful example of a type of proof which has been put forward in one shape or another by a number of writers. We have initially, according to this view, a certain number of hypothetical possibilities, all equally probable, some favourable and some unfavourable to our conclusion. Experience, or rather knowledge that the event has happened, rules out a number of these alternatives, and we are left with a field of possibilities narrower than that with which we started. Only part of the original field or *Spielraum* of possibility is now admissible (*zulässig*). Causes have *à posteriori* probabilities which are proportional to the extent of their occurrence in the now restricted field of possibility.

There is much in this which seems to be true, but it hardly amounts to a proof. The whole discussion is in reality an appeal to intuition. For how do we know that the possibilities admissible *à posteriori* are still, as they were assumed to be *à priori*, equal possibilities? Von Kries himself notices that there is a difficulty; and I do not see how he is to avoid it, except by the introduction of an axiom.

This was in fact the course taken by Professor Donkin in 1851, in an article which aroused some interest in the *Philosophical*

[1] "Sixth Paper on the Calculus of Equivalent Statements," *Proc. Lond. Math. Soc.*, 1897, vol. xxviii. p. 567. See also p. 155 above.

[2] *Wahrscheinlichkeitsrechnung*, p. 178.

[3] *Die Principien der Wahrscheinlichkeitsrechnung*, pp. 117-121. The above account of Von Kries's argument is much condensed.

Magazine at the time, but which has since been forgotten. Donkin's theory is, however, of considerable interest. He laid down as one of the fundamental principles of probability the following: [1]

"If there be any number of mutually exclusive hypotheses $h_1h_2h_3$, . . . of which the probabilities relative to a particular state of information are $p_1p_2p_3$ . . ., and if new information be gained which changes the probabilities of some of them, suppose of h_{m+1} and all that follow, without having otherwise any reference to the rest, then the probabilities of these latter have the *same ratios* to one another, *after* the new information, that they had before." [2]

Donkin goes on to say that the most important case is where the new information consists in the knowledge that some of the hypotheses must be rejected, without any further information as to those of the original set which are retained. This is the proposition which Von Kries requires.

As it stands, the phrase "without having otherwise any reference to the rest" obviously lacks precision. An interpretation, however, *can* be put upon it, with which the principle is true. If, given the old information and the truth of one of the hypotheses $h_1 \ldots h_m$ to the exclusion of the rest, the probability of what is conveyed by the new information is the *same* whichever of the hypotheses $h_1 \ldots h_m$ has been taken, then Donkin's principle is valid. For let a be the old information, a' the new, and let $h_r/a = p_r$, $h_r/aa' = p_r'$; then

$$p_r' = h_r/aa' = \frac{h_r a'/a}{a'/a} = \frac{a'/h_r a \cdot p_r}{a'/a},$$

$\therefore \frac{p_r'}{p_r} = \frac{p_s'}{p_s}$, etc., if $a'/h_r a = a'/h_s a$, which is the condition already explained.

14. Difficulties connected with the Inverse Principle have arisen, however, not so much in attempts to *prove* the principle as in those to *enunciate* it—though it may have been the lack

[1] "On certain Questions relating to the Theory of Probabilities," *Phil. Mag.* 4th series, vol. i., 1851.

[2] It is interesting to notice that an axiom, practically equivalent to this, has been laid down more lately by A. A. Markoff (*Wahrscheinlichkeitsrechnung*, p. 8) under the title 'Unabhängigkeitsaxiom.'

of a rigorous proof that has been responsible for the frequent enunciation of an inaccurate principle.

It will be noticed that in the formula (38·2) the *à priori* probabilities of the hypotheses a_1 and a_2 drop out if $p_1 = p_2$, and the results can then be expressed in a much simpler shape. This is the shape in which the principle is enunciated by Laplace for the *general* case,[1] and represents the uninstructed view expressed with great clearness by De Morgan: [2] "Causes are likely or unlikely, just in the same proportion that it is likely or unlikely that observed events should follow from them. The most probable cause is that from which the observed event could most easily have arisen." If this were true the principle of Inverse Probability would certainly be a most powerful weapon of proof, even equal, perhaps, to the heavy burdens which have been laid on it. But the proof given in Chapter XIV. makes plain the necessity in general of taking into account the *à priori* probabilities of the possible causes. Apart from formal proof this necessity commends itself to careful reflection. If a cause is very improbable in itself, the occurrence of an event, which might very easily follow from it, is not necessarily, so long as there are other possible causes, strong evidence in its favour. Amongst the many writers who, forgetting the theoretic qualification, have been led into actual error, are philosophers as diverse as Laplace, De Morgan, Jevons, and Sigwart, Jevons [3] going so far as to maintain that the fallacious principle he enunciates is "that which common sense leads us to adopt almost instinctively."

15. The theory of the combination of premisses dealt with in §§ 7, 8 of Chapter XIV. has not often been discussed, and the history of it is meagre. Archbishop Whately [4] was led astray

[1] See the passage quoted above, p. 175.

[2] "Essay on Probabilities," in the *Cabinet Encyclopædia*, p. 27.

[3] *Principles of Science*, vol. i. p. 280.

[4] *Logic*, 8th ed. p. 211: "As in the case of two probable premisses, the conclusion is not established except upon the supposition of their being *both* true, so in the case of two distinct and independent indications of the truth of some proposition, unless *both* of them fail, the proposition must be true: we therefore multiply together the fractions indicating the probability of the failure of each—the chances against it—and, the result being the total chances against the establishment of the conclusion by these arguments, this fraction being deducted from unity, the remainder gives the probability *for* it. *E.g.* a certain book is conjectured to be by such and such an author, partly, 1st, from its resemblance in style to his known works; partly, 2nd, from its being attri-

by a superficial error, and De Morgan, adopting the same mistaken rule, pushed it to the point of absurdity.[1] Bishop Terrot[2] approached the question more critically. Boole's[3] last and most considered contribution to the subject of probability dealt with the same topic. I know of no discussion of it during the past sixty years.

Boole's treatment is full and detailed. He states the problem as follows: "Required the probability of an event z, when two circumstances x and y are known to be present,—the probability of the event z, when we know only of the existence of the circumstances x, being p, and the probability, when we only know of the existence of y, being q."[4] His solution, however, is vitiated by the fundamental error examined in § 6 above. Two of his conclusions may be mentioned for their plausibility, but neither is valid.

"If the causes in operation, or the testimonies borne," he

---

buted to him by some one likely to be pretty well informed. Let the probability of the conclusion, as deduced from one of these arguments by itself, be supposed $\frac{2}{5}$, and in the other case $\frac{3}{7}$; then the *opposite* probabilities will be $\frac{3}{5}$ and $\frac{4}{7}$, which multiplied together give $\frac{12}{35}$ as the probability against the conclusion. . . ."

The Archbishop's error, in that a negative can always be turned into an affirmative by a change of verbal expression, was first pointed out by a mere diocesan, Bishop Terrot, in the *Edin. Phil. Trans.* vol. xxi. The mistake is well explained by Boole in the same volume of the *Edin. Phil. Trans.*: "A confusion may here be noted between the probability that a conclusion is proved, and the probability in favour of a conclusion furnished by evidence which does not prove it. In the proof and statement of his rule, Archbishop Whately adopts the former view of the nature of the probabilities concerned in the data. In the exemplification of it, he adopts the latter."

[1] "Theory of Probabilities," *Encyclopædia Metropolitana*, p. 400. He shows by means of it that "if any assertion appear neither likely nor unlikely in itself, then any logical argument in favour of it, however weak the premisses, makes it in some degree more likely than not—a theorem which will be readily admitted on its own evidence." He then gives an example: "*à priori* vegetation on the planets is neither likely nor unlikely; suppose argument from analogy makes it $\frac{3}{10}$; then the total probability is $\frac{1}{2}+\frac{1}{2} \cdot \frac{3}{10}$ or $\frac{13}{20}$." De Morgan seems to accept without hesitation the conclusion to be derived from this, that everything which is not impossible is as probable as not.

[2] "On the Possibility of combining two or more Probabilities of the same Event, so as to form one definite Probability," *Edin. Phil. Trans.*, 1856, vol. xxi.

[3] "On the Application of the Theory of Probabilities to the Question of the Combination of Testimonies or Judgments," *Edin. Phil. Trans.*, 1857, vol. xxi.

[4] *Loc. cit.* p. 631. Boole's principle (*loc. cit.* p. 620) that "the mean strength of any probabilities of an event which are founded upon different judgments or observations is to be measured by that supposed probability of the event *à priori* which those judgments or observations following thereupon would not tend to alter," is not correct if it means more than that the mean strength of z/x and z/y is to be measured by z/xy.

argues, "are, separately, such as to leave the mind in a state of equipoise as respects the event whose probability is sought, united they will but produce the same effect." If, that is to say, $a/h_1 = \frac{1}{2}$ and $a/h_2 = \frac{1}{2}$, he concludes that $a/h_1h_2 = \frac{1}{2}$. The plausibility of this is superficial. Consider, for example, the following instance: h_1 = A is black and B is black or white, h_2 = B is black and A is black or white, a = both A and B are black. Boole also concluded without valid reason that a/h_1h_2 increases, the greater the *à priori* improbability of the combination h_1h_2.

16. The theory of "Testimony" itself, the theory, that is to say, of the combination of the evidence of witnesses, has occupied so considerable a space in the traditional treatment of Probability that it will be worth while to examine it briefly. It may, however, be safely said that the principal conclusions on the subject set out by Condorcet, Laplace, Poisson, Cournot, and Boole, are demonstrably false. The interest of the discussion is chiefly due to the memory of these distinguished failures.

It seems to have been generally believed by these and other logicians and mathematicians[1] that the probability of two witnesses speaking the truth, who are independent in the sense that there is no collusion between them, is always the product of the probabilities that each of them separately will speak the truth.[2] On this basis conclusions such as the following, for example, are arrived at:

X and Y are independent witnesses (*i.e.* there is no collusion between them). The probability that X will speak the truth is x, that Y will speak the truth is y. X and Y agree in a particular statement. The chance that this statement is true is

$$\frac{xy}{xy+(1-x)(1-y)}.$$

For the chance that they both speak the truth is xy, and the chance that they both speak falsely is $(1-x)(1-y)$. As, in this

[1] Perhaps M. Bertrand should be registered as an honourable exception. At least he points out a precisely analogous fallacy in an example where two meteorologists prophesy the weather, *Calcul des Probabilités*, p. 31.

[2] *E.g.*, Boole, *Laws of Thought*, p. 279.
De Morgan, *Formal Logic*, p. 195.
Condorcet, *Essai*, p. 4.
Lacroix, *Traité*, p. 248.
Cournot, *Exposition*, p. 354.
Poisson, *Recherches*, p. 323.
This list could be greatly extended.

case, our hypothesis is that they agree, these two alternatives are exhaustive; whence the above result, which may be found in almost every discussion of the subject.

The fallacy of such reasoning is easily exposed by a more exact statement of the problem. For let a_1 stand for "X_1 asserts a," and let $a/a_1h = x_1$, where h, our general data, is by itself irrelevant to a, *i.e.*, x_1 is the probability that a statement is true of which we only know that X_1 has asserted it. Similarly let us write $b/b_2h = x_2$, where b_2 stands for "X_2 asserts b." The above argument then assumes that, if X_1 and X_2 are witnesses who are causally independent in the sense there is no collusion between them direct or indirect, $ab/a_1b_2h = a/a_1h \,.\, b/b_2h = x_1x_2$.

But $ab/a_1b_2h = a/a_1bb_2h \,.\, b/a_1b_2h$, and this is not equal to x_1x_2 unless $a/a_1bb_2h = a/a_1h$ and $b/a_1b_2h = b/b_2h$. It is not a sufficient condition for this, as seems usually to be supposed, that X_1 and X_2 should be witnesses causally independent of one another. It is also necessary that a and b, *i.e.* the propositions asserted by the witnesses, should be irrelevant to one another and also each of them irrelevant to the fact of the assertion of the other by a witness. If a knowledge of a affects the probability either of b or of b_1, it is evident that the formula breaks down. In the one extreme case, where the assertions of the two contradict one another, $ab/a_1b_2h = 0$. In the other extreme, where the two agree in the same assertion, *i.e.* where $a \equiv b$, $a/a_1bb_2h = 1$ and not $= a/a_1h$.

17. The special problem of the agreement of witnesses, who make the same statement, can be best attacked as follows, a certain amount of simplification being introduced. Let the general data h of the problem include the hypothesis that X_1 and X_2 are each asked and reply to a question to which there is *only one* correct answer. Let $a_i =$ "X_i asserts a in reply to the question," and $m_i =$ "X_i gives the correct answer to the question." Then

$$m_1/a_1h = x_1 \text{ and } m_2/a_2h = x_2,$$

x_1 and x_2 being, in the conventional language of this problem, the "credibilities" of the witnesses. We have, since the witnesses agree and since a follows from m_ia_i and m_i follows from aa_i,

$$m_1/a_1a_2h = m_1m_2/a_1a_2h = m_2/a_1a_2h\,;$$
$$a/a_ih = m_i/a_ih\,;$$
$$a/a_im_ih = 1\,;\; m_i/aa_ih = 1.$$

Also, since the witnesses are, in the ordinary sense, "independent" witnesses, $a_2/a_1ah = a_2/ah$ and $a_2/a_1\bar{a}h = a_2/\bar{a}h$; that is to say, the probability of X_2's asserting a is independent of the fact of X_1's having asserted a, given we know that a is, in fact, true or false, as the case may be.

The probability that, if the witnesses agree, their assertion is true is

$$a/a_1a_2h = m_1/a_1a_2h = \frac{m_1a_2/a_1h}{a_2/a_1h}$$

$$= \frac{a_2/a_1m_1h \,.\, m_1/a_1h}{a_2a/a_1h + a_2\bar{a}/a_1h} = \frac{a_2/a_1ah \,.\, x_1}{a_2/a_1ah \,.\, x_1 + a_2/a_1\bar{a}h \,.\, (1-x_2)}.$$

If this is to be equal to $\dfrac{x_1x_2}{x_1x_2 + (1-x_1)(1-x_2)}$, we must have

$$\frac{a_2/a_1ah}{a_2/a_1\bar{a}h} = \frac{x_2}{1-x_2}.$$

Now $\dfrac{a_2/a_1ah}{a_2/a_1\bar{a}h} = \dfrac{a_2/ah}{a_2/\bar{a}h}$ by the hypothesis of "independence"

$$= \frac{aa_2/h}{\bar{a}a_2/h} \cdot \frac{\bar{a}/h}{a/h} = \frac{a/a_2h}{\bar{a}/a_2h} \cdot \frac{\bar{a}/h}{a/h}$$

$$= \frac{x_2}{1-x_2} \cdot \frac{\bar{a}/h}{a/h}.$$

This then is the assumption which has tacitly slipped into the conventional formula,—that $a/h = \bar{a}/h = \frac{1}{2}$. It is assumed, that is to say, that any proposition taken at random is as likely as not to be true, so that *any* answer to a given question is, *à priori*, as likely as not to be correct. Thus the conventional formula ought to be employed only in those cases where the answer which the "independent" witnesses agree in giving is, *à priori* and apart from their agreement, as likely as not.

18. A somewhat similar confusion has led to the controversy as to whether and in what manner the *à priori* improbability of a statement modifies its credibility in the mouth of a witness whose degree of reliability is known. The fallacy of attaching the same weight to a testimony regardless of the character of what is asserted, is pointed out, of course, by Hume in the *Essay on Miracles*, and his argument, that the great *à priori* improbability of some assertions outweighs the force of testimony otherwise reliable, depends on the avoidance of it. The correct view is also taken by Laplace in his *Essai philosophique* (pp.

98-102), where he argues that a witness is less to be believed when he asserts an extraordinary fact, declaring the opposite view (taken by Diderot in the article on "Certitude" in the *Encyclopédie*) to be inconceivable before "le simple bon sens."

The manner in which the resultant probability is affected depends upon the precise meaning we attach to "degree of reliability" or "coefficient of credibility." If a witness's credibility is represented by x, do we mean that, if a is the true answer, the probability of his giving it is x, or do we mean that if he answers a the probability of a's being true is x? These two things are not equivalent.

Let a_1 stand for "a is asserted by the witness"; h_1 for our evidence bearing on the witness's veracity; and h_2 for other evidence bearing on the truth of a. Let a/h_1h_2, *i.e.* the *à priori* probability of a apart from our knowledge of the fact that the witness has asserted it, be represented by p.

Let $a/a_1h_1 = x_1$ and $a_1/ah_1 = x_2$; so that $x_1 = \frac{a/h_1}{a_1/h_1} \cdot x_2$. In general $a/h_1 \neq a_1/h_1$. Do we mean by the witness's credibility x_1 or x_2?

We require $a/a_1h_2h_2$.

Let $a_1/\bar{a}h_1 = r$, *i.e.* the probability, apart from our special knowledge concerning a, that, if a is false, the witness will hit on that particular falsehood.

$$a/a_1h_1h_2 = \frac{a_1/ah_1h_2 \cdot a/h_1h_2}{a_1/h_1h_2} = \frac{x_2p}{a_1a/h_1h_2 + a_1\bar{a}/h_1h_2}$$

$$= \frac{x_2p}{x_2p + a_1/\bar{a}h_1h_2 \cdot (1-p)} = \frac{x_2p}{x_2p + r(1-p)};$$

for $a_1/ah_1h_2 = a_1/ah_1$ and $a_1/\bar{a}h_1h_2 = a_1/\bar{a}h_1$, since, given *certain* knowledge concerning a, h_2 is irrelevant to the probability of a_1.

19. Generally speaking, all problems, in regard to the combination of testimonies or to the combination of evidence derived from testimony with evidence derived from other sources, may be treated as special instances of the general problem of the combination of arguments. Beyond pointing out the above plausible fallacies, there is little to add. Mr. W. E. Johnson, however, has proposed a method of defining credibility, which is sometimes valuable, because it regards the witness's credibility not absolutely, but with reference to a given type of question, so that it enables us to measure the force of the witness's testimony under special circumstances. If a represents the fact of A's testimony regarding x, then we may define A's credibility for x as a, where a is given by the equation

$$x/ah = x/h + a\sqrt{x/h \cdot \bar{x}/h};$$

so that $a\sqrt{x/h \cdot \bar{x}/h}$ measures the amount by which A's assertion of x increases its probability.

20. One of the most ancient problems in probability is concerned with the gradual diminution of the probability of a past event, as the length of the tradition increases by which it is established. Perhaps the most famous solution of it is that propounded by Craig in his *Theologiae Christianae Principia Mathematica*, published in 1699.[1] He proves that suspicions of any history vary in the duplicate ratio of the times taken from the beginning of the history in a manner which has been described as a kind of parody of Newton's *Principia*. "Craig," says Todhunter, "concluded that faith in the Gospel so far as it depended on oral tradition expired about the year 880, and that so far as it depended on written tradition it would expire in the year 3150. Peterson by adopting a different law of diminution concluded that faith would expire in 1789."[2] About the same time Locke raised the matter in chap. xvi. bk. iv. of the *Essay Concerning Human Understanding*: "Traditional testimonies the farther removed, the less their proof. . . . No Probability can rise higher than its first original." This is evidently intended to combat the view that the long acceptance by the human race of a reputed fact is an additional argument

[1] See Todhunter's *History*, p. 54. It has been suggested that the anonymous essay in the *Phil. Trans.* for 1699 entitled "A Calculation of the Credibility of Human Testimony" is due to Craig. In this it is argued that, if the credibilities of a set of witnesses are $p_1 \ldots p_n$, then if they are successive the resulting probability is the product $p_1 p_2 \ldots p_n$; if they are concurrent, it is:

$$1-(1-p_1)(1-p_2) \ldots (1-p_n).$$

This last result follows from the supposition that the first witness leaves an amount of doubt represented by $1-p_1$; of this the second removes the fraction p_2, and so on. See also Lacroix, *Traité élémentaire*, p. 262. The above theory was actually adopted by Bicquilley.

[2] In the *Budget of Paradoxes* De Morgan quotes Lee, the Cambridge Orientalist, to the effect that Mahometan writers, in reply to the argument that the Koran has not the evidence derived from Christian miracles, contend that, as evidence of Christian miracles is daily weaker, a time must at last arrive when it will fail of affording assurance that they were miracles at all: whence the necessity of another prophet and other miracles.

in its favour and that a long tradition increases rather than diminishes the strength of an assertion. "This is certain," says Locke, "that what in one age was affirmed upon slight grounds, can never after come to be more valid in future ages, by being often repeated." In this connection he calls attention to "a rule observed in the law of England, which is, that though the attested copy of a record be good proof, yet the copy of a copy never so well attested, and by never so credible witnesses, will not be admitted as a proof in Judicature." If this is still a good rule of law, it seems to indicate an excessive subservience to the principle of the decay of evidence.

But, although Locke affirms sound maxims, he gives no theory that can afford a basis for calculation. Craig, however, was the more typical professor of probability, and in attempting an algebraic formula he was the first of a considerable family. The last grand discussion of the problem took place in the columns of the *Educational Times*.[1] Macfarlane[2] mentions that four different solutions have been put forward by mathematicians of the problem: "A says that B says that a certain event took place; required the probability that the event did take place, p_1 and p_2 being A's and B's respective probabilities of speaking the truth." Of these solutions only Cayley's is correct.

[1] Reprinted in *Mathematics from the Educational Times*, vol. xxvii.

[2] *Algebra of Logic*, p. 151. Macfarlane attempts a solution of the general problem without success. Its solution is not difficult, if enough unknowns are introduced, but of very little interest.

CHAPTER XVII

SOME PROBLEMS IN INVERSE PROBABILITY, INCLUDING AVERAGES

1. THE present chapter deals with 'problems'—that is to say, with applications to particular abstract questions of some of the fundamental theorems demonstrated in Chapter XIV. It is without philosophical interest and should probably be omitted by most readers. I introduce it here in order to show the analytical power of the method developed above and its advantage in ease and especially in accuracy over other methods which have been employed.[1] § **2** is mainly based upon some problems discussed by Boole. §§ **3-7** deal with the fundamental theory connecting averages and laws of error. §§ **8-11** treat discursively the Arithmetic Average, the Method of Least Squares, and Weighting.

2. In the following paragraph solutions are given of some problems posed by Boole in chapter xx. of his *Laws of Thought*. Boole's own method of solving them is constantly erroneous,[2] and the difficulty of his method is so great that I do not know of any one but himself who has ever attempted to use it. The term 'cause' is frequently used in these examples where it might have been better to use the term 'hypothesis.' For by a possible cause of an event no more is here meant than an antecedent occurrence, the knowledge of which is relevant to our anticipation of the event; it does not mean an antecedent from which the event in question *must* follow.

(56) The *à priori* probabilities of two causes A_1 and A_2 are c_1 and c_2 respectively. The probability that if the cause A_1

[1] Such examples as these might sometimes be set to test the wits of students. The problems on Probability usually given are simply problems on mathematical combinations. These, on the other hand, are really problems in logic.

[2] For the reason given in § 6 of Chapter XVI. The solutions of problems I.–VI., for example, in the *Laws of Thought*, chap. xx., are all erroneous.

occur, an event E will accompany it (whether as a consequence of A_1 or not), is p_1, and the probability that E will accompany A_2, if A_2 present itself, is p_2. Moreover, the event E cannot appear in the absence of both the causes A_1 and A_2. Required the probability of the event E.

This problem is of great historical interest and has been called Boole's 'Challenge Problem.' Boole originally proposed it for solution to mathematicians in 1851 in the *Cambridge and Dublin Mathematical Journal*. A result was given by Cayley[1] in the *Philosophical Magazine*, which Boole declared to be erroneous.[2] He then entered the field with his own solution.[3] "Several attempts at its solution," he says, "have been forwarded to me, all of them by mathematicians of great eminence, all of them admitting of particular verification, yet differing from each other and from the truth."[4] After calculations of considerable length and great difficulty he arrives at the conclusion that u is the probability of the event E where u is that root of the equation

$$\frac{[1-c_1(1-p_1)-u][1-c_2(1-p_2)-u]}{1-u}=\frac{(u-c_1p_1)(u-c_2p_2)}{c_1p_1+c_2p_2-u}$$

which is not less than c_1p_1 and c_2p_2 and not greater than $1-c_1(1-p_1)$, $1-c_2(1-p_2)$, or $c_1p_1+c_2p_2$.

This solution can easily be seen to be wrong. For in the case where A_1 and A_2 cannot both occur, the solution is $u=c_1p_1+c_2p_2$; whereas Boole's equations do not reduce to

[1] *Phil. Mag.* 4th series, vol. vi.

[2] Cayley's solution was defended against Boole by Dedekind (*Crelle's Journal*, vol. l. p. 268). The difference arises out of the extreme ambiguity as to the meaning of the terms as employed by Cayley.

[3] "Solution of a Question in the Theory of Probabilities," *Phil. Mag.* 4th series, vol. vii., 1854. This solution is the same as that printed by Boole shortly afterwards in the *Laws of Thought*, pp. 321-326. In the *Phil. Mag.* Wilbraham gave as the solution $u=c_1p_1+c_2p_2-z$, where z is necessarily less than either c_1p_1 or c_2p_2. This solution is correct so far as it goes, but is not complete. The problem is also discussed by Macfarlane, *Algebra of Logic*, p. 154.

[4] In proposing the problem Boole had said: "The motives which have led me, after much consideration, to adopt, with reference to this question, a course unusual in the present day, and not upon slight grounds to be revived, are the following: First, I propose the question as a test of the sufficiency of received methods. Secondly, I anticipate that its discussion will in some measure add to our knowledge of an important branch of pure analysis." When printing his own solution in the *Laws of Thought*, he adds, that the above "led to some interesting private correspondence, but did not elicit a solution."

this simplified form. The mistake which Boole has made is the one general to his system, referred to in Chapter XVI., § 6.[1]

The correct solution, which is very simple, can be reached as follows:

Let a_1, a_2, e assert the occurrences of the two causes and the event respectively, and let h be the *data* of the problem.

Then we have $a_1/h = c_1$, $a_2/h = c_2$, $e/a_1h = p_1$, $e/a_2h = p_2$: we require e/h. Let $e/h = u$, and let $a_1a_2/eh = z$. Since the event cannot occur in the absence of both the causes,

$$e/\bar{a}_1\bar{a}_2h = 0.$$

It follows from this that $\bar{a}_1\bar{a}_2/eh = 0$, unless $e/h = 0$,

i.e.
$$(a_1 + a_2)/eh = 1,$$

whence
$$a_1/eh + a_2/eh = 1 + a_1a_2/eh \qquad \text{by (24).}$$

Now
$$a_1/eh = \frac{c_1p_1}{u} \text{ and } a_2/eh = \frac{c_2p_2}{u},$$

$$\therefore\ u = \frac{c_1p_1 + c_2p_2}{1 + z},$$

where z is the probability after the event that *both* the causes were present.

If we write $ea_1a_2/h = y$,

$$y = a_1a_2/eh\ .\ e/h = uz,$$

so that
$$u = (c_1p_1 + c_2p_2) - y.$$

Boole's solution fails by attempting to be independent of y or z.

(56.1). Suppose that we wish to find limits for the solution which are independent of y and z: then, since $y \geqslant 0$, $u \leqslant c_1p_1 + c_2p_2$.

Again

$$e/h = e\bar{a}_1/h + ea_1/h \leqslant \bar{a}_1/h + ea_1/h \leqslant 1 - c_1 + c_1p_1 \text{ by (24.2) and (4).}$$

Similarly $e/h \leqslant 1 - c_2 + c_2p_2$. From the same equations it appears that $e/h \geqslant c_1p_1$ and $\geqslant c_2p_2$.

[1] Boole's error is pointed out and a correct solution given in Mr. McColl's "Sixth Article on the Calculus of Equivalent Statements" (*Proc. Lond. Math. Soc.* vol. xxviii. p. 562).

$\therefore$ u lies between

$$\text{the greatest of} \begin{cases} c_1p_1 \\ c_2p_2 \end{cases} \text{and the least of} \begin{cases} c_1p_1 + c_2p_2 \\ 1 - c_1(1-p_1) \\ 1 - c_2(1-p_2). \end{cases}$$

It will be seen that these numerical limits are the same as the limits obtained by Boole for the roots of his equations.

(56.2) Suppose that the *à priori* probabilities of the causes c_1 and c_2 are to be eliminated. The only limit we then have is $u<p_1+p_2$.

(56.3) Suppose that one of the *à priori* probabilities c_2 is to be eliminated. We then have limits $c_1p_1 \lessgtr u \lessgtr 1 - c_1 + c_1p_1$. If, therefore, c_1 is large, u does not differ widely from c_1p_1.

(56.4) Suppose p_2 is to be eliminated. We then have

$$\begin{aligned} c_1p_1 \lessgtr u \lessgtr c_1p_1 + c_2 \\ \lessgtr c_1p_1 + 1 - c_1. \end{aligned}$$

If therefore c_1 is large or c_2 small, u does not differ widely from c_1p_1.

(56.5) If $a_1/a_2h = a_1/h$, *i.e.* if our knowledge of each of the causes is independent, we have a closer approximation. For

$$\begin{gathered} y = ea_1a_2/h = e/a_1a_2h \,.\, a_1/a_2h \,.\, a_2/h = e/a_1a_2h \,.\, c_1c_2, \\ \therefore\ u = c_1p_1 + c_2p_2 - c_1c_2 \,.\, e/a_1a_2h, \\ \therefore\ u > c_1p_1 + c_2p_2 - c_1c_2. \end{gathered}$$

(57) We may now generalise (56) and discuss the case of n causes. If an event can only happen as a consequence of one *or more* of certain causes $A_1, A_2, \ldots A_n$, and if c_1 is the *à priori* probability of the cause A_1 and p_1 the probability that, if the cause A_1 be known to exist, the event E will occur : required the probability of E.

This is Boole's problem VI. (*Laws of Thought*, p. 336). As the result of ten pages of mathematics, he finds the solution to be the root lying between certain limits of an equation of the nth degree which he cannot solve. I know no other discussion of the problem. The solution is as follows :

$$e/h = e\bar{a}_1/h + ea_1/h = e\bar{a}_1/h + e/a_1h \,.\, a_1/h = e\bar{a}_1/h + c_1p_1 \quad \text{(i.)}$$
$$e\bar{a}_1/h = e\bar{a}_1\bar{a}_2/h + e\bar{a}_1/a_2h \,.\, a_2/h = e\bar{a}_1\bar{a}_2/h + c_2 \,.\, e\bar{a}_1/a_2h,$$
$$e\bar{a}_1/a_2h = e/a_2h - ea_1/a_2h = p_2 - \frac{1}{c_2} \,.\, ea_1a_2/h,$$

$$\therefore\ e/h = e\bar{a}_1\bar{a}_2/h + c_1p_1 + c_2p_2 - ea_1a_2/h,$$

$$e\bar{a}_1\bar{a}_2/h = e\bar{a}_1\bar{a}_2\bar{a}_3/h + e\bar{a}_1\bar{a}_2a_3/h,$$

and

$$e\bar{a}_1\bar{a}_2a_3/h = e\bar{a}_1\bar{a}_2/a_3h \,.\, c_3 = c_3\{e/a_3h - e\overline{\bar{a}_1\bar{a}_2}/a_3h\}$$

$$= c_3p_3 - e\overline{\bar{a}_1\bar{a}_2}a_3/h,$$

$$\therefore\ e/h = e\bar{a}_1\bar{a}_2\bar{a}_3/h + c_1p_1 + c_2p_2 + c_3p_3 - e\bar{\bar{a}}_1a_2/h - e\overline{\bar{a}_1\bar{a}_2}a_3/h.$$

In general

$$e\bar{a}_1\bar{a}_2 \ldots \bar{a}_{r-1}/h = e\bar{a}_1\bar{a}_2 \ldots \bar{a}_{r-1}\bar{a}_r/h + e\bar{a}_1\bar{a}_2 \ldots \bar{a}_{r-1}a_r/h$$

$$= e\bar{a}_1 \ldots \bar{a}_r/h + e\bar{a}_1 \ldots \bar{a}_{r-1}/a_rh \,.\, c_r$$

$$= e\bar{a}_1 \ldots \bar{a}_r/h + c_r\{e/a_rh - e\overline{\bar{a}_1 \ldots \bar{a}_{r-1}}/a_rh\}$$

$$= e\bar{a}_1 \ldots \bar{a}_r/h + c_rp_r - e\overline{\bar{a}_1 \ldots \bar{a}_{r-1}}a_r/h,$$

$\therefore$ finally we have $e/h = e\bar{a}_1 \ldots \bar{a}_n/h + \sum_1^n c_rp_r - \sum_2^n e\overline{\bar{a} \ldots \bar{a}_{r-1}}a_r/h.$

But since the n causes are supposed to be exhaustive

$$e\bar{a}_1 \ldots \bar{a}_n/h = 0,$$

$$\therefore\ e/h = \sum_1^n c_rp_r - \sum_2^n e\overline{\bar{a}_1 \ldots \bar{a}_{r-1}}a_r/h \qquad \text{(ii.)}.$$

Let

$$\overline{e\bar{a}_1 \ldots \bar{a}_{r-1}}a_r/h = n_r\,;$$

then

$$e/h = \sum_1^n c_rp_r - \sum_2^n n_r \qquad \text{(iii.)}.$$

(57.1) If our knowledge of the several causes is independent, if, that is to say, our knowledge of the existence of any one of them is not relevant to the probability of the existence of any other, so that $a_r/a_sh = a_r/h = c_r$, then

$$\overline{e\bar{a}_1 \ldots \bar{a}_{r-1}}a_r/h = e\overline{\bar{a}_1 \ldots \bar{a}_{r-1}}/a_rh \,.\, c_r$$

$$= c_r \,.\, e/\overline{\bar{a}_1 \ldots \bar{a}_{r-1}}a_rh\{1 - \bar{a}_1 \ldots \bar{a}_{r-1}/a_rh\}$$

$$= c_r[1 - \prod_1^{r-1}(1 - c_1) \ldots (1 - c_{r-1})]e/\overline{\bar{a}_1 \ldots \bar{a}_{r-1}}a_rh.$$

Let

$$e/\overline{\bar{a}_1 \ldots \bar{a}_{r-1}}a_rh = m_r,$$

then

$$e/h = \sum_{r=1}^{r=n} c_rp_r - \sum_{r=2}^{r=n} c_r[1 - \prod_{s=1}^{s=r-1}(1 - c_s)]m_r.$$

These results do not look very promising as they stand, but they lead to some useful approximations on the elimination of m_r and n_r and to some interesting special cases.

(57.2) From equation (i.) it follows that $e/h > c_1p_1$ and $e/h < 1 - c_1(1-p_1)$; and from equation (ii.) that $e/h < \sum_1^n c_r p_r$;

$\therefore$ e/h lies between

$$\text{the greatest of} \begin{cases} c_1p_1 \\ \vdots \\ c_np_n \end{cases} \text{and the least of} \begin{cases} \sum_1^n c_rp_r \\ 1-c_1(1-p_1) \\ \vdots \\ 1-c_n(1-p_n). \end{cases}$$

(57.3) Further, if the causes are independent it follows from (57.1) that

$$e/h > \sum_1^n c_rp_r - \sum_2^n c_r[1 - \prod_1^{r-1}(1-c_s)],$$

so that e/h lies between

$$\text{the greatest of} \begin{cases} \sum_1^n c_rp_r - \sum_2^n c_r[1-\prod_1^{r-1}(1-c_s)] \\ c_1p_1 \\ \vdots \\ c_np_n \end{cases} \text{and the least of} \begin{cases} \sum_1^n c_rp_r \\ 1-c_1(1-p_1) \\ \vdots \\ 1-c_n(1-p_n). \end{cases}$$

(57.4) Now consider the case in which $p_1 = p_2 = \ldots = p_n = 1$, *i.e.* in which any of the causes would be sufficient, and in which the causes are independent. Then $m_r = 1$; so that

$$e/h = \sum_{r=1}^{r=n} c_r - \sum_{r=2}^{r=n} c_r[1 - \prod_{s=1}^{s=r-1}(1-c_s)]$$
$$= 1 - (1-c_1)(1-c_2)\ldots(1-c_n).$$

(57.5) Let $c_1, c_2 \ldots c_n$ be small quantities so that their squares and products may be neglected.

Then
$$e/h = \Sigma c_rp_r,$$

i.e. the smaller the probabilities of the causes the more do they approach the condition of being mutually exclusive.[1]

(57.6) The *à posteriori* probability of a particular cause a_r after the event has been observed is

$$a_r/eh = \frac{e/a_rh \,.\, a_r/h}{e/h}$$
$$= \frac{p_rc_r}{e/h}.$$

(This is Boole's problem IX., p. 357).

[1] Boole arrives at this result, *Laws of Thought*, p. 345, but I doubt his proof.

(58) The probability of the occurrence of a certain natural phenomenon under given circumstances is p. There is also a probability a of a permanent cause of the phenomenon, *i.e.* of a cause which would always produce the event under the circumstances supposed. What is the probability that the phenomenon, being observed n times, will occur the $n+1^{th}$?

This is Boole's problem X. (*Laws of Thought*, p. 358). Boole arrives by his own method at the same result as that given below. It is necessary first of all to state the assumption somewhat more precisely. If x_r asserts the occurrence of the event at the rth trial and t the existence of the 'permanent cause' we have

$$x_r/h = p,\ t/h = a,\ x_r/th = 1,$$

and we require $$x_{n+1}/x_1 \ldots x_n h = y_{n+1}.$$

It is also assumed that if there is *no* permanent cause the probability of x_s is not affected by the observations x_r, etc., *i.e.*

$$x_s/x_r \ldots x_t \bar{t} h = x_s/\bar{t}h,$$ [1]

$$x_s/\bar{t}h = \frac{x_s\bar{t}/h}{\bar{t}/h} = \frac{x_s/h - x_s t/h}{\bar{t}/h} = \frac{p-a}{1-a},$$

$$x_r/x_1 \ldots x_{r-1}h = x_r t/x_1 \ldots x_{r-1}h + x_r\bar{t}/x_1 \ldots x_{r-1}h$$

$$= t/x_1 \ldots x_{r-1}h + x_r/\bar{t}x_1 \ldots x_{r-1}h \,.\, \bar{t}/x_1 \ldots x_{r-1}h$$

$$= \frac{x_1 \ldots x_{r-1}t/h}{x_1 \ldots x_{r-1}/h} + \frac{p-a}{1-a} \cdot \frac{x_1 \ldots x_{r-1}/\bar{t}h \,.\, \bar{t}/h}{x_1 \ldots x_{r-1}/h}$$

$$= \frac{a}{y_1 y_2 \ldots y_{r-1}} + \frac{p-a}{1-a}\,\frac{\left(\frac{p-a}{1-a}\right)^{r-1}(1-a)}{y_1 y_2 \ldots y_{r-1}},$$

i.e. $$y_r = \frac{a + (p-a)\left(\frac{p-a}{1-a}\right)^{r-1}}{y_1 y_2 \ldots y_{r-1}}.$$

Also $$y_1 = p \text{ and } y_2 = \frac{a + (p-a)\frac{p-a}{1-a}}{y_1},$$

---

[1] This assumption, which is tacitly introduced by Boole, is not generally justifiable. I use it here, as my main purpose is to illustrate a method. The same problem, *without* this assumption, will be discussed in dealing with Pure Induction.

so that
$$y_{n+1}=\frac{a+(p-a)\left(\dfrac{p-a}{1-a}\right)^{n}}{a+(p-a)\left(\dfrac{p-a}{1-a}\right)^{n-1}}.$$

(58.1) If $p=a$, $y_n=1$; for if an event can only occur as the result of a permanent cause, a single occurrence makes future occurrences certain under similar conditions.

(58.2)
$$y_{n+1}-y_n=\frac{a(p-a)\left(\dfrac{p-a}{1-a}\right)^{n-2}\left(1-\dfrac{p-a}{1-a}\right)}{\left[a+(p-a)\left(\dfrac{p-a}{1-a}\right)^{n-1}\right]\left[a+(p-a)\left(\dfrac{p-a}{1-a}\right)^{n-2}\right]}$$

(by easy algebra);

and p is always $>a$ and <1.

So that $(p-a)\left(\dfrac{p-a}{1-a}\right)^{r}$ is positive and decreases as r increases,
$$\therefore\ y_{n+1}>y_n.$$

As n increases $y_n=1-\epsilon$, where
$$\epsilon=(p-a)\left[1-\left(\frac{p-a}{1-a}\right)\right]\frac{\left(\dfrac{p-a}{1-a}\right)^{n-2}}{a+(p-a)\left(\dfrac{p-a}{1-a}\right)^{n-2}},$$

so that for any value of η however small a value of n can be found such that $\epsilon<\eta$ so long as a is not zero.

(58.3) t_n the *à posteriori* probability of a permanent cause after n successful observations is
$$t/x_1\ldots x_nh=\frac{x_1\ldots x_n/th\,.\,t/h}{x_1\ldots x_n/h}=\frac{a}{y_1y_2\ldots y_n},$$

i.e.
$$t_n=\frac{a}{a+(p-a)\left(\dfrac{p-a}{1-a}\right)^{n}},$$

$$t_n=1-\epsilon',\text{ where }\epsilon'=\frac{(p-a)\left(\dfrac{p-a}{1-a}\right)^{n}}{a+(p-a)\left(\dfrac{p-a}{1-a}\right)^{n}}.$$

So that t_n approaches the limit unity as n increases, so long as a is not zero.

3. The following is a common type of statistical problem.[1] We are given a series of measurements, or observations, or estimates of the true value of a given quantity; and we wish to determine what function of these measurements will yield us the *most probable* value of the quantity, on the basis of this evidence. The problem is not determinate unless we have some good ground for making an assumption as to how likely we are in each case to make errors of given magnitudes. But such an assumption, with or without justification, is frequently made.

The functions of the original measurements which we commonly employ, in order to yield us approximations to the most probable value of the quantity measured, are the various kinds of means or averages—the arithmetic mean, for example, or the median. The relation, which we assume, between errors of different magnitudes and the probabilities that we have made errors of those magnitudes, is called a *law of error*. Corresponding to each law of error which we might assume, there is some function of the measurements which represents the most probable value of the quantity. The object of the following paragraphs is to discover what laws of error, if we assume them, correspond to each of the simple types of average, and to discover this by means of a systematic method.

4. Let us assume that the real value of the quantity is either $b_1, \ldots b_r \ldots b_n$, and let a_r represent the conclusion that the value is, in fact, b_r. Further let x_r represent the evidence that a measurement has been made of magnitude y_r.

If a measurement y_p has been made, what is the probability that the real value is b_s? The application of the theorem of inverse probability yields the following result:

$$a_s/x_p h = \frac{x_p/a_s h. \quad a_s/h}{\sum_{r=1}^{r=n} x_p/a_r h. \quad a_r/h}$$

(the number of possible values of the quantity being n), where h stands for any other relevant evidence which we may have, in addition to the fact that a measurement x_p has been made.

Next, let us suppose that a number of measurements $y_1 \ldots y_m$

[1] The substance of §§ 3-7 has been printed in the *Journal of the Royal Statistical Society*, vol. lxxiv. p. 323 (February 1911).

have been made; what is now the probability that the real value is b_s? We require the value of $a_s/x_1x_2 \ldots x_mh$. As before,

$$a_s/x_1x_2 \ldots x_mh = \frac{x_1 \ldots\ldots x_m/a_sh.\ \ a_s/h}{\sum_{r=1}^{r=x} x_1 \ldots x_m/a_rh.\ \ a_r/h}.$$

At this point we must introduce the simplifying assumption that, if we knew the real value of the quantity, the different measurements of it would be *independent*, in the sense that a knowledge of what errors have actually been made in some of the measurements would not affect in any way our estimate of what errors are likely to be made in the others. We assume, in fact, that $x_r/x_p \ldots x_sa_rh = x_r/a_rh$. This assumption is exceedingly important. It is tantamount to the assumption that our law of error is unchanged throughout the series of observations in question. The general evidence h, that is to say, which justifies our assumption of the particular law of error which we do assume, is of such a character that a knowledge of the actual errors made in a number of measurements, not more numerous than those in question, are absolutely or approximately irrelevant to the question of what form of law we ought to assume. The law of error which we assume will be based, presumably, on an experience of the relative frequency with which errors of different magnitudes have been made under analogous circumstances in the past. The above assumption will *not* be justified if the additional experience, which a knowledge of the errors in the new measurements would supply, is sufficiently comprehensive, relatively to our former experience, to be capable of modifying our assumption as to the shape of the law of error, or if it suggests that the circumstances, in which the measurements are being carried out, are not so closely analogous as was originally supposed.

With this assumption, *i.e.* that x_1, etc., are independent of one another relatively to evidence a_rh, etc., it follows from the ordinary rule for the multiplication of independent probabilities that

$$x_1 \ldots\ldots x_m/a_sh = \prod_{q=1}^{q=m} x_q/a_sh.$$

Hence

$$a_s/x_1x_2 \ldots x_mh = \frac{a_s/h.\ \ \prod_{q=1}^{q=m} x_q/a_sh}{\sum_{r=1}^{r=n}\left[\prod_{q=1}^{q=m} x_q/a_rh.\ \ a_r/h\right]}.$$

The *most probable* value of the quantity under measurement, given the m measurements y_1, etc.—which is our *quaesitum*—is therefore that value which makes the above expression a maximum. Since the denominator is the same for all values of b_s, we must find the value which makes the numerator a maximum. Let us assume that $a_1/h = a_2/h = \ldots = a_n/h$. We assume, that is to say, that we have no reason *à priori* (*i.e.* before any measurements have been made) for thinking any one of the possible values of the quantity more likely than any other. We require, therefore, the value of b_s, which makes the expression $\prod_{q=1}^{q=m} x_q/a_s h$ a maximum. Let us denote this value by y.

We can make no further progress without a further assumption. Let us assume that $x_q/a_s h$—namely, the probability of a measurement y_q assuming the real value to be b_s—is an algebraic function f of y_q and b_s, the same function for all values of y_q and b_s within the limits of the problem.[1] We assume, that is to say, $x_q/a_s h = f(y_q, b_s)$, and we have to find the value of b_s, namely y, which makes $\prod_{q=1}^{q=m} f(y_q, y)$ a maximum. Equating to zero the differential coefficient of this expression with respect to y, we have $\sum_{q=1}^{q=m} \frac{f'(y_q, y)}{f(y_q, y)} = 0$,[2] where $f' = \frac{df}{dy}$. This equation may be written for brevity in the form $\Sigma \frac{f'_q}{f_q} = 0$.

If we solve this equation for y, the result gives us the value of the quantity under observation, which is most probable relatively to the measurements we have made.

The act of differentiation assumes that the possible values of y are so numerous and so uniformly distributed within the range in question, that we may, without sensible error, regard them as continuous.

5. This completes the *prolegomena* of the inquiry. We are

[1] Gauss, in obtaining the normal law of error, made, in effect, the more special assumption that $x_q/a_s h$ is a function of e_q only, where e_q is the error and $e_q = b_s - y_q$. We shall find in the sequel that all symmetrical laws of error, such that positive and negative errors of the same absolute magnitude are equally likely, satisfy this condition—the normal law, for example, and the simplest median law. But other laws, such as those which lead to the geometric mean, do not satisfy it.

[2] Since none of the measurements actually made can be impossible, none of the expressions $f(y_q, y)$ can vanish.

now in a position to discover what laws of error correspond to given assumptions respecting the algebraic relation between the measurements and the most probable value of the quantity, and *vice versa*. For the law of error determines the form of $f(y_q,y)$. And the form of $f(y_q,y)$ determines the algebraic relation $\Sigma\frac{f'_q}{f_q}=0$ between the measurements and the most probable value. It may be well to repeat that $f(y_q,y)$ denotes the probability to us that an observer will make a measurement y_q in observing a quantity whose true value we know to be y. A law of error tells us what this probability is for all possible values of y_q and y within the limits of the problem.

(i.) If the most probable value of the quantity is equal to the arithmetic mean of the measurements, what law of error does this imply?

$\Sigma\frac{f'_q}{f_q}=0$ must be equivalent to $\Sigma(y-y_q)=0$, since the most probable value y must equal $\frac{1}{m}\sum_{q=1}^{q=m}y_q$.

$\therefore \frac{f'_q}{f_q}=\phi''(y)(y-y_q)$ where $\phi''(y)$ is some function which is not zero and is independent of y_q.

Integrating,

$\log f_q=\int\phi''(y)(y-y_q)dy+\psi(y_q)$ where $\psi(y_q)$ is some function independent of y.

$$=\phi'(y)(y-y_q)-\phi(y)+\psi(y_q).$$

So that $\quad f_q=e^{\phi'(y)(y-y_q)-\phi(y)+\psi(y_q)}.$

Any law of error of this type, therefore, leads to the arithmetic mean of the measurements as the most probable value of the quantity measured.

If we put $\phi(y)=-k^2y^2$ and $\psi(y_q)=-k^2y_q^{\,2}+\log A$, we obtain $f_q=Ae^{-k^2(y-y_q)^2}$, the form normally assumed.

$=Ae^{-k^2z_q^2}$, where z_q is the absolute magnitude of the error in the measurement y_q.

This is, clearly, only one amongst a number of possible solutions. But with one additional assumption we can prove that this is the only law of error which leads to the arithmetic mean.

Let us assume that negative and positive errors of the same absolute amount are equally likely.

In this case f_q must be of the form $Be^{\theta(y-y_q)^2}$,

$$\therefore \phi'(y)(y-y_q)-\phi(y)+\psi(y_q)=\theta(y-y_q)^2.$$

Differentiating with respect to y,

$$\phi''(y)=2\frac{d}{d(y-y_q)^2}\theta(y-y_q)^2.$$

But $\phi''(y)$ is, by hypothesis, independent of y_q.

$\therefore \quad \frac{d}{d(y-y_q)^2}\theta(y-y_q)^2=-k^2$ where k is constant; integrating,

$\theta(y-y_q)^2=-k^2(y-y_q)^2+\log C$ and we have $f_q=Ae^{-k^2(y-y_q)^2}$ (where $A=BC$).

(ii.) What is the law of error, if the geometric mean of the measurements leads to the most probable value of the quantity?

In this case $\Sigma\frac{f'_q}{f_q}=0$ must be equivalent to $\prod_{q=1}^{q=m} y_q=y^m$, *i.e.* to

$$\Sigma\log\frac{y_q}{y}=0.$$

Proceeding as before, we find that the law of error is

$$f_q=A\,e^{\phi'(y)\log\frac{y_q}{y}+\int\frac{\phi'(y)}{y}dy+\psi(y_q)}.$$

There is no solution of this which satisfies the condition that negative and positive errors of the same absolute magnitude are equally likely. For we must have

$$\phi'(y)\log\frac{y_q}{y}+\int\frac{\phi'(y)}{y}dy+\psi(y_q)=\phi(y-y_q)^2$$

$$\text{or }\phi''(y)\log\frac{y_q}{y}=\frac{d}{dy}\phi(y-y_q)^2,$$

which is impossible.

The simplest law of error, which leads to the geometric mean, seems to be obtained by putting $\phi'(y)=-ky$, $\psi(y_q)=0$. This gives $f_q=A\left(\frac{y}{y_q}\right)^{ky}e^{-ky}$.

A law of error, which leads to the geometric mean of the observations as the most probable value of the quantity, has been previously discussed by Sir Donald McAlister (*Proceedings of the Royal Society*, vol. xxix. (1879) p. 365). His investigation depends upon the obvious fact that, if the geometric mean of the observations yields the most probable value of the quantity, the arithmetic mean of the logarithms of the observations must yield the most probable value of the logarithm of the quantity. Hence, if we suppose that the logarithms of the observations obey the normal law of error (which leads to their arithmetic mean as the most probable value of the logarithms of the quantity), we can by substitution find a law of error for the observations themselves which must lead to the geometric mean of them as the most probable value of the quantity itself.

If, as before, the observations are denoted by y_q, etc., and the quantity by y, let their logarithms be denoted by l_q, etc., and by l. Then, if l_q, etc., obey the normal law of error, $f(l_q,l) = \mathrm{A}e^{-k^2(l_q-l)^2}$. Hence the law of error for y_q, etc., is determined by

$$f(y_q,y) = \mathrm{A}e^{-k^2(\log y_q - \log y)^2}$$
$$= \mathrm{A}e^{-k^2\left(\log \frac{y_q}{y}\right)^2},$$

and the most probable value of y must, clearly, be the geometric mean of y_q, etc.

This is the law of error which was arrived at by Sir Donald McAlister. It can easily be shown that it is a special case of the generalised form which I have given above of all laws of error leading to the geometric mean. For if we put $\psi(y_q) = -k^2(\log y_q)^2$, and $\phi'(y) = 2k^2 \log y$, we have

$$f_q = \mathrm{A}e^{2k^2 \log y \log \frac{y_q}{y} + \int 2k^2 \frac{\log y}{y} dy - k^2(\log y_q)^2}$$
$$= \mathrm{A}e^{2k^2 \log y \log y_q - 2k^2(\log y)^2 + k^2(\log y)^2 - k^2(\log y_q)^2}$$
$$= \mathrm{A}e^{-k^2\left(\log \frac{y_q}{y}\right)^2}.$$

A similar result has been obtained by Professor J. C. Kapteyn.[1] But he is investigating frequency curves, not laws of error, and this result is merely incidental to his main discussion. His method, however, is not unlike a more generalised form of Sir Donald McAlister's. In order to discover the frequency curve of certain quantities y, he supposes that there are certain other quantities z, functions of the quantities y, which are given by $z = \mathrm{F}(y)$, and that the frequency curve of these quantities z is *normal*. By this device he is enabled in the investigation of a type of skew frequency curve, which is likely to be met with often, to utilise certain statistical constants corresponding to

[1] *Skew Frequency Curves*, p. 22, published by the Astronomical Laboratory at Groningen (1903).

those which have been already calculated for the normal curve.

In fact the main advantage both of Sir Donald McAlister's law of error and of Professor Kapteyn's frequency curves lies in the possibility of adapting without much trouble to unsymmetrical phenomena numerous expressions which have been already calculated for the normal law of error and the normal curve of frequency.[1]

This method of proceeding from arithmetic to geometric laws of error is clearly capable of generalisation. We have dealt with the geometric law which can be derived from the normal arithmetic law. Similarly if we start from the simplest geometric law of error, namely, $f_q = A\left(\frac{y}{y_q}\right)^{k^2y} e^{-k^2y}$, we can easily find, by writing $\log y = l$ and $\log y_q = l_q$, the corresponding arithmetic law, namely, $f_q = Ae^{k^2e^l(l-l_q)-k^2e^l}$, which is obtained from the generalised arithmetic law by putting $\phi(l) = k^2e^l$ and $\psi(l_q) = 0$. And, in general, corresponding to the arithmetic law

$$f_q = Ae^{\phi'(y)(y-y_q)-\phi(y)+\psi(y_q)},$$

we have the geometric law

$$f_q = Ae^{\phi'_1(z)\log\frac{z_q}{z} + \int\frac{\phi_1(z)}{z}dz + \psi_1(z_q)},$$

where

$$y = \log z,\ y_q = \log z_q,\ \int\frac{\phi_1'(z)}{z}dz = \phi(\log z) \text{ and } \psi_1(z_q) = \psi(\log z_q).$$

(iii.) What law of error does the harmonic mean imply?

In this case, $\Sigma\frac{f'_q}{f_q} = 0$ must be equivalent to $\Sigma\left(\frac{1}{y_q} - \frac{1}{y}\right) = 0$.

Proceeding as before, we find that $f_q = Ae^{\phi'(y)\left[\frac{1}{y_q}-\frac{1}{y}\right] - \int\frac{\phi'(y)}{y^2}dy + \psi(y_q)}$. A simple form of this is obtained by putting $\phi'(y) = -k^2y^2$ and $\psi(y_q) = -k^2y_q$. Then $f_q = Ae^{\frac{k^2}{y_q}(y-y_q)^2} = Ae^{-k^2\frac{z_q^2}{y_q}}$. With this law, positive and negative errors of the same absolute magnitude are not equally likely.

(iv.) If the most probable value of the quantity is equal to the median of the measurements, what is the law of error?

The median is usually defined as the measurement which

[1] It may be added that Professor Kapteyn's monograph brings forward considerations which would be extremely valuable in determining the types of phenomena to which geometric laws of error are likely to be applicable.

occupies the middle position when the measurements are ranged in order of magnitude. If the number of measurements m is odd, the most probable value of the quantity is the $\frac{m+1}{2}$th, and, if the number is even, all values between the $\frac{m}{2}$th and the $\left(\frac{m}{2}+1\right)$th are equally probable amongst themselves and more probable than any other. For the present purpose, however, it is necessary to make use of another property of the median, which was known to Fechner (who first introduced the median into use) but which seldom receives as much attention as it deserves. *If y is the median of a number of magnitudes, the sum of the absolute differences* (i.e. *the difference always reckoned positive*) *between y and each of the magnitudes is a minimum.* The median y of $y_1\, y_2\, .\, .\, .\, y_m$ is found, that is to say, by making $\sum_1^m |y_q - y|$ a minimum where $|y_q - y|$ is the difference always reckoned positive between y_q and y.

We can now return to the investigation of the law of error corresponding to the median.

Write $|y - y_q| = z_q$. Then since $\sum_1^m z_q$ is to be a minimum we must have $\sum_1^m \frac{y-y_q}{z_q} = 0$. Whence, proceeding as before, we have

$$f_q = \mathrm{A}e^{\int \frac{y-y_q}{z_q}\phi''(y)dy + \psi(y_q)}.$$

The simplest case of this is obtained by putting

$$\phi''(y) = -k^2,$$

$$\psi(y_q) = \frac{y-y_q}{z_q}k^2 y_q,$$

whence
$$f_q = \mathrm{A}e^{-k^2|y-y_q|} = \mathrm{A}e^{-k^2 z_q}.$$

This satisfies the additional condition that positive and negative errors of equal magnitude are equally likely. Thus in this important respect the median is as satisfactory as the arithmetic mean, and the law of error which leads to it is as simple. It also resembles the normal law in that it is a function of the error *only*, and not of the magnitude of the measurement as well.

The median law of error, $f_q = \mathrm{A}e^{-k^2 z_q}$, where z_q is the absolute amount of the error always reckoned positive, is of some historical interest, because it was the earliest law of error to be formulated. The first attempt to bring the doctrine of averages into definite relation with the theory of probability and with laws of error was published by Laplace in 1774 in a memoir "sur la probabilité des causes par les événemens."[1] This memoir was not subsequently incorporated in his *Théorie analytique*, and does not represent his more mature view. In the *Théorie* he drops altogether the law tentatively adopted in the memoir, and lays down the main lines of investigation for the next hundred years by the introduction of the *normal* law of error. The popularity of the normal law, with the arithmetic mean and the method of least squares as its corollaries, has been very largely due to its overwhelming advantages, in comparison with all other laws of error, for the purposes of mathematical development and manipulation. And in addition to these technical advantages, it is probably applicable as a first approximation to a larger and more manageable group of phenomena than any other single law. So powerful a hold indeed did the normal law obtain on the minds of statisticians, that until quite recent times only a few pioneers have seriously considered the possibility of preferring in certain circumstances other means to the arithmetic and other laws of error to the normal. Laplace's earlier memoir fell, therefore, out of remembrance. But it remains interesting, if only for the fact that a law of error there makes its appearance for the first time.

Laplace sets himself the problem in a somewhat simplified form: "Déterminer le milieu que l'on doit prendre entre trois observations données d'un même phénomène." He begins by assuming a law of error $z=\phi(y)$, where z is the probability of an error y; and finally, by means of a number of somewhat arbitrary assumptions, arrives at the result $\phi(z)=\frac{m}{2}e^{-my}$. If this formula is to follow from his arguments, y must denote the *absolute* error, always taken positive. It is not unlikely that Laplace was led to this result by considerations other than those by which he attempts to justify it.

Laplace, however, did not notice that his law of error led to the median. For, instead of finding the most probable value, which would have led him straight to it, he seeks the "mean of error"—the value, that is to say, which the true value is as likely

[1] *Mémoires présentés à l'Académie des Sciences*, vol. vi.

to fall short of as to exceed. This value is, for the median law, laborious to find and awkward in the result. Laplace works it out correctly for the case where the observations are no more than three.

6. I do not think that it is possible to find by this method a law of error which leads to the mode. But the following general formulae are easily obtained:

(v.) If $\Sigma\theta(y_q,y)=0$ is the law of relation between the measurements and the most probable value of the quantity, then the law of error $f_q(y_q,y)$ is given by $f_q=\mathrm{A}e^{\int\theta(y_qy)\phi''(y)dy+\psi(y_q)}$. Since f_q lies between 0 and 1, $\int\theta(y_qy)\phi''(y)dy+\psi(y_q)+\log \mathrm{A}$ must be negative for all values of y_q and y that are physically possible; and, since the values of y_q are between them exhaustive,

$$\Sigma \mathrm{A}e^{\int\theta(y_qy)\phi''(y)dy+\psi(y_q)}=1,$$

where the summation is for all terms that can be formed by giving y_q every value *à priori* possible.

(vi.) The most general form of the law of error, when it is assumed that positive and negative errors of the same magnitude are equally probable, is $\mathrm{A}e^{-k^2f(y-y_q)^2}$, where the most probable value of the quantity is given by the equation

$$\Sigma(y-y_q)f'(y-y_q)^2=0,\text{ where } f'(y-y_q)^2=\frac{d}{d(y-y_q)^2}f(y-y_q)^2.$$

The arithmetic mean is a special case of this obtained by putting $f(y-y_q)^2=(y-y_q)^2$; and the median is a special case obtained by putting $f(y-y_q)^2=+\sqrt{(y-y_q)^2}$.

We can obtain other special cases by putting

$$f(y-y_q)^2=(y-y_q)^4,$$

when the law of error is $\mathrm{A}e^{-k^2(y-y_q)^4}$ and the most probable values are the roots of $my^3-3y^2\Sigma y_q+3y\Sigma y^2_q-\Sigma y_q{}^3=0$; and by putting $f(y-y_q)^2=\log(y-y_q)^2$, when the law of error is $\dfrac{\mathrm{A}}{(y-y_q)^{2k^2}}$ and the most probable values the roots of $\Sigma\dfrac{1}{y-y_q}=0$. In all these cases the law is a function of the error only.

7. These results may be summarised thus. We have assumed:

(*a*) That we have no reason, before making measurements, for supposing that the quantity we measure is more likely to have any one of its possible values than any other.

(*b*) That the errors are independent, in the sense that a *knowledge* of how great an error has been made in one case does not affect our expectation of the probable magnitude of the error in the next.

(*c*) That the probability of a measurement of given magnitude, when in addition to the *à priori* evidence the real value of the quantity is supposed known, is an algebraic function of this given magnitude of the measurement and of the real value of the quantity.

(*d*) That we may regard the series of possible values as continuous, without sensible error.

(*e*) That the *à priori* evidence permits us to assume a law of error of the type specified in (*c*); *i.e.* that the algebraic function referred to in (*c*) is known to us *à priori*.

Subject to these assumptions, we have reached the following conclusions:

(1) The most general form of the law of error is

$$f_q = \mathrm{A}e^{\int \phi''(y)\,\theta(y_q y)dy + \psi(y_q)},$$

leading to the equation $\Sigma\,\theta(y_q y) = 0$, connecting the most probable value and the actual measurements, where y is the most probable value and y_q, etc., the measurements.

(2) Assuming that positive and negative errors of the same absolute magnitude are equally likely, the most general form is $f_q = \mathrm{A}e^{-k^2 f(y-y_q)^2}$, leading to the equation $\Sigma(y-y_q)f'(y-y_q)^2 = 0$, where $f'z = \dfrac{d}{dz}fz$. Of the special cases to which this form gives rise, the most interesting were

(3) $f_q = \mathrm{A}e^{-k^2(y-y_q)^2} = \mathrm{A}e^{-k^2 z_q^2}$, where $z_q = |y - y_q|$, leading to the arithmetic mean of the measurements as the most probable value of the quantity; and

(4) $f_q = \mathrm{A}e^{-k^2 z_q}$, leading to the median.

(5) The most general form leading to the arithmetic mean is $f_q = \mathrm{A}e^{\phi'(y)(y-y_q) - \phi(y) + \psi(y_q)}$, with the special cases (3), and

(6) $f_q = \mathrm{A}e^{k^2 e^y (y-y_q) - k^2 e^y}$.

(7) The most general form leading to the geometric mean is $f_q = \mathrm{A}e^{\phi'(y)\log\frac{y_q}{y} + \int\frac{\phi'(y)}{y}dy + \psi(y_q)}$, with the special cases:

(8) $f_q = \mathrm{A}\left(\frac{y}{y_q}\right)^{k^2 y} e^{-k^2 y}$, and

(9) $f_q = \mathrm{A}e^{-k^2\left(\log \frac{y_q}{y}\right)^2}$.

(10) The most general form leading to the harmonic mean is $f_q = \mathrm{A}\, e^{\phi'(y)\left[\frac{1}{y_q}-\frac{1}{y}\right]-\int\frac{\phi'(y)}{y^2}+\psi(y_q)}$, with the special case

(11) $f_q = \mathrm{A}e^{-k^2\frac{(y-y_q)^2}{y_q}} = \mathrm{A}e^{-k^2\frac{z_q^2}{y_q}}$.

(12) The most general form leading to the median is

$$f_q = \mathrm{A}e^{\phi'(y)\frac{y-y_q}{z_q}+\psi(y_q)},$$

with the special case (4).

In each of these expressions, f_q is the probability of a measurement y_q, given that the true value is y.

8. The doctrine of Means and the allied theory of Least Squares comprise so extensive a subject-matter that they cannot be adequately treated except in a volume primarily devoted to them. As, however, they are one of the important practical applications of the theory of probability, I am unwilling to pass them by entirely; and the following discursive observations, chiefly relating to the Normal Law of Error, will serve, taken in conjunction with the paragraphs immediately preceding, to illustrate the connection between the theories of this treatise and the general treatment of averages.

9. *The Claims of the Arithmetic Average.*—By definition the arithmetic average of a number of quantities is nothing more than their arithmetic sum divided by their number. But the utility of an average generally consists in our supposed right to substitute, in certain cases, this single measure for the varying measures of which it is a function. Sometimes this requires no justification; the word "average" is in these cases used for the sake of shortness, and merely to summarise a set of facts: as, for instance, when we say that the birth-rate in England is greater than the birth-rate in France.

But there are other cases in which the average makes a more substantial claim to add to our knowledge. After a number of examiners of equal capacity have given varying marks to a candidate for the same paper, it may be thought fair to allow the candidate the average of the different marks allotted: and in general if several estimates of a magnitude have been made, between the accuracy of which we have no reason to discriminate, we often think it reasonable to act as if the true magnitude were the average of the several measurements. Perhaps De Witt, in his report on Annuities to the States General in 1671,[1] was the first to use it scientifically. But as Leibniz points out: "Our peasants have made use of it for a long time according to their natural mathematics. For example, when some inheritance or land is to be sold, they form three bodies of appraisers; these bodies are called *Schurzen* in Low Saxon, and each body makes an estimate of the property in question. Suppose, then, that the first estimates its value to be 1000 crowns, the second, 1400, the third, 1500; the sum of these three estimates is taken, viz. 3900, and because they were three bodies, the third, *i.e.* 1300, is taken as the mean value asked for. This is the axiom: *aequalibus aequalia*, equal suppositions must have equal consideration."[2]

But this is a very inadequate axiom. Equal suppositions would have equal consideration, if the three estimates had been multiplied together instead of being added. The truth is that at all times the arithmetic mean has had *simplicity* to recommend it. It is always easier to add than to multiply. But simplicity is a dangerous criterion: "La nature," says Fresnel, "ne s'est pas embarassée des difficultés d'analyse, elle n'a évité que la complication des moyens."

With Laplace and Gauss there began a series of attempts to *prove* the worth of the arithmetic mean. It was discovered that its use involved the assumption of a particular type of law of error for the *à priori* probabilities of given errors. It was also found that the assumption of this law led on to a more complicated rule, known as the Method of Least Squares, for combining the results of observations which contain more than one doubtful quantity. In spite of a popular belief that, whilst the Arithmetic Mean is intuitively obvious, the Method of Least Squares depends upon doubtful and arbitrary assumptions, it can be demonstrated that the two stand and fall together.[3]

[1] *De vardye van de lif-renten na proportie van de losrenten.* The Hague, 1671.

[2] *Nouveaux Essais.* Engl. transl. p. 540.

[3] Venn (*Logic of Chance*, p. 40) thinks that the Normal Law of Error and the Method of Least Squares "are not only totally distinct things, but they have scarcely even any necessary connection with each other. The Law of Error is the statement of a physical fact. . . . The Method of Least Squares, on the other hand, is not a law at all in the scientific sense of the term. It is simply a rule or direction. . . ."

The analytical theorems of Laplace and Gauss are complicated, but the special assumptions upon which they are based are easily stated.[1] Gauss supposes (*a*) that the probability of a given error is a function of the error only and not also of the magnitude of the observation, (*b*) that the errors are so small that their cubes and higher powers may be neglected. Assumption (*a*) is arbitrary,[2] and Gauss did not state it explicitly. These two assumptions, together with certain others, lead us to the result. For let $\phi(z)$ be the law of error where z is the error, and let us assume, as it always is assumed in these proofs, that $\phi(z)$ can be expanded by Maclaurin's Theorem. Then $\phi(x) = \phi(0) + z\phi'(0) + \frac{z^2}{2!}\phi''(0) + \frac{z^3}{3!}\phi'''(0) + \ldots$ It is also supposed that positive and negative errors are equally probable, *i.e.* $\phi(z) = \phi(-z)$, so that $\phi'(0)$ and $\phi'''(0)$ vanish. Since we may neglect z^4 in comparison with z^2, $\phi(z) = \phi(0) + \frac{1}{2}z^2\phi''(0)$. But (neglecting z^4 and higher powers) $a + bz^2 = ae^{\frac{bz^2}{a}}$, so that $\phi(z) = ae^{\frac{bz^2}{a}}$.

Gauss's proof looks much more complicated than this, but he obtains the form $ae^{\frac{bz^2}{a}}$ by neglecting higher powers of z, so that this expression is really equivalent to $a + bz^2$. By this approximation he has reduced all the possible laws to an equivalent form.[3] It is true, therefore, that the normal law of error is, to the second power of the error, equivalent to any law of error, *which is a function of the error only, and for which positive and negative errors are equally probable.* Laplace also introduces assumptions equivalent to these.

While mathematicians have endeavoured to establish the normal law of error and the arithmetic mean as a law of logic,

[1] For an account of the three principal methods of arriving at the Method of Least Squares and the Arithmetic Mean, see Ellis, *Least Squares.* Gauss's first method is in the *Theoria Motus*, and his second in the *Theoria Combinationis Observationum.* Laplace's investigations are in chap. iv. of the second Book of the *Théorie analytique.* Laplace's method was improved by Poisson in the *Connaissance des temps* for 1827 and 1832.

[2] It does not follow, as G. Hagen argues (*Grundzüge der Wahrscheinlichkeitsrechnung*, p. 29), that, because a larger error is less probable than a smaller, *therefore* the probability of a given error is a function of its magnitude only.

[3] This is pointed out by Bertrand, *Calcul des probabilités*, p 267.

others have claimed for it the testimony of experience and have deemed it a law of nature.[1]

That this cannot be so, is evident. For suppose that $x_1x_2 \ldots x_n$ are a set of observations of an unknown quantity x. Then, by this principle, $x = \frac{1}{n}\Sigma x_r$ gives the most probable value of x. But suppose we had wished to determine x^2, our observations, assuming that we can multiply correctly, would be $x_1^2, x_2^2 \ldots x_n^2$, and the most probable value of $x^2 = \frac{1}{n}\Sigma x_r^2$. But $(\frac{1}{n}\Sigma x_r)^2 \neq \frac{1}{n}\Sigma x_r^2$. And in general, $\frac{1}{n}\Sigma f(x_r) \neq f(\frac{1}{n}\Sigma x_r)$. Nor is this a consideration which can safely be ignored in practice. For our "observations" are often the result of some manipulation, and the particular shape in which we get them is not necessarily fixed for us. It is not easy to say what the *direct* observation is. In particular if any such law of sensation, as that enunciated by Fechner, is true (*i.e.* that sensation varies as the logarithm of the stimulus), the arithmetic mean must break down as a *practical* rule in all cases where human sensation is part of the instrument by means of which the observations are recorded.[2]

Apart, however, from theoretical refutations, statisticians now recognise that the arithmetic mean and the normal law of error can only be applied to certain special classes of phenomena. Quetelet[3] was, I think, the first to point this out. In England, Galton drew attention to the fact many years ago, and Professor Pearson[4] has shown "that the Gaussian-Laplace normal distribution is very far from being a general law of frequency distribution either for errors of observation or for the distribution of deviations from type such as occur in organic populations. . . . It is not even approximately correct, for example, in the distribution of barometric variations, of grades of fertility and incidence of disease."

[1] This is, of course, a very common point of view indeed. Cf. Bertrand, *op. cit.* p. 183: "Malgré les objections précédentes, la formule de Gauss doit être adoptée. L'observation la confirme: cela doit suffire dans les applications."

[2] This was noticed by Galton.

[3] E.g. *Letters on the Theory of Probabilities*, p. 114.

[4] On "Errors of Judgment, etc.," *Phil. Trans.* A, vol. cxcviii. pp. 235-299. The following quotation is from his memoir *On the General Theory of Skew Correlation and Nonlinear Regression*, where further references are given.

The Arithmetic Mean occupies, therefore, no unique position; and it is worth while, from the point of view of probability, to discuss the properties of other possible means and laws of error, as, for example, on the lines indicated in the earlier part of this chapter.

10. *The Method of Least Squares.*—The problem, to which this method is applied, is no more than the application of the same considerations, as those which we have just been discussing, to cases where the relation between the observed measurements and the quantity whose most probable value we require, involves more than one unknown.

Owing to the surprising character of its conclusions, if they could be accepted as universally valid, and to the obscurity of the mathematical fabric that has been reared on and about it, this method has been surrounded by an unnecessary air of mystery. It is true that in recent times scepticism has grown at the expense of mystery. It is also true that just views have been held by individuals for sixty years past, notably by Leslie Ellis. But the old mistakes are not always corrected in the current text-books, and even so useful and generally used a treatise on Least Squares, as Professor Mansfield Merriman's, opens with a series of very fallacious statements.

The controversial side of the Method of Least Squares is purely logical; in the later developments there is much elaborate mathematics of whose correctness no one is in doubt. What it is important to state with the utmost possible clearness is the precise assumptions on which the mathematics is based; when these assumptions have been set forth, it remains to determine their applicability in particular cases.

In dealing with averages we supposed ourselves to be presented with a number of direct observations of some quantity which it is desired to determine. But it is obvious that direct observations will be in many cases either impracticable or inconvenient; and our natural course will be to measure certain other quantities which we know to bear fixed and invariable relations to the unknowns we wish to determine. In surveying, for instance, or in astronomy, we constantly prefer to take measurements of angles or distances in which we are not interested for their own sakes, but which bear known geometrical relationships to the set of ultimate unknowns.

If we wish to determine the most probable values of a set of unknowns x_1, x_2, x_3 . . . x_r, instead of obtaining a number of sets of direct observations of each, we may obtain a number of equations of observation of the following type:

$$\begin{aligned} a_1x_1 + a_2x_2 + \ldots + a_rx_r &= V_1, \\ b_1x_1 + b_2x_2 + \ldots + b_rx_r &= V_2, \\ \cdot \quad \cdot \quad \cdot \quad &\cdot \quad \cdot \\ k_1x_1 + k_2x_2 + \ldots + k_rx_r &= V_n, \end{aligned}$$

where V_1, etc., are the quantities *directly* observed, and the a's, b's, etc., are supposed known ($n > r$).

We have in such a case n equations to determine r unknowns, and since the observations are likely to be inexact, there may be no precise solution whatever. In these circumstances we wish to know the most probable set of values of the x's warranted by these observations.

The problem is precisely similar in kind to that dealt with by averages and differs only in the degree of its complexity. It is the problem of finding the most probable solution of such a set of discrepant equations of observation that the Method of Least Squares claims to solve.

By 1750 the astronomers were obtaining such equations of observation in the course of their investigations, and the question arose as to the proper manner of their solution. Boscovich in Italy, Mayer and Lambert in Germany, Laplace in France, Euler in Russia, and Simpson in England proposed different methods of solution. Simpson, in 1757, was the first to introduce, by way of simplification, the assumption or axiom that positive and negative errors are equally probable.[1] The Method of Least Squares was first definitely stated by Legendre in 1805, who proposed it as an advantageous method of adjusting observations. This was soon followed by the 'proofs' of Laplace and Gauss. But it is easily shown that these proofs involve the normal law of error $y = ke^{-h^2x^2}$, and the theory of Least Squares simply develops the mathematical results of applying to equations of observation, which involve more than one unknown, that law

[1] See Merriman's *Method of Least Squares*, p. 181, for an historical sketch, from which the above is taken. In 1877 Merriman published in the *Transactions of the Connecticut Academy* a list of writings relating to the Method of Least Squares and the theory of accidental errors of observation, which comprised 408 titles—classified as 313 memoirs, 72 books, 23 parts of books.

of error which leads to the Arithmetic Mean in the case of a single unknown.

11. *The Weighting of Averages.*—It is necessary to recur to the distinction made at the beginning of § 9 between the two types to which our average, or, as it is generally termed in social inquiries, our index number, may belong. The average or index number may simply summarise a set of facts and give us the actual value of a composite quantity, as, for example, the index number of the cost of living. In such cases the composite quantity, in which we are interested, need not contain precisely the same number of units of each of the elementary quantities of which it is composed, so that the 'weights,' which denote the numbers of each elementary quantity appropriate to the composite quantity, are part of the definition of the composite quantity, and can no more be dispensed with than the magnitudes of the elementary quantities themselves. Nor in such cases is the rejection of discordant observations permissible; if, that is to say, some of the elementary quantities are subject to much wider variation, or to variations of a different type than the majority, that is no reason for rejecting them.

On the other hand, the individual items, out of which the average is composed, may each be *indications* or approximate estimates of some *one single* quantity; and the average, instead of representing the measure of a composite quantity, may be selected as furnishing the most probable value of the single quantity, given, as evidence of its magnitude, the values of the various terms which make up the average.

If this is the character of our average, the problem of weighting depends upon what we know about the individual observations or samples or indications, out of which our average is to be built up. The units in question may be *known* to differ in respects relevant to the probable value of the *quaesitum*. Thus there may be reasons, quite apart from the actual results of the individual observations or samples, for trusting some of them more than others. Our knowledge may indicate to us, in fact, that the constants of the laws of error appropriate to the several instances, even if the type of the law can be assumed to be constant, should be varied according to the data we possess about each. It may also indicate to us that the condition of *independence* between the instances, which the method of averages presumes, is imperfectly satisfied, and consequently that our mode of combining the instances in an average must be modified accordingly.

Some modern statisticians, who, really influenced perhaps by practical considerations, have been inclined to deprecate the importance of weighting on theoretical grounds, have not always been quite clear what kind of average they supposed themselves to be dealing with. In particular, discussions of the question of weighting in connection with index numbers of the value of money have suffered from this confusion. It has not been clear whether such index numbers really represent measures of a composite quantity or whether they are probable estimates of the value of a single quantity formed by combining a number of independent approximations towards the value of this quantity. The original Jevonian conception of an index number of the value of money was decidedly of the latter type. Modern work on the subject has been increasingly dominated by the other conception. A discussion of where the truth lies would lead me too far into the field of a subject-matter alien to that of this treatise.

Theoretical arguments against weighting have sometimes been based on the fact that to weight the items of the average in an irrelevant manner, or, as it is generally expressed, in a random manner, is not likely, provided the variations between the weights are small compared with the variations between the items, to affect the result very much. But why should any one wish to weight an average "at random"? Such observations overlook the real meaning and significance of weights. They are probably inspired by the fact that a superficial treatment of statistics would sometimes lead to the introduction of weights which are irrelevant. In drawing a conclusion, for example, from the vital statistics of various towns, the figures of population for the different towns may or may not be relevant to our conclusion. It depends on the character of the argument. If they are relevant, it may be right to employ them as weights. If they are irrelevant, it must be wrong and unnecessary to do so. The fact that wheat is a more important article of consumption than pins *may*, on certain assumptions, be irrelevant to the usefulness of variations in the price of each article as indications of variation in the value of money. With other assumptions, it may be extremely relevant. Or again, we may know that observations with a particular instrument tend to be too large and must, therefore, be weighted down. It is contrary both to theory and to common sense to suppose that the possession of information as to the relative reliability of different statistics is not useful. There is no place, therefore, in my judgment, for a *generalised* argument as to the propriety or impropriety of weighting an average.

It should be added that, where we seek to build up an index number of a conception, which is quantitative but is not itself numerically measurable in any defined or unambiguous sense, by combining a number of numerical quantities, which, while they do not measure our *quaesitum* are nevertheless indications of its quantitative variations and tend to fluctuate in the same sense, as, for example, by means of what are sometimes called *economic barometers* of the state of business, or the prosperity of the country or the like, some very confusing questions can arise both as to what sort of a thing our resulting index really is, and as to the mode of compilation appropriate to it.

These confusing questions always arise when, instead of measuring a quantity directly, we seek an index to fluctuations in its magnitude by combining in an average the fluctuations of a series of magnitudes, which are, each of them in a different way, to some extent (but only to some extent), correlated with fluctuations in our *quaesitum*. I must not burden this book with a discussion of the problems of Index Numbers. But I venture to think that they would be sooner cleared up if the natures and purposes of differing index numbers were more sharply distinguished—those, namely, which are simply descriptive of a composite commodity, those which seek to combine results differing from one another in a way analogous to the variations of an instrument of precision, and those which combine results, not of the *quaesitum* itself, but of various other quantities, variations in which are partly due to variations in the *quaesitum*, but which we well know to be also due to other distinguishable influences. Index numbers of the third type are often treated by methods and arguments only appropriate to those of the second type.

12. *The Rejection of Discordant Observations.*—This differs from the problem just discussed, because we have supposed so far that our system of weighting is determined by data which we possess prior to and apart from our knowledge of the actual magnitude of the items of our average. The principle of the rejection of discordant observations comes in when it is argued that, if one or more of our observations show great discrepancies from the results of the greater number, these ought to be partly or entirely neglected in striking the average, even if there is no reason, except their discrepancy from the rest, for attributing less weight to them than to the others. By some this practice has been thought to be in accordance with the dictates of common sense; by others it is denounced as savouring even of forgery.[1]

This controversy, like so many others in Probability, is due to a failure to understand the meaning of 'independence.' The mathematics of the orthodox theory of Averages and Least Squares depend, as we have seen, upon the assumption that the observations are 'independent'; but this has sometimes been interpreted to mean a *physical* independence. In point of fact, the theory requires that the observations shall be independent, in the sense that a *knowledge* of the result of some does not affect the probability that the others, when known, involve given errors.

Clearly there may be initial data in relation to which this supposition is entirely or approximately accurate. But in many cases the assumption will be inadmissible. A knowledge of the results of a number of observations may lead us to modify our opinion as to the relative reliabilities of others.

The question, whether or not discordant observations should be specially weighted down, turns, therefore, upon the nature of the preliminary data by which we have been guided in initially adopting a particular law of error as appropriate to the observations. If the observations are, relevant to these data, strictly 'independent,' in the sense required for probability, then rejection is not permissible. But if this condition is not fulfilled, a bias against discordant observations may be well justified.

[1] *E.g.* G. Hagen's *Grundzüge der Wahrscheinlichkeitsrechnung*, p. 63: "Die Täuschung, die man durch Verschweigen von Messungen begeht, lässt sich eben so wenig entschuldigen, als wenn man Messungen fälschen oder fingiren wollte."

PART III

INDUCTION AND ANALOGY

CHAPTER XVIII

INTRODUCTION

Nothing so like as eggs ; yet no one, on account of this apparent similarity, expects the same taste and relish in all of them. 'Tis only after a long course of uniform experiments in any kind, that we attain a firm reliance and security with regard to a particular event. Now where is that process of reasoning, which from one instance draws a conclusion, so different from that which it infers from a hundred instances, that are no way different from that single instance ? This question I propose as much for the sake of information, as with any intention of raising difficulties. I cannot find, I cannot imagine any such reasoning. But I keep my mind still open to instruction, if any one will vouchsafe to bestow it on me.—HUME.[1]

1. I HAVE described Probability as comprising that part of logic which deals with arguments which are rational but not conclusive. By far the most important types of such arguments are those which are based on the methods of Induction and Analogy. Almost all empirical science rests on these. And the decisions dictated by experience in the ordinary conduct of life generally depend on them. To the analysis and logical justification of these methods the following chapters are directed.

Inductive processes have formed, of course, at all times a vital, habitual part of the mind's machinery. Whenever we learn by experience, we are using them. But in the logic of the schools they have taken their proper place slowly. No clear or satisfactory account of them is to be found anywhere. Within and yet beyond the scope of formal logic, on the line, apparently, between mental and natural philosophy, Induction has been admitted into the organon of scientific proof, without much help from the logicians, no one quite knows when.

2. What are its distinguishing characteristics ? What are the qualities which in ordinary discourse seem to afford strength to an inductive argument ?

[1] *Philosophical Essays concerning Human Understanding.*

I shall try to answer these questions before I proceed to the more fundamental problem—What ground have we for regarding such arguments as rational?

Let the reader remember, therefore, that in the first of the succeeding chapters my main purpose is no more than to state in precise language what elements are commonly regarded as adding weight to an empirical or inductive argument. This requires some patience and a good deal of definition and special terminology. But I do not think that the work is controversial. At any rate, I am satisfied myself that the analysis of Chapter XIX. is fairly adequate.

In the next section, Chapters XX. and XXI., I continue in part the same task, but also try to elucidate what sort of assumptions, *if* we could adopt them, lie behind and are required by the methods just analysed. In Chapter XXII. the nature of these assumptions is discussed further, and their possible justification is debated.

3. The passage quoted from Hume at the head of this chapter is a good introduction to our subject. Nothing so *like* as eggs, and after a *long* course of uniform experiments we can expect with a firm reliance and security the same taste and relish in all of them. The eggs must be like eggs, and we must have tasted many of them. This argument is based partly upon *Analogy* and partly upon what may be termed *Pure Induction*. We argue from Analogy in so far as we depend upon the *likeness* of the eggs, and from Pure Induction when we trust the *number* of the experiments.

It will be useful to call arguments *inductive* which depend in any way on the methods of Analogy and Pure Induction. But I do not mean to suggest by the use of the term *inductive* that these methods are necessarily confined to the objects of phenomenal experience and to what are sometimes called empirical questions; or to preclude from the outset the possibility of their use in abstract and metaphysical inquiries. While the term *inductive* will be employed in this general sense, the expression *Pure Induction* must be kept for that part of the argument which arises out of the repetition of instances.

4. Hume's account, however, is incomplete. His argument could have been improved. His experiments should not have been too uniform, and ought to have differed from one another

as much as possible in all respects save that of the likeness of the eggs. He should have tried eggs in the town and in the country, in January and in June. He might then have discovered that eggs could be good or bad, however like they looked.

This principle of varying those of the characteristics of the instances, which we regard in the conditions of our generalisation as non-essential, may be termed *Negative Analogy*.

It will be argued later on that an increase in the *number* of experiments is *only* valuable in so far as, by increasing, or possibly increasing, the variety found amongst the non-essential characteristics of the instances, it strengthens the Negative Analogy. If Hume's experiments had been *absolutely* uniform, he would have been right to raise doubts about the conclusion. There is no process of reasoning, which from one instance draws a conclusion different from that which it infers from a hundred instances, if the latter are known to be in *no* way different from the former. Hume has unconsciously misrepresented the typical inductive argument.

When our control of the experiments is fairly complete, and the conditions in which they take place are well known, there is not much room for assistance from Pure Induction. If the Negative Analogies are known, there is no need to count the instances. But where our control is incomplete, and we do not know accurately in what ways the instances differ from one another, then an increase in the mere number of the instances helps the argument. For unless we know for certain that the instances are perfectly uniform, each new instance *may* possibly add to the Negative Analogy.

Hume might also have weakened his argument. He expects no more than the same taste and relish from his eggs. He attempts no conclusion as to whether his stomach will always draw from them the same nourishment. He has conserved the force of his generalisation by keeping it narrow.

5. In an inductive argument, therefore, we start with a number of instances similar in some respects AB, dissimilar in others C. We pick out one or more respects A in which the instances are similar, and argue that some of the other respects B in which they are also similar are likely to be associated with the characteristics A in other unexamined cases. The more comprehensive the essential characteristics A, the greater the

variety amongst the non-essential characteristics C, and the less comprehensive the characteristics B which we seek to associate with A, the stronger is the likelihood or probability of the generalisation we seek to establish.

These are the three ultimate logical elements on which the probability of an empirical argument depends,—the Positive and the Negative Analogies and the scope of the generalisation.

6. Amongst the generalisations arising out of empirical argument we can distinguish two separate types. The first of these may be termed *universal induction.* Although such inductions are themselves susceptible of any degree of probability, they affirm *invariable* relations. The generalisations which they assert, that is to say, claim universality, and are upset if a single exception to them can be discovered. Only in the more exact sciences, however, do we aim at establishing universal inductions. In the majority of cases we are content with that other kind of induction which leads up to laws upon which we can generally depend, but which does not claim, however adequately established, to assert a law of more than probable connection.[1] This second type may be termed *Inductive Correlation.* If, for instance, we base upon the data, that this and that and those swans are white, the conclusion that *all* swans are white, we are endeavouring to establish a universal induction. But if we base upon the data that this and those swans are white and that swan is black, the conclusion that *most* swans are white, or that the probability of a swan's being white is such and such, then we are establishing an inductive correlation.

Of these two types, the former—universal induction—presents both the simpler and the more fundamental problem. In this part of my treatise I shall confine myself to it almost entirely. In Part V., on the Foundations of Statistical Inference, I shall discuss, so far as I can, the logical basis of inductive correlation.

7. The fundamental connection between Inductive Method and Probability deserves all the emphasis I can give it. Many writers, it is true, have recognised that the conclusions which we reach by inductive argument are probable and inconclusive. Jevons, for instance, endeavoured to justify inductive processes by means of the principles of inverse probability. And it is true also that much of the work of Laplace and his followers was

[1] What Mill calls 'approximate generalisations.'

directed to the solution of essentially inductive problems. But it has been seldom apprehended clearly, either by these writers or by others, that the validity of every induction, strictly interpreted, depends, not on a matter of fact, but on the existence of a relation of probability. An inductive argument affirms, not that a certain matter of fact *is* so, but that *relative to certain evidence* there is a probability in its favour. The validity of the induction, relative to the original evidence, is not upset, therefore, if, as a fact, the truth turns out to be otherwise.

The clear apprehension of this truth profoundly modifies our attitude towards the solution of the inductive problem. The validity of the inductive method does *not* depend on the success of its predictions. Its repeated failure in the past may, of course, supply us with new evidence, the inclusion of which will modify the force of subsequent inductions. But the force of the old induction *relative to the old evidence* is untouched. The evidence with which our experience has supplied us in the past may have proved misleading, but this is entirely irrelevant to the question of what conclusion we ought reasonably to have drawn from the evidence then before us. The validity and reasonable nature of inductive generalisation is, therefore, a question of logic and not of experience, of formal and not of material laws. The actual constitution of the phenomenal universe determines the character of our evidence ; but it cannot determine what conclusions *given* evidence *rationally* supports.

CHAPTER XIX

THE NATURE OF ARGUMENT BY ANALOGY

All kinds of reasoning from causes or effects are founded on two particulars, viz. the constant conjunction of any two objects in all past experience, and the resemblance of a present object to any of them. Without some degree of resemblance, as well as union, 'tis impossible there can be any reasoning.—HUME.[1]

1. HUME rightly maintains that some degree of resemblance must always exist between the various instances upon which a generalisation is based. For they must have this, at least, in common, that they are instances of the proposition which generalises them. Some element of analogy must, therefore, lie at the base of every inductive argument. In this chapter I shall try to explain with precision the meaning of Analogy, and to analyse the reasons, for which, rightly or wrongly, we usually regard analogies as strong or weak, without considering at present whether it is possible to find a *good* reason for our instinctive principle that likeness breeds the expectation of likeness.

2. There are a few technical terms to be defined. We mean by a *generalisation* a statement that all of a certain definable class of propositions are true. It is convenient to specify this class in the following way. If $f(x)$ is true for all those values of x for which $\phi(x)$ is true, then we have a generalisation about ϕ and f which we may write $g(\phi, f)$. If, for example, we are dealing with the generalisation, "All swans are white," this is equivalent to the statement, "'x is white' is true for all those values of x for which 'x is a swan' is true." The proposition $\phi(a).f(a)$ is an *instance* of the generalisation $g(\phi, f)$.

By thus defining a generalisation in terms of propositional functions, it becomes possible to deal with all kinds of generalisa-

[1] *A Treatise of Human Nature.*

tions in a uniform way; and also to bring generalisation into convenient connection with our definition of Analogy.

If some one thing is true about both of two objects, if, that is to say, they both satisfy the same propositional function, then to this extent there is an *analogy* between them. Every generalisation $g(\phi, f)$, therefore, asserts that one analogy is always accompanied by another, namely, that between all objects having the analogy ϕ there is also the analogy f. The set of propositional functions, which are satisfied by both of the two objects, constitute the *positive analogy*. The analogies, which would be disclosed by complete knowledge, may be termed the *total positive analogy*; those which are relative to partial knowledge, the *known positive analogy*.

As the positive analogy measures the resemblances, so the negative analogy measures the differences between the two objects. The set of functions, such that each is satisfied by one and not by the other of the objects, constitutes the *negative analogy*. We have, as before, the distinction between the *total negative analogy* and the *known negative analogy*.

This set of definitions is soon extended to the cases in which the number of instances exceeds two. The functions which are true of *all* of the instances constitute the positive analogy of the set of instances, and those which are true of *some only*, and are false of others, constitute the negative analogy. It is clear that a function, which represents positive analogy for a group of instances taken out of the set, may be a negative analogy for the set as a whole. Analogies of this kind, which are positive for a sub-class of the instances, but negative for the whole class, we may term *sub-analogies*. By this it is meant that there are resemblances which are common to some of the instances, but not to all.

A simple notation, in accordance with these definitions, will be useful. If there is a positive analogy ϕ between a set of instances $a_1 \ldots a_n$, whether or not this is the total analogy between them, let us write this—

$$\underset{a_1 \ldots a_n}{\mathrm{A}}(\phi).$$[1]

[1] Hence $\underset{a_1 \ldots a_n}{\mathrm{A}}(\phi) \equiv \phi(a_1) \cdot \phi(a_2) \ldots \phi(a_n) \equiv \prod_{x=a_1}^{x=a_n} \phi(x)$.

And if there is a negative analogy ϕ', let us write this—

$$\bar{A}_{a_1 \ldots a_n}(\phi').$$[1]

Thus $A_{a_1 \ldots a_n}(\phi)$ expresses the fact that there is a set of characteristics ϕ which are common to all the instances, and $\bar{A}_{a_1 \ldots a_n}(\phi')$ that there is a set of characteristics ϕ' which is true of at least one of the instances and false of at least one.

3. In the typical argument from analogy we wish to generalise from one part to another of the total analogy which experience has shown to exist between certain selected instances. In all the cases where one characteristic ϕ has been found to exist, another characteristic f has been found to be associated with it. We argue from this that any instance, which is known to share the first analogy ϕ, is likely to share also the second analogy f. We have found in certain cases, that is to say, that both ϕ and f are true of them; and we wish to assert f as true of other cases in which we have only observed ϕ. We seek to establish the generalisation $g(\phi, f)$, on the ground that ϕ and f constitute between them an observed positive analogy in a given set of experiences.

But while the argument is of this character, the grounds, upon which we attribute more or less weight to it, are often rather complex; and we must discuss them, therefore, in a systematic manner.

4. According to the view suggested in the last chapter, the value of such an argument depends partly upon the nature of the conclusion which we seek to draw, partly upon the evidence which supports it. If Hume had expected the same degree of nourishment as well as the same taste and relish from all of the eggs, he would have drawn a conclusion of weaker probability. Let us consider, then, this dependence of the probability upon the *scope* of the generalisation $g(\phi, f)$,—upon the comprehensiveness, that is to say, of the condition ϕ and the conclusion f respectively.

The more comprehensive the condition ϕ and the less comprehensive the conclusion f, the greater *à priori* probability do we attribute to the generalisation g. With every increase in ϕ this probability increases, and with every increase in f it will diminish.

[1] Hence $\bar{A}_{a_1 \ldots a_n}(\phi') \equiv \sum_{x=a_r}^{x=a_s} \phi'(x) \,.\, \sum_{x=a_r'}^{x=a_s'} \overline{\phi'(x)}.$

The condition $\phi(\equiv\phi_1\phi_2)$ is more comprehensive than the condition ϕ_1, relative to the general evidence h, if ϕ_2 is a condition independent of ϕ_1 relative to h, ϕ_2 being independent of ϕ_1, if $g(\phi_1, \phi_2)/h \neq 1$, *i.e.* if, relative to h, the satisfaction of ϕ_2 is not inferrible from that of ϕ_1.

Similarly the conclusion $f(\equiv f_1 f_2)$ is more comprehensive than the conclusion f_1, relative to the general evidence h, if f_2 is a conclusion independent of f_1, relative to h, *i.e.* if $g(f_1, f_2)/h \neq 1$.

If $\phi \equiv \phi_1\phi_2$ and $f \equiv f_1 f_2$, where ϕ_1 and ϕ_2 are independent and f_1 and f_2 are independent relative to h, we have—

$$g(\phi_1, f)/h = g(\phi_1\phi_2, f) \cdot g(\phi_1\bar{\phi}_2, f)/h$$
$$\not< g(\phi, f)/h,$$

and
$$g(\phi, f)/h = g(\phi, f_1 f_2)/h$$
$$= g(\phi f_1, f_2)/h \cdot g(\phi, f_1)/h$$
$$\leq g(\phi, f_1)/h,$$

so that
$$g(\phi, f_1)/h \geq g(\phi, f)/h \geq g(\phi_1, f)/h.$$

This proves the statement made above. It will be noticed that we cannot necessarily compare the *à priori* probabilities of two generalisations in respect of more and less, unless the condition of the first is included in the condition of the second, and the conclusion of the second is included in that of the first.

We see, therefore, that some generalisations stand *initially* in a stronger position than others. In order to attain a given degree of probability, generalisations require, according to their scope, different amounts of favourable evidence to support them.

5. Let us now pass from the character of the generalisation *à priori* to the evidence by which we support it. Since, whenever the conclusion f is complex, *i.e.* resolvable into the form $f_1 f_2$ where $g(f_1, f_2)/h \neq 1$, we can express the probability of the generalisation $g(\phi, f)$ as the product of the probabilities of the two generalisations $g(\phi f_1, f_2)$ and $g(\phi, f_1)$, we may assume in what follows, that the conclusion f is simple and not capable of further analysis, without diminishing the generality of our argument.

We will begin with the simplest case, namely, that which arises in the following conditions. First, let us assume that our knowledge of the examined instances is complete, so that we know of every statement, which is about the examined instances, whether it is true or false of each.[1] Second, let us assume that

[1] If $\psi(a)$ is a proposition and $\psi(a) = h \cdot \theta(a)$, where h is a proposition not involving a, then we must regard $\theta(a)$, not $\psi(a)$ as the statement *about* a.

all the instances which are known to satisfy the condition ϕ, are also known to satisfy the conclusion f of the generalisation. And third let us assume that there is nothing which is true of *all* the examined instances and yet not included either in ϕ or in f, *i.e.* that the positive analogy between the instances is exactly co-extensive with the analogy ϕf which is covered by the generalisation.

Such evidence as this constitutes what we may term a perfect analogy. The argument in favour of the generalisation cannot be further improved by a knowledge of additional instances. Since the positive analogy between the instances is exactly coextensive with the analogy covered by the generalisation, and since our knowledge of the examined instances is complete, there is no need to take account of the negative analogy.

An analogy of this kind, however, is not likely to have much practical utility; for if the analogy covered by the generalisation, covers the *whole* of the positive analogy between the instances it is difficult to see to what *other* instances the generalisation can be applicable. Any instance, about which everything is true which is true of all of a set of instances, must be identical with one of them. Indeed, an argument from perfect analogy can only have practical utility, if, as will be argued later on, there are some distinctions between instances which are *irrelevant* for the purposes of analogy, and if, in a perfect analogy, the positive analogy, of which we must take account, need cover only those distinctions which are relevant. In this case a generalisation based on perfect analogy might cover instances numerically distinct from those of the original set.

The law of the Uniformity of Nature appears to me to amount to an assertion that an analogy which is perfect, except that mere differences of position in time and space are treated as irrelevant, is a valid basis for a generalisation, two total causes being regarded as the *same* if they only differ in their positions in time or space. This, I think, is the whole of the importance which this law has for the theory of inductive argument. It involves the assertion of a generalised judgment of irrelevance, namely, of the irrelevance of mere position in time and space to generalisations which have no reference to particular positions in time and space. It is in respect of such position in time or space that 'nature' is supposed 'uniform.' The significance of the law and the nature of its justification, if any, are further discussed in Chapter XXII.

6. Let us now pass to the type which is next in order of simplicity. We will relax the first condition and no longer assume that the *whole* of the positive analogy between the instances is covered by the generalisation, though retaining the assumption that our knowledge of the examined instances is complete. We know, that is to say, that there are some respects in which the examined instances are all alike, and yet which are not covered by the generalisation. If ϕ_1 is the part of the positive analogy between the instances which is *not* covered by the generalisation, then the probability of this type of argument from analogy can be written—

$$g(\phi, f) \Big/ \underset{a_1 \ldots a_n}{\mathrm{A}} (\phi\phi_1 f).$$

The value of this probability turns on the comprehensiveness of ϕ_1. There are some characteristics ϕ_1 common to all the instances, which the generalisation treats as unessential, but the less comprehensive these are the better. ϕ_1 stands for the characteristics in which all the instances resemble one another outside those covered by the generalisation. To reduce these resemblances between the instances is the same thing as to increase the differences between them. And hence any increase in the Negative Analogy involves a reduction in the comprehensiveness of ϕ_1. When, however, our knowledge of the instances is complete, it is not necessary to make separate mention of the negative analogy $\underset{a_1 \ldots a_n}{\bar{\mathrm{A}}} (\phi')$ in the above formula. For ϕ' simply includes all those functions about the instances, which are not included in $\phi\phi_1 f$, and of which the contradictories are not included in them; so that in stating $\underset{a_1 \ldots a_n}{\mathrm{A}} (\phi\phi_1 f)$, we state by implication $\underset{a_1 \ldots a_n}{\bar{\mathrm{A}}} (\phi')$ also.

The whole process of strengthening the argument in favour of the generalisation $g(\phi, f)$ by the accumulation of further experience appears to me to consist in making the argument approximate as nearly as possible to the conditions of a perfect analogy, by steadily reducing the comprehensiveness of those resemblances ϕ_1 between the instances which our generalisation disregards. Thus the advantage of additional instances, derived from experience, arises not out of their number as such, but out of their tendency to limit and reduce the comprehensiveness of ϕ_1, or, in other words, out of their tendency to increase the negative analogy ϕ', since $\phi_1\phi'$ comprise between them whatever is not covered by ϕf. The more numerous the instances, the less comprehensive are their superfluous resemblances likely to be. But a single additional instance which greatly reduced ϕ_1 would increase the probability of the argument more than a large number of instances which affected ϕ_1 less.

7. The nature of the argument examined so far is, then, that the instances all have some characteristics in common which we have ignored in framing our generalisation; but it is still assumed that our knowledge about the examined instances is complete. We will next dispense with this latter assumption, and deal with the case in which our knowledge of the characteristics of the examined instances themselves is or may be incomplete.

It is now necessary to take explicit account of the known negative analogy. For when the known positive analogy falls short of the total positive analogy, it is not possible to infer the negative analogy from it. Differences may be known between the instances which cannot be inferred from the known positive analogy. The probability of the argument must, therefore, be written—

$$g(\phi, f) \Big/ \underset{a_1 \ldots a_n}{\mathrm{A}} (\phi\phi_1 f) \underset{a_1 \ldots a_n}{\bar{\mathrm{A}}} (\phi'),$$

where $\phi\phi_1 f$ stands for the characteristics in which all n instances $a_1 \ldots a_n$ are *known* to be alike, and ϕ' stands for the characteristics in which they are *known* to differ.

This argument is strengthened by any additional instance or by any additional knowledge about the former instances which diminishes the known superfluous resemblances ϕ_1 or increases the negative analogy ϕ'. The object of the accumulation of further experience is still the same as before, namely, to make the form of the argument approximate more and more closely to that of perfect analogy. Now, however, that our knowledge of the instances is no longer assumed to be complete, we must take account of the mere *number* n of the instances, as well as of our specific knowledge in regard to them; for the more numerous the instances are, the greater the opportunity for the *total* negative analogy to exceed the *known* negative analogy. But the more complete our knowledge of the instances, the less attention need we pay to their mere number, and the more imperfect our knowledge the greater the stress which must be laid upon the argument from number. This part of the argument will be discussed in detail in the following chapter on Pure Induction.

8. When our knowledge of the instances is incomplete, there may exist analogies which are known to be true of some of the instances and are not known to be false of any. These sub-analogies (see § 2) are not so dangerous as the positive analogies ϕ_1, which are known to be true of *all* the instances, but their existence is, evidently, an element of weakness, which we must endeavour to eliminate by the growth of knowledge and the multiplication of instances. A sub-analogy of this kind between the instances $a_r \dots a_s$ may be written $\underset{a_r \dots a_s}{\mathrm{A}}(\psi_k)$; and the formula, if it is to take account of all the relevant information, ought, therefore, to be written—

$$g(\phi, f) \Big/ \underset{a_1 \dots a_n}{\mathrm{A}}(\phi\phi_1 f)\ \underset{a_1 \dots a_n}{\bar{\mathrm{A}}}(\phi')\Pi\left\{\underset{a_r \dots a_s}{\mathrm{A}}(\psi_k)\right\},$$

where the terms of $\Pi\left\{\underset{a_r \dots a_s}{\mathrm{A}}(\psi_k)\right\}$ stand for the various sub-analogies between sub-classes of the instances, which are not included in $\phi\phi_1 f$ or in ϕ'.

9. There is now another complexity to be introduced. We must dispense with the assumption that the whole of the analogy covered by the generalisation is known to exist in all the instances. For there may be some instances within our experience, about which our knowledge is incomplete, but which show *part* of the analogy required by the generalisation and nothing which contradicts it; and such instances afford some support to the generalisation. Suppose that ${}_b\phi$ and ${}_bf$ are *part* of ϕ and f respectively, then we may have a set of instances $b_1 \dots b_m$ which show the following analogies:

$$\underset{b_1 \dots b_m}{\mathrm{A}}({}_b\phi{}_b\phi_{1b}f)\ \underset{b_1 \dots b_m}{\bar{\mathrm{A}}}({}_b\phi')\Pi\left\{\underset{b_r \dots b_s}{\mathrm{A}}({}_b\psi_k)\right\},$$

where ${}_b\phi_1$ is the analogy not covered by the generalisation, and so on, as before.

The formula, therefore, is now as follows:

$$g(\phi, f) \Big/ \prod_{a,b\ldots} \left\{ \underset{a_1 \ldots a_n}{\mathrm{A}} ({}_a\phi_a\phi_{1a}f) \underset{a_1 \ldots a_n}{\bar{\mathrm{A}}} ({}_a\phi^1) \right\} \prod \left\{ \underset{a_r b_s \ldots}{\mathrm{A}} (\psi_k) \right\}.$$

In this expression ${}_a\phi$, ${}_af$ are the whole *or* part of ϕ, f; the product $\prod_{a,b,\ldots}$ is composed of the positive and negative analogies for each of the sets of instances $a_1 \ldots a_n$, $b_1 \ldots b_m$, etc.; and the product Π contains the various sub-analogies of different sub-classes of all the instances $a_1 \ldots a_n$, $b_1 \ldots b_m$, etc., regarded as *one* set.[1]

10. This completes our classification of the *positive* evidence which supports a generalisation; but the probability may also be affected by a consideration of the negative evidence. We have taken account so far of that part of the evidence only which shows the whole or part of the analogy we require, and we have neglected those instances of which ϕ, the condition of the generalisation, or f, its conclusion, or part of ϕ or of f is *known to be false*. Suppose that there are instances of which ϕ is true and f false, it is clear that the generalisation is ruined. But cases in which we know *part* of ϕ to be true and f to be false, and are ignorant as to the truth or falsity of the rest of ϕ, weaken it to some extent. We must take account, therefore, of analogies

$$\underset{a_1' \ldots a_n'}{\mathrm{A}} ({}_{a'}\phi_{a'}\bar{f}),$$

where ${}_{a'}\phi$, part of ϕ, is true of all the set, and ${}_{a'}f$, part of f, is false of all the set, while the truth or falsity of some part of ϕ and f is unknown. The negative evidence, however, can strengthen as well as weaken the evidence. We deem instances favourably relevant in which ϕ and f are both false together.[2]

Our final formula, therefore, must include terms, similar to those in the formula which concludes § 9, not only for sets of instances which show analogies ${}_a\phi_af$, where ${}_a\phi$ and ${}_af$ are parts of ϕ and f, but also for sets which show analogies ${}_a\bar{\phi}_af$,

[1] Even if we want to distinguish between the sub-analogies of the a set and the sub-analogies of the b set, this information can be gathered from the product Π.

[2] I am disposed to think that we need not pay attention to instances for which part of ϕ is known to be false, and part of f to be true. But the question is a little perplexing.

or analogies ${}_a\bar{\phi}_a\bar{f}$, where ${}_a\phi$ and ${}_af$ are the whole or part of ϕ and f, and $\bar{\phi}\,\bar{f}$ are the contradictories of ϕ and f.[1]

It should be added, perhaps, that the theoretical classification of most empirical arguments in daily use is complicated by the account which we reasonably take of generalisations previously established. We often take account indirectly, therefore, of evidence which supports in some degree other generalisations than that which we are concerned to establish or refute at the moment, but the probability of which is relevant to the problem under investigation.

11. The argument will be rendered unnecessarily complex, without much benefit to its theoretical interest, if we deal with the most general case of all. What follows, therefore, will deal with the formula of the third degree of generality, namely—

$$g(\phi, f) \Big/ \underset{a_1 \ldots a_n}{\mathrm{A}} (\phi\phi_1 f) \underset{a_1 \ldots a_n}{\bar{\mathrm{A}}} (\phi')\Pi\left\{\underset{a_r \ldots a_s}{\mathrm{A}} (\psi_k)\right\},$$

in which no *partial* instances occur, *i.e.* no instances in which part only of the analogy, required by the generalisation, is known to exist. In this third degree of generality, it will be remembered, our knowledge of the characteristics of the instances is incomplete, there is more analogy between the instances than is covered by the generalisation, and there are some sub-analogies to be reckoned with. In the above formula the incompleteness of our knowledge is implicitly recognised in that $\phi\phi_1 f\phi'$ are not between them entirely comprehensive. It is also supposed that all the evidence we have is positive, no knowledge is assumed, that is to say, of instances characterised by the conjunctions ${}_a\bar{\phi}_a f$, ${}_a\phi_a\bar{f}$, or ${}_a\bar{\phi}_a\bar{f}$, where ${}_a\phi$ and ${}_af$ are part of ϕ and f.

An argument, therefore, from experience, in which, on the basis of examined instances, we establish a generalisation applicable beyond these instances, can be strengthened, if we restrict our attention to the simpler type of case, by the following means:

(1) By reducing the resemblances ϕ_1 known to be common to all the instances, but ignored as unessential by the generalisation.

(2) By increasing the differences ϕ' known to exist between the instances.

[1] Where the conclusion f is simple and not complex (see § 5), some of these complications cannot, of course, arise.

(3) By diminishing the sub-analogies or unessential resemblances ψ_k known to be common to some of the instances and not known to be false of any.

These results can generally be obtained in two ways, either by increasing the number of our instances or by increasing our knowledge of those we have.

The reasons why these methods seem to common sense to strengthen the argument are fairly obvious. The object of (1) is to avoid the possibility that ϕ_1 as well as ϕ is a necessary condition of f. The object of (2) is to avoid the possibility that there may be some resemblances additional to ϕ, common to all the instances, which have escaped our notice. The object of (3) is to get rid of indications that the total value of ϕ_1 may be greater than the known value. When $\phi\phi_1 f$ is the *total* positive analogy between the instances, so that the known value of ϕ_1 is its total value, it is (1) which is fundamental; and we need take account of (2) and (3) only when our knowledge of the instances is incomplete. But when our knowledge of the instances is incomplete, so that ϕ_1 falls short of its total value and we cannot infer ϕ' from it, it is better to regard (2) as fundamental; in any case every reduction of ϕ_1 must increase ϕ'.

12. I have now attempted to analyse the various ways in which common practice seems to assume that considerations of Analogy can yield us presumptive evidence in favour of a generalisation.

It has been my object, in making a classification of empirical arguments, not so much to put my results in forms closely similar to those in which problems of generalisation commonly present themselves to scientific investigators, as to inquire whether ultimate uniformities of method can be found beneath the innumerable modes, superficially differing from another, in which we do in fact argue.

I have not yet attempted to justify this way of arguing. After turning aside to discuss in more detail the method of Pure Induction, I shall make this attempt; or rather I shall try to see *what sort* of assumptions are capable of justifying empirical reasoning of this kind.

CHAPTER XX

THE VALUE OF MULTIPLICATION OF INSTANCES, OR PURE INDUCTION

1. It has often been thought that the essence of inductive argument lies in the multiplication of instances. "Where is that process of reasoning," Hume inquired, "which from one instance draws a conclusion, so different from that which it infers from a hundred instances, that are no way different from that single instance?" I repeat that by emphasising the number of the instances Hume obscured the real object of the method. If it were strictly true that the hundred instances are *no* way different from the single instance, Hume would be right to wonder in what manner they can strengthen the argument. The object of increasing the number of instances arises out of the fact that we are nearly always aware of *some* difference between the instances, and that even where the known difference is insignificant we may suspect, especially when our knowledge of the instances is very incomplete, that there may be more. Every new instance *may* diminish the unessential resemblances between the instances and by introducing a new difference increase the Negative Analogy. For this reason, and for this reason only, new instances are valuable.

If our premisses comprise the body of memory and tradition which has been originally derived from direct experience, and the conclusion which we seek to establish is the Newtonian theory of the Solar System, our argument is one of Pure Induction, in so far as we support the Newtonian theory by pointing to the great number of consequences which it has in common with the facts of experience. The predictions of the Nautical Almanack are a consequence of the Newtonian theory, and these predictions are verified many thousand times a day. But even here the force of the argument largely depends, not on the mere number of these predictions, but on the knowledge that the circumstances in which they are fulfilled differ widely from one another in a vast number of important respects. The *variety* of the circumstances, in which the Newtonian generalisation is fulfilled, rather than the number of them, is what seems to impress our reasonable faculties.

2. I hold, then, that our object is always to increase the Negative Analogy, or, which is the same thing, to diminish the characteristics common to all the examined instances and yet not taken account of by our generalisation. Our method, however, may be one which certainly achieves this object, or it may be one which possibly achieves it. The former of these, which is obviously the more satisfactory, may consist either in increasing our definite knowledge respecting instances examined already, or in finding additional instances respecting which definite knowledge is obtainable. The second of them consists in finding additional instances of the generalisation, about which, however, our definite knowledge may be meagre; such further instances, if our knowledge about them were more complete, would either increase or leave unchanged the Negative Analogy; in the former case they would strengthen the argument and in the latter case they would not weaken it; and they must, therefore, be allowed some weight. The two methods are not entirely distinct, because new instances, about which we have some knowledge but not much, may be known to increase the Negative Analogy a little by the first method, and suspected of increasing it further by the second.

It is characteristic of advanced scientific method to depend on the former, and of the crude unregulated induction of ordinary experience to depend on the latter. It is when our definite knowledge about the instances is limited, that we must pay attention to their number rather than to the specific differences between them, and must fall back on what I term Pure Induction.

In this chapter I investigate the conditions and the manner in which the mere repetition of instances can add to the force of the argument. The chief value of the chapter, in my judgment, is negative, and consists in showing that a line of advance, which might have seemed promising, turns out to be a blind alley, and that we are thrown back on known Analogy. Pure

Induction will not give us any very substantial assistance in getting to the bottom of the general inductive problem.

3. The problem of generalisation[1] by Pure Induction can be stated in the following symbolic form:

Let h represent the general *à priori data* of the investigation; let g represent the generalisation which we seek to establish; let $x_1x_2 \ldots x_n$ represent instances of g.

Then $x_1/gh=1$, $x_2/gh=1 \ldots x_n/gh=1$; given g, that is to say, the truth of each of its instances follows. The problem is to determine the probability $g/hx_1x_2 \ldots x_n$, *i.e.* the probability of the generalisation when n instances of it are given. Our analysis will be simplified, and nothing of fundamental importance will be lost, if we introduce the assumption that there is nothing in our *à priori data* which leads us to distinguish between the *à priori* likelihood of the different instances; we assume, that is to say, that there is no reason *à priori* for expecting the occurrence of any one instance with greater reliance than any other, *i.e.*

$$x_1/h = x_2/h = \ldots = x_n/h.$$

Write
$$g/hx_1x_2 \ldots x_n = p_n$$
and
$$x_{n+1}/hx_1x_2 \ldots x_n = y_{n+1};$$
then
$$\frac{p_n}{p_{n-1}} = \frac{g/hx_1 \ldots x_n}{g/hx_1 \ldots x_{n-1}} = \frac{x_n/hx_1 \ldots x_{n-1}}{g/hx_1 \ldots x_{n-1} \cdot x_n/hx_1 \ldots x_{n-1}}$$
$$= \frac{x_n/ghx_1 \ldots x_{n-1}}{x_n/hx_1 \ldots x_{n-1}}$$
$$= \frac{1}{y_n}.$$

$\therefore \dfrac{p_n}{p_{n-1}} = \dfrac{1}{y_n}$, and hence $p_n = \dfrac{1}{y_1y_2 \cdots y_n} \cdot p_0$, where $p_0 = g/h$, *i.e.* p_0 is the *à priori* probability of the generalisation.

[1] In the most general sense we can regard any proposition as the generalisation of all the propositions which follow from it. For if h is any proposition, and we put $\phi(x) \equiv$ 'x can be inferred from h' and $f(x) \equiv x$, then $g(\phi, f) \equiv h$. Since Pure Induction consists in finding as many instances of a generalisation as possible, it is, in the widest sense, the process of strengthening the probability of any proposition by adducing numerous instances of known truths which follow from it. The argument is one of Pure Induction, therefore, in so far as the probability of a conclusion is based upon the number of independent consequences which the conclusion and the premisses have in common.

It follows, therefore, that $p_n > p_{n-1}$ so long as $y_n \neq 1$.
Further,

$$\begin{aligned} x_1 x_2 \ldots x_n/h &= x_n/h x_1 x_2 \ldots x_{n-1} \,.\, x_1 x_2 \ldots x_{n-1}/h \\ &= y_n \,.\, x_1 x_2 \ldots x_{n-1}/h \\ &= y_n y_{n-1} \ldots y_1. \end{aligned}$$

$$\begin{aligned} \therefore\ p_n &= \frac{p_0}{y_1 y_2 \ldots y_n} = \frac{p_0}{x_1 x_2 \ldots x_n/h} \\ &= \frac{p_0}{x_1 x_2 \ldots x_n g/h + x_1 x_2 \ldots x_n \bar{g}/h} \\ &= \frac{p_0}{g/h + x_1 x_2 \ldots x_n/\bar{g}h \,.\, \bar{g}/h} \\ &= \frac{p_0}{p_0 + x_1 x_2 \ldots x_n/\bar{g}h(1-p_0)}. \end{aligned}$$

This approaches unity as a limit, if $x_1 x_2 \ldots x_n/\bar{g}h \,.\, \frac{1}{p_0}$ approaches zero as a limit, when n increases.

4. We may now stop to consider how much this argument has proved. We have shown that if each of the instances necessarily follows from the generalisation, then each additional instance increases the probability of the generalisation, so long as the new instance could not have been predicted with certainty from a knowledge of the former instances.[1] This condition is the same as that which came to light when we were discussing Analogy. If the new instance were identical with one of the former instances, a knowledge of the latter would enable us to predict it. If it differs or may differ in analogy, then the condition required above is satisfied.

The common notion, that each successive verification of a doubtful principle strengthens it, is formally proved, therefore, without any appeal to conceptions of law or of causality. *But we have not proved* that this probability approaches certainty as a limit, or even that our conclusion becomes more likely than not, as the number of verifications or instances is indefinitely increased.

5. What are the conditions which must be satisfied in order that the rate, at which the probability of the generalisation increases, may be such that it will approach certainty as a

[1] Since $p_n > p_{n-1}$ so long as $y_n \neq 1$.

limit when the number of independent instances of it are indefinitely increased? We have already shown, as a basis for this investigation, that p_n approaches the limit of certainty for a generalisation g, if, as n increases, $x_1x_2 \ldots x_n/\bar{g}h$ becomes small compared with p_0, *i.e.* if the *à priori* probability of so many instances, assuming the falsehood of the generalisation, is small compared with the generalisation's *à priori* probability. It follows, therefore, that the probability of an induction tends towards certainty as a limit, when the number of instances is increased, provided that

$$x_r/x_1x_2 \ldots x_{r-1}\bar{g}h < 1 - \epsilon$$

for all values of r, and $p_0 > \eta$, where ϵ and η are finite probabilities, separated, that is to say, from impossibility by a value of some finite amount, however small. These conditions appear simple, but the meaning of a 'finite probability' requires a word of explanation.[1]

I argued in Chapter III. that not all probabilities have an exact numerical value, and that, in the case of some, one can say no more about their relation to certainty and impossibility than that they fall short of the former and exceed the latter. There is one class of probabilities, however, which I called the numerical class, the ratio of each of whose members to certainty can be expressed by some number less than unity; and we can sometimes compare a non-numerical probability in respect of more and less with one of these numerical probabilities. This enables us to give a definition of 'finite probability' which is capable of application to non-numerical as well as to numerical probabilities. I define a 'finite probability' as one which *exceeds* some numerical probability, the ratio of which to certainty can be expressed by a finite number.[2] The principal method, in which a probability can be proved finite by a process of argument, arises either when

[1] The proof of these conditions, which is obvious, is as follows:

$$x_1x_2 \ldots x_n/\bar{g}h = x_n/x_1x_2 \ldots x_{n-1}\bar{g}h \,.\, x_1x_2 \ldots x_{n-1}/\bar{g}h < (1-\epsilon)^n,$$

where ϵ is finite and $p_0 > \eta$ where η is finite. There is always, under these conditions, some finite value of n such that both $(1-\epsilon)^n$ and $\frac{(1-\epsilon)^n}{\eta}$ are less than any given finite quantity, however small.

[2] Hence a series of probabilities $p_1p_2 \ldots p_r$ approaches a limit L, if, given any positive finite number ϵ however small, a positive integer n can always be found such that for all values of r greater than n the difference between L and p_r is less than $\epsilon . \gamma$, where γ is the measure of certainty.

its conclusion can be shown to be one of a finite number of alternatives, which are between them exhaustive or, at any rate, have a finite probability, and to which the Principle of Indifference is applicable; or (more usually), when its conclusion is *more* probable than some hypothesis which satisfies this first condition.

6. The conditions, which we have now established in order that the probability of a pure induction may tend towards certainty as the number of instances is increased, are (1) that $x_r/x_1x_2 \ldots x_{r-1}\bar{g}h$ falls short of certainty by a finite amount for all values of r, and (2) that p_0, the *à priori* probability of our generalisation, exceeds impossibility by a finite amount. It is easy to see that we can show by an exactly similar argument that the following more general conditions are equally satisfactory:

(1) That $x_r/x_1x_2 \ldots x_{r-1}\bar{g}h$ falls short of certainty by a finite amount for all values of r beyond a specified value s.

(2) That p_s, the probability of the generalisation relative to a knowledge of these first s instances, exceeds impossibility by a finite amount.

In other words Pure Induction can be usefully employed to strengthen an argument if, after a certain number of instances have been examined, we have, from some other source, a finite probability in favour of the generalisation, and, assuming the generalisation is false, a finite uncertainty as to its conclusion being satisfied by the next hitherto unexamined instance which satisfies its premiss. To take an example, Pure Induction can be used to support the generalisation that the sun will rise every morning for the next million years, provided that with the experience we have actually had there are finite probabilities, however small, *derived from some other source*, first, in favour of the generalisation, and, second, in favour of the sun's *not* rising to-morrow assuming the generalisation to be false. Given these finite probabilities, obtained otherwise, however small, then the probability can be strengthened and can tend to increase towards certainty by the mere multiplication of instances provided that these instances are so far distinct that they are not inferrible one from another.

7. Those supposed proofs of the Inductive Principle, which are based openly or implicitly on an argument in inverse probability, are all vitiated by unjustifiable assumptions relating to the magnitude of the *à priori* probability p_0. Jevons, for instance, avowedly assumes that we may, in the absence of special information, suppose any unexamined hypothesis to be as likely as not. It is difficult to see how such a belief, if even its most immediate implications had been properly apprehended, could have remained plausible to a mind of so sound a practical judgment as his. The arguments against it and the contradictions to which it leads have been dealt with in Chapter IV. The demonstration of Laplace, which depends upon the Rule of Succession, will be discussed in Chapter XXX.

8. The prior probability, which must always be found, before the method of pure induction can be usefully employed to support a substantial argument, is derived, I think, in most ordinary cases—with what justification it remains to discuss—from considerations of Analogy. But the conditions of valid induction as they have been enunciated above, are quite independent of analogy, and might be applicable to other types of argument. In certain cases we might feel justified in assuming *directly* that the necessary conditions are satisfied.

Our belief, for instance, in the validity of a logical scheme is based partly upon inductive grounds—on the *number* of conclusions, each seemingly true on its own account, which can be derived from the axioms—and partly on a degree of self-evidence in the axioms themselves sufficient to give them the initial probability upon which induction can build. We depend upon the initial presumption that, if a proposition appears to us to be true, this is by itself, in the absence of opposing evidence, *some reason* for its *being* as well as appearing true. We cannot deny that what appears true is sometimes false, but, unless we can assume some substantial relation of probability between the appearance and the reality of truth, the possibility of even probable knowledge is at an end.

The conception of our having *some* reason, though not a conclusive one, for certain beliefs, arising out of direct inspection, may prove important to the theory of epistemology. The old metaphysics has been greatly hindered by reason of its having always demanded demonstrative certainty. Much of the cogency of Hume's criticism arises out of the assumption of methods of certainty on the part of those systems against which it was directed. The earlier realists were hampered by their not perceiving that lesser claims in the beginning might yield them what they wanted in the end. And transcendental philosophy has partly arisen, I believe, through the belief that there is no knowledge on these matters short of certain knowledge, being combined with the belief that such certain knowledge of metaphysical questions is beyond the power of ordinary methods.

When we allow that probable knowledge is, nevertheless, real, a new method of argument can be introduced into metaphysical discussions. The demonstrative method can be laid on one side, and we may attempt to advance the argument by taking account of circumstances which seem to give *some* reason for preferring one alternative to another. Great progress may follow if the nature and reality of objects of perception,[1] for instance, can be usefully investigated by methods not altogether dissimilar from those employed in science and with the prospect of obtaining as high a degree of certainty as that which belongs to some scientific conclusions; and it may conceivably be shown that a belief in the conclusions of science, enunciated in any reasonable manner however restricted, involves a preference for some metaphysical conclusions over others.

9. Apart from analysis, careful reflection would hardly lead us to expect that a conclusion which is based on no other than grounds of pure induction, defined as I have defined them as consisting of repetition of instances merely, could attain in this way to a high degree of probability. To this extent we ought all of us to agree with Hume. We have found that the suggestions of common sense are supported by more precise methods. Moreover, we constantly distinguish between arguments, which we call inductive, upon other grounds than the number of instances upon which they are based; and under certain conditions we regard as crucial an insignificant number of experiments. The method of pure induction may be a useful means of strengthening a probability based on some other ground. In the case, however, of most scientific arguments, which would commonly be called inductive, the probability that we are right, when we make predictions on the basis of past experience, depends not so much on the number of past experiences upon which we rely, as on the degree in which the circumstances of these experiences

[1] A paper by Mr. G. E. Moore entitled, "The Nature and Reality of Objects of Perception," which was published in the *Proceedings of the Aristotelian Society for 1906*, seems to me to apply for the first time a method somewhat resembling that which is described above.

resemble the known circumstances in which the prediction is to take effect. Scientific method, indeed, is mainly devoted to discovering means of so heightening the known analogy that we may dispense as far as possible with the methods of pure induction.

When, therefore, our previous knowledge is considerable and the analogy is good, the purely inductive part of the argument may take a very subsidiary place. But when our knowledge of the instances is slight, we may have to depend upon pure induction a good deal. In an advanced science it is a last resort,—the least satisfactory of the methods. But sometimes it must be our first resort, the method upon which we must depend in the dawn of knowledge and in fundamental inquiries where we must presuppose nothing.

CHAPTER XXI

THE NATURE OF INDUCTIVE ARGUMENT CONTINUED

1. In the enunciation, given in the two preceding chapters, of the Principles of Analogy and Pure Induction there has been no reference to experience or causality or law. So far, the argument has been perfectly formal and might relate to a set of propositions of any type. But these methods are most commonly employed in physical arguments where material objects or experiences are the terms of the generalisation. We must consider, therefore, whether there is any good ground, as some logicians seem to have supposed, for restricting them to this kind of inquiry.

I am inclined to think that, whether reasonably or not, we naturally apply them to all kinds of argument alike, including formal arguments as, for example, about numbers. When we are told that Fermat's formula for a prime, namely, $2^{2^a}+1$ for all values of a, has been verified in every case in which verification is not excessively laborious—namely, for $a=1$, 2, 3, and 4, we feel that this is *some* reason for accepting it, or, at least, that it raises a sufficient presumption to justify a further examination of the formula.[1] Yet there can be no reference here to the uniformity of nature or physical causation. If inductive methods are limited to natural objects, there can no more be an appreciable ground for thinking that $2^{2^a}+1$ is a true formula for primes, because empirical methods show that it yields primes up to $a=4$, or *even if they showed that it yielded primes for every number up to a million million*, than there is to think that any formula which I may choose to write down

[1] This formula has, in fact, been disproved in recent times, *e.g.* $2^{2^5}+1=$ 4, 294, 967, 297 $=641\times6$, 700, 417. Thus it is no longer so good an illustration as it would have been a hundred years ago.

at random is a true source of primes. To maintain that there is no appreciable ground in such a case is paradoxical. If, on the other hand, a partial verification does raise some just appreciable presumption in the formula's favour, then we must include numbers, at any rate, as well as material objects amongst the proper subjects of the inductive method. The conclusion of the previous chapter indicates, however, that, if arguments of this kind have force, it can only be in virtue of there being some finite *à priori* probability for the formula based on other than inductive grounds.

There are some illustrations in Jevons's *Principles of Science*,[1] which are relevant to this discussion. We find it to be true of the following six numbers:

5, 15, 35, 45, 65, 95

that they all end in five, and are all divisible by five without remainder. Would this fact, by itself, raise any kind of presumption that all numbers ending in five are divisible by five without remainder? Let us also consider the six numbers,

7, 17, 37, 47, 67, 97.

They all end in seven and also agree in being primes. Would this raise a presumption in favour of the generalisation that all numbers are prime, which end in seven? We might be prejudiced in favour of the first argument, because it would lead us to a true conclusion; but we ought not to be prejudiced against the second because it would lead us to a false one; for the validity of empirical arguments as the foundation of a probability cannot be affected by the actual truth or falsity of their conclusions. If, on the evidence, the analogy is similar and equal, and if the scope of the generalisation and its conclusion is similar, then the value of the two arguments must be equal also.

Whether or not the use of empirical argument appears plausible to us in these particular examples, it is certainly true that many mathematical theorems have actually been discovered by such methods. Generalisations have been suggested nearly as often, perhaps, in the logical and mathematical sciences, as in the

[1] Pp. 229-231 (one volume edition). Jevons uses these illustrations, not for the purpose to which I am here putting them, but to demonstrate the fallibility of empirical laws.

physical, by the recognition of particular instances, even where formal proof has been forthcoming subsequently. Yet if the suggestions of analogy have no appreciable probability in the formal sciences, and should be permitted only in the material, it must be unreasonable for us to pursue them. If no finite probability exists that a formula, for which we have empirical verification, is in fact universally true, Newton was acting fortunately, but not reasonably, when he hit on the Binomial Theorem by methods of empiricism.[1]

2. I am inclined to believe, therefore, that, if we trust the promptings of common sense, we have the same kind of ground for trusting analogy in mathematics that we have in physics, and that we ought to be able to apply any justification of the method, which suits the latter case, to the former also. This does not mean that the *à priori* probabilities, from some other source than induction, which the inductive method requires as its foundation, may not be sought and found differently in the two types of inquiry. A reason why it has been thought that analogy ought to be confined to natural laws may be, perhaps, that in most of those cases, in which we could support a mathematical theorem by a very strong analogy, the existence of a formal proof has done away with the necessity for the limping methods of empiricism; and because in most mathematical investigations, while in our earliest thoughts we are not ashamed to consult analogy, our later work will be more profitably spent in searching for a formal proof than in establishing analogies which must, at the best, be relatively weak. As the modern scientist discards, as a rule, the method of pure induction, in favour of experimental analogy, where, if he takes account of his previous knowledge, one or two cases may prove immensely significant; so the modern mathematician prefers the resources of his analysis, which may yield him certainty, to the doubtful promises of empiricism.

3. The main reason, however, why it has often been held that we ought to limit inductive methods to the content of the particular material universe in which we live, is, most probably, the fact that we can easily imagine a universe so constructed that such methods would be useless. This suggests that analogy and induction, while they happen to be useful to us in this world,

[1] See Jevons, *loc. cit.* p. 231.

cannot be universal principles of logic, on the same footing, for instance, as the syllogism.

In one sense this opinion may be well founded. I do not deny or affirm at present that it may be necessary to confine inductive methods to arguments about certain kinds of objects or certain kinds of experiences. It may be true that in every useful argument from analogy our premisses must contain fundamental assumptions, obtained directly and not inductively, which some possible experiences might preclude. Moreover, the success of induction in the past can certainly affect its probable usefulness for the future. We may discover something about the nature of the universe—we may even discover it by means of induction itself—the knowledge of which has the effect of destroying the further utility of induction. I shall argue later on that the confidence with which we ourselves use the method does in fact depend upon the nature of our past experience.

But this empirical attitude towards induction may, on the other hand, arise out of either one of two possible confusions. It may confuse, first, the reasonable character of arguments with their practical usefulness. The usefulness of induction depends, no doubt, upon the actual content of experience. If there were no repetition of detail in the universe, induction would have no utility. If there were only a single object in the universe, the laws of addition would have no utility. But the processes of induction and addition would remain reasonable. It may confuse, secondly, the validity of attributing probability to the conclusion of an argument with the question of the actual truth of the conclusion. Induction tells us that, on the basis of certain evidence, a certain conclusion is reasonable, *not* that it is true. If the sun does not rise to-morrow, if Queen Anne still lives, this will not prove that it was foolish or unreasonable of us to have believed the contrary.

4. It will be worth while to say a little more in this connection about the not infrequent failure to distinguish the rational from the true. The excessive ridicule, which this mistake has visited on the supposed irrationality of barbarous and primitive peoples, affords some good examples. "Reflection and enquiry should satisfy us," says Dr. Frazer in the *Golden Bough*, "that to our predecessors we are indebted for much of what we thought most our own, and that their errors were not wilful extravagances or the ravings of insanity, but simply hypotheses, justifiable as such at the time when they were propounded, but which a fuller experience has proved to be inadequate. . . . Therefore, in reviewing the opinions and practices of ruder ages and races we shall do well to look with leniency upon their errors as inevitable slips made in the search for truth. . . ." The first introduction of iron ploughshares into Poland, he tells in another passage, having been followed by a succession of bad harvests, the farmers attributed the badness of the crops to the iron ploughshares, and discarded them for the old wooden ones. The method of reasoning of the farmers is not different from that of science, and may, surely, have had for them some appreciable probability in its favour. "It is a curious superstition," says a recent pioneer in Borneo, "this of the Dusuns, to attribute anything—whether good or bad, lucky or unlucky—that happens to them to something novel which has arrived in their country. For instance, my living in Kindram has caused the intensely hot weather we have experienced of late."[1] What is this curious superstition but the Method of Difference?

The following passage from Jevons's *Principles of Science* well illustrates the tendency, to which he himself yielded, to depreciate the favourite analogies of one age, because the experience of their successors has confuted them. Between things which are the same in number, he points out, there is a certain resemblance, namely in number; and in the infancy of science men could not be persuaded that there was not a deeper resemblance implied in that of number. "Seven days are mentioned in Genesis; infants acquire their teeth at the end of seven months; they change them at the end of seven years; seven feet was the limit of man's height; every seventh year was a climacteric or critical year, at which a change of disposition took place. In natural science there were not only the seven planets, and the seven metals, but also the seven primitive colours, and the seven tones of music. So deep a hold did this doctrine take that we still have its results in many customs, not only in the seven days of the week, but the seven years' apprenticeship, puberty at fourteen years, the second climacteric, and legal majority at twenty-one years, the third climacteric." Religious systems from Pythagoras to Comte have sought to derive strength from the virtue of seven.

[1] *Golden Bough*, p. 174.

"And even in scientific matters the loftiest intellects have occasionally yielded, as when Newton was misled by the analogy between the seven tones of music and the seven colours of his spectrum. . . . Even the genius of Huyghens did not prevent him from inferring that but one satellite could belong to Saturn, because, with those of Jupiter and the earth, it completed the perfect number of six." But is it certain that Newton and Huyghens were only reasonable when their theories were true, and that their mistakes were the fruit of a disordered fancy? Or that the savages, from whom we have inherited the most fundamental inductions of our knowledge, were always superstitious when they believed what we now know to be preposterous?

It is important to understand that the common sense of the race has been impressed by very weak analogies and has attributed to them an appreciable probability, and that a logical theory, which is to justify common sense, need not be afraid of including these marginal cases. Even our belief in the real existence of other people, which we all hold to be well established, may require for its justification the combination of experience with a just appreciable *à priori* possibility for Animism generally.[1] If we actually possess evidence which renders some conclusion absurd, it is very difficult for us to appreciate the relation of this conclusion to data which are different and less complete; but it is essential that we should realise arguments from analogy as *relative to premisses*, if we are to approach the logical theory of Induction without prejudice.

5. While we depreciate the former probability of beliefs which we no longer hold, we tend, I think, to exaggerate the present degree of certainty of what we still believe. The preceding paragraph is not intended to deny that savages often greatly

[1] "This is animism, or that sense of something in Nature which to the enlightened or civilised man is not there, and in the civilised man's child, if it be admitted that he has it at all, is but a faint survival of a phase of the primitive mind. And by animism I do not mean the theory of a soul in nature, but the tendency or impulse or instinct, in which all myth originates, to *animate* all things; the projection of ourselves into nature; the sense and apprehension of an intelligence like our own, but more powerful in all visible things" (Hudson, *Far Away and Long Ago*, pp. 224-5). This 'tendency or impulse or instinct,' refined by reason and enlarged by experience, may be required, in the shape of an intuitive *à priori* probability, if some of those universal conclusions of common sense, which the most sceptical do not kick away, are to be supported with rational foundations.

overestimate the value of their crude inductions, and are to this extent irrational. It is not easy to distinguish between a belief's being the most reasonable of those which it is open to us to believe, and its being more probable than not. In the same way we, perhaps, put an excessive confidence in those conclusions—the existence of other people, for instance, the law of gravity, or to-morrow's sunrise—of which, in comparison with many other beliefs, we are very well assured. We may sometimes confuse the practical certainty, attaching to the class of beliefs upon which it is rational to act with the utmost confidence, with the more wholly objective certainty of logic. We might rashly assert, for instance, that to-morrow's sunrise is as likely to us as failure, and the special virtue of the number seven as unlikely, even to Pythagoras, as success, in an attempt to throw heads a hundred times in succession with an unbiassed coin.[1]

6. As it has often been held upon various grounds, with reason or without, that the validity of Induction and Analogy depends in some way upon the character of the actual world, logicians have sought for material laws upon which these methods can be founded. The Laws of Universal Causation and the Uniformity of Nature, namely, that all events have *some* cause and that the same total cause always produces the same effect, are those which commonly do service. But these principles merely assert that there are *some* data from which events posterior to them in time could be inferred. They do not seem to yield us much assistance in solving the inductive problem proper, or in determining how we can infer with probability from *partial* data. It has been suggested in the previous chapter that the Principle of the Uniformity of Nature amounts to an assertion that an argument from perfect analogy (defined as I have defined it) is valid when applied to events only differing in their positions in time or space.[2] It has also been pointed out that ordinary inductive arguments appear to be strengthened by any evidence which makes them approximate more closely in character to a perfect analogy. But this, I think, is the whole extent to which this principle, even if its truth could be assumed, would help us.

[1] Yet if every inhabitant of the world, Grimsehl has calculated, were to toss a coin every second, day and night, this latter event would only occur once on the average in every twenty billion years.

[2] Is this interpretation of the Principle of the Uniformity of Nature affected by the Doctrine of Relativity?

States of the universe, identical in every particular, may never recur, and, even if identical states were to recur, we should not know it.

The kind of fundamental assumption about the character of material laws, on which scientists appear commonly to act, seems to me to be much less simple than the bare principle of Uniformity. They appear to assume something much more like what mathematicians call the principle of the superposition of small effects, or, as I prefer to call it, in this connection, the *atomic* character of natural law. The system of the material universe must consist, if this kind of assumption is warranted, of bodies which we may term (without any implication as to their size being conveyed thereby) *legal atoms*, such that each of them exercises its own separate, independent, and invariable effect, a change of the total state being compounded of a number of separate changes each of which is solely due to a separate portion of the preceding state. We do not have an invariable relation between particular bodies, but nevertheless each has on the others its own separate and invariable effect, which does not change with changing circumstances, although, of course, the total effect may be changed to almost any extent if all the other accompanying causes are different. Each atom can, according to this theory, be treated as a separate cause and does not enter into different organic combinations in each of which it is regulated by different laws.

Perhaps it has not always been realised that this atomic uniformity is in no way implied by the principle of the Uniformity of Nature. Yet there might well be quite different laws for wholes of different degrees of complexity, and laws of connection between complexes which could not be stated in terms of laws connecting individual parts. In this case natural law would be organic and not, as it is generally supposed, atomic. If every configuration of the Universe were subject to a separate and independent law, or if very small differences between bodies—in their shape or size, for instance,—led to their obeying quite different laws, prediction would be impossible and the inductive method useless. Yet nature might still be uniform, causation sovereign, and laws timeless and absolute.

The scientist wishes, in fact, to assume that the occurrence of a phenomenon which has appeared as part of a more complex phenomenon, may be *some* reason for expecting it to be associated on another occasion with part of the same complex. Yet if different wholes were subject to different laws *quâ* wholes and not simply on account of and in proportion to the differences of their parts, knowledge of a part could not lead, it would seem, even to presumptive or probable knowledge as to its association with other parts. Given, on the other hand, a number of legally atomic units and the laws connecting them, it would be possible to deduce their effects *pro tanto* without an exhaustive knowledge of all the coexisting circumstances.

We do habitually assume, I think, that the size of the atomic unit is for mental events an individual consciousness, and for material events an object small in relation to our perceptions. These considerations do not show us a way by which we can justify Induction. But they help to elucidate the kind of assumptions which we do actually make, and may serve as an introduction to what follows.

CHAPTER XXII

THE JUSTIFICATION OF THESE METHODS

1. The general line of thought to be followed in this chapter may be indicated, briefly, at the outset.

A system of facts or propositions, as we ordinarily conceive it, may comprise an indefinite number of members. But the ultimate constituents or indefinables of the system, which all the members of it are about, are less in number than these members themselves. Further, there are certain laws of necessary connection between the members, by which it is meant (I do not stop to consider whether *more* than this is meant) that the truth or falsity of every member can be inferred from a knowledge of the laws of necessary connection together with a knowledge of the truth or falsity of some (but not all) of the members.

The ultimate constituents together with the laws of necessary connection make up what I shall term the *independent variety* of the system. The more numerous the ultimate constituents and the necessary laws, the greater is the system's independent variety. It is not necessary for my present purpose, which is merely to bring before the reader's mind the sort of conception which is in mine, that I should attempt a complete definition of what I mean by a system.

Now it is characteristic of a system, as distinguished from a collection of heterogeneous and independent facts or propositions, that the number of its premisses, or, in other words, the amount of independent variety in it, should be less than the number of its members. But it is not an obviously essential characteristic of a system that its premisses or its independent variety should be actually finite. We must distinguish, therefore, between systems which may be termed finite and infinite respectively, the terms *finite* and *infinite* referring not to the number of members in the system but to the amount of independent variety in it.

The purpose of the discussion, which occupies the greater part of this chapter, is to maintain that, if the premisses of our argument permit us to assume that the facts or propositions, with which the argument is concerned, belong to a *finite* system, then probable knowledge can be validly obtained by means of an inductive argument. I now proceed to approach the question from a slightly different standpoint, the controlling idea, however, being that which is outlined above.

2. What is our actual course of procedure in an inductive argument? We have before us, let us suppose, a set of n instances which have r known qualities, $a_1a_2 \ldots a_r$ in common, these r qualities constituting the known positive analogy. From these qualities three (say) are picked out, namely, a_1, a_2, a_3, and we inquire with what probability *all* objects having these three qualities have also certain other qualities which we have picked out, namely, a_{r-1}, a_r. We wish to determine, that is to say, whether the qualities a_{r-1}, a_r are *bound up* with the qualities a_1, a_2, a_3. In thus approaching this question we seem to suppose that the qualities of an object are bound together in a limited number of *groups*, a sub-class of each group being an infallible symptom of the coexistence of certain other members of it also.

Three possibilities are open, any of which would prove destructive to our generalisation. It may be the case (1) that a_{r-1} or a_r is independent of all the other qualities of the instances —they may not overlap, that is to say, with any other groups; or (2) that $a_1a_2a_3$ do not belong to the same groups as $a_{r-1}a_r$; or (3) that $a_1a_2a_3$, while they belong to the same group as $a_{r-1}a_r$, are not sufficient to specify this group uniquely—they belong, that is to say, to other groups also which do not include a_{r-1} and a_r. The precautions we take are directed towards reducing the likelihood, so far as we can, of each of these possibilities. We distrust the generalisation if the terms typified by $a_{r-1}a_r$ are numerous and comprehensive, because this increases the likelihood that some at least of them fall under heading (1), and also because it increases the likelihood of (3). We trust it if the terms typified by $a_1a_2a_3$ are numerous and comprehensive, because this decreases the likelihood both of (2) and of (3). If we find a new instance which agrees with the former instances in $a_1 a_2 a_3 a_{r-1} a_r$ but not in a_4, we welcome it, because this disposes of the possibility that it is a_4, alone or in combination, that is bound up with $a_{r-1} a_r$. We desire to increase our knowledge of the properties, lest there be some positive analogy which is escaping us, and when our knowledge is incomplete we multiply instances, which we do not know to increase the negative analogy for certain, in the hope that they may do so.

If we sum up the various methods of Analogy, we find, I think, that they are all capable of arising out of an underlying assumption, that if we find two sets of qualities in coexistence there is a finite probability that they belong to the same group, and a finite probability also that the first set specifies this group uniquely. Starting from this assumption, the object of the methods is to increase the finite probability and make it large. Whether or not anything of this sort is explicitly present to our minds when we reason scientifically, it seems clear to me that we do act exactly as we should act, if this were the assumption from which we set out.

In most cases, of course, the field is greatly simplified from the first by the use of our pre-existing knowledge. Of the properties before us we generally have good reason, derived from prior analogies, for supposing some to belong to the same group and others to belong to different groups. But this does not affect the theoretical problem confronting us.

3. What kind of ground could justify us in assuming the existence of these finite probabilities which we seem to require? If we are to obtain them, not directly, but by means of argument, we must somehow base them upon a finite number of exhaustive alternatives.

The following line of argument seems to me to represent, on the whole, the kind of assumption which is obscurely present to our minds. We suppose, I think, that the almost innumerable apparent properties of any given object all arise out of a finite number of generator properties, which we may call $\phi_1 \phi_2 \phi_3 \ldots$. Some arise out of ϕ_1 alone, some out of ϕ_1 in conjunction with ϕ_2, and so on. The properties which arise out of ϕ_1 alone form one group; those which arise out of $\phi_1 \phi_2$ in conjunction form another group, and so on. Since the number of generator properties is finite, the number of groups also is finite. If a set of apparent properties arise (say) out of three generator properties $\phi_1\phi_2\phi_3$, then this set of properties may be said to specify the group $\phi_1\phi_2\phi_3$. Since the total number of apparent properties is assumed to be greater than that of the generator properties, and since the number of groups is finite, it follows that, if two sets of apparent properties are taken, there is, in the absence of evidence to the contrary, a finite probability that the second set will belong to the group specified by the first set.

There is, however, the possibility of a plurality of generators. The first set of apparent properties may specify more than one group,—there is more than one group of generators, that is to say, which are competent to produce it; and some only of these groups may contain the second set of properties. Let us, for the moment, rule out this possibility.

When we argue from an analogy, and the instances have two groups of characters in common, namely ϕ and f, either f belongs to the group ϕ or it arises out of generators partly distinct from those out of which ϕ arises. For the reason already explained there is a finite probability that f and ϕ belong to the same group. If this is the case, *i.e.* if the generalisation $g(\phi f)$ is valid, then f will certainly be true of all other cases in which ϕ is true; if this is not the case, then f will not always be true when ϕ is true. We have, therefore, the preliminary conditions necessary for the application of pure induction. If x_r, etc., are the instances,

$$g/h = p_0, \text{ where } p_0 \text{ is finite,}$$

$$x_r/gh = 1, \text{ etc.,}$$

and $$x_r/x_1x_2 \ldots x_{r-1}\bar{g}h = 1 - \epsilon, \text{ where } \epsilon \text{ is finite.}$$

And hence, by the argument of Chapter XX., the probability of a generalisation, based on such evidence as this, is capable, under suitable conditions, of tending towards certainty as a limit, when the number of instances is increased.

If ϕ is complex and includes a number of characters which are not always found together, it must include a number of separate generator properties and specify a large group; hence the initial probability that f belongs to this group is relatively large. If, on the other hand, f is complex, there will be, for the same reasons *mutatis mutandis*, a relatively smaller initial probability than otherwise that f belongs to any other given group.

When the argument is mainly by analogy, we endeavour to obtain evidence which makes the initial probability p_0 relatively high; when the analogy is weak and the argument depends for its strength upon pure induction, p_0 is small and p_m, which is based upon numerous instances, depends for its magnitude upon their number. But an argument from induction must always involve some element of analogy, and, on the other hand, few arguments from analogy can afford to ignore altogether the strengthening influence of pure induction.

4. Let us consider the manner in which the methods of analogy increase the initial likelihood that two characters belong to the same group. The numerous characters of an object which are known to us may be represented by $a_1 a_2 \ldots a_n$. We select two sets of these, a_r and a_s, and seek to determine whether a_s always belongs to the group specified by a_r. Our previous knowledge will enable us, in general, to rule out many of the object's characters as being irrelevant to the groups specified by a_r and a_s, although this will not be possible in the most fundamental inquiries. We may also know that certain characters are always associated with a_r or with a_s. But there will be left a residuum of whose connection with a_r or a_s we are ignorant. These characters, whose relevance is in doubt, may be represented by $a_{r+1} \ldots a_{s-1}$. If the analogy is perfect, these characters are eliminated altogether. Otherwise, the argument is weakened in proportion to the comprehensiveness of these doubtful characters. For it may be the case that some of $a_{r+1} \ldots a_{s-1}$ are necessary as well as a_r, in order to specify all the generators which are required to produce a_s.

5. We may possibly be justified in neglecting certain of the characters $a_{r+1} \ldots a_{s-1}$ by *direct* judgments of irrelevance. There are certain properties of objects which we rule out from the beginning as wholly or largely independent and irrelevant to all, or to some, other properties. The principal judgments of this kind, and those alone about which we seem to feel much confidence, are concerned with absolute position in time and space, this class of judgments of irrelevance being summed up, I have suggested, in the Principle of the Uniformity of Nature. We judge that *mere* position in time and space cannot possibly affect, as a determining cause, any other characters; and this belief appears so strong and certain, although it is hard to see

how it can be based on experience, that the judgment by which we arrive at it seems perhaps to be direct. A further type of instance in which some philosophers seem to have trusted direct judgments of relevance in these matters arises out of the relation between mind and matter. They have believed that no mental event can possibly be a *necessary* condition for the occurrence of a material event.

The Principle of the Uniformity of Nature, as I interpret it, supplies the answer, if it is correct, to the criticism that the instances, on which generalisations are based, are all alike in being past, and that any generalisation, which is applicable to the future, must be based, for this reason, upon imperfect analogy. We judge directly that the resemblance between instances, which consists in their being past, is in itself irrelevant, and does not supply a valid ground for impugning a generalisation.

But these judgments of irrelevance are not free from difficulty, and we must be suspicious of using them. When I say that position is irrelevant, I do not mean to deny that a generalisation, the premiss of which specifies position, may be true, and that the same generalisation without this limitation might be false. But this is because the generalisation is incompletely stated; it happens that objects so specified have the required characters, and hence their position supplies a sufficient criterion. Position may be relevant as a sufficient condition but never as a *necessary* condition, and the inclusion of it can only affect the truth of a generalisation when we have left out some other essential condition. A generalisation which is true of one instance must be true of another which *only* differs from the former by reason of its position in time or space.

6. Excluding, therefore, the possibility of a plurality of generators, we can justify the method of perfect analogy, and other inductive methods in so far as they can be made to approximate to this, by means of the assumption that the objects in the field, over which our generalisations extend, do not have an infinite number of independent qualities; that, in other words, their characteristics, however numerous, cohere together in groups of invariable connection, which are finite in number. This does not limit the number of entities which are only *numerically* distinct. In the language used at the beginning of this chapter, the use of inductive methods can be justified if they are applied to what we have reason to suppose a finite system.[1]

7. Let us now take account of a possible plurality of generators. I mean by this the possibility that a given character can arise in more than one way, can belong to more than one distinct group, and can arise out of more than one generator. ϕ might, for instance, be sometimes due to a generator a_1, and a_1 might invariably produce f. But we could not generalise from ϕ to f, if ϕ might be due in other cases to a different generator a_2 which would *not* be competent to produce f.

If we were dealing with inductive correlation, where we do not claim universality for our conclusions, it would be sufficient for us to assume that the number of distinct generators, to which a given property ϕ can be due, is always finite. To obtain validity for universal generalisations it seems necessary to make the more comprehensive and less plausible assumption that a finite probability always exists that there is *not*, in any given case, a plurality of causes. With this assumption we have a valid argument from pure induction on the same lines, nearly, as before.

8. We have thus two distinct difficulties to deal with, and we require for the solution of each a separate assumption. The point may be illustrated by an example in which only one of the difficulties is present. There are few arguments from analogy of which we are better assured than the existence of other people. We feel indeed so well assured of their existence that it has been thought sometimes that our knowledge of them must be in some way direct. But analogy does not seem to me unequal to the proof. We have numerous experiences in our own person of acts which are associated with states of consciousness, and we infer that similar acts in others are likely to be associated with similar states of consciousness. But this argument from analogy is superior in one respect to nearly all other empirical arguments, and this superiority may possibly explain the great confidence which we feel in it. We do seem in this case to have direct knowledge, such as we have in no other case, that our states of consciousness are, sometimes at least, causally connected with some of our acts. We do not, as in other cases,

[1] Mr. C. D. Broad, in two articles "On the Relation between Induction and Probability" (*Mind*, 1918 and 1920), has been following a similar line of thought.

merely observe invariable sequence or coexistence between consciousness and act; and we do believe it to be vastly improbable in the case of some at least of our own physical acts that they could have occurred without a mental act to support them. Thus, we seem to have a special assurance of a kind not usually available for believing that there is *sometimes* a necessary connection between the conclusion and the condition of the generalisation; we doubt it only from the possibility of a plurality of causes.

The objection to this argument on the ground that the analogy is always imperfect, in that all the observed connections of consciousness and act are alike in being *mine*, seems to me to be invalid on the same ground as that on which I have put on one side objections to future generalisations, which are based on the fact that the instances which support them are all alike in being *past*. If direct judgments of irrelevance are ever permissible, there seems some ground for admitting one here.

9. As a logical foundation for Analogy, therefore, we seem to need some such assumption as that the amount of variety in the universe is limited in such a way that there is no one object so complex that its qualities fall into an infinite number of independent groups (*i.e.* groups which might exist independently as well as in conjunction); or rather that none of the objects about which we generalise are as complex as this; or at least that, though some objects may be infinitely complex, we sometimes have a finite probability that an object about which we seek to generalise is not infinitely complex.

To meet a possible plurality of causes some further assumption is necessary. If we were content with Inductive Correlations and sought to prove merely that there was a probability in favour of *any* instance of the generalisation in question, without inquiring whether there was a probability in favour of *every* instance, it would be sufficient to suppose that, while there may be more than one sufficient cause of a character, there is not an infinite number of distinct causes competent to produce it. And this involves no new assumption; for if the aggregate variety of the system is finite, the possible plurality of causes must also be finite. If, however, our generalisation is to be universal, so that it breaks down if there is a single exception to it, we must obtain, by some means or other, a finite probability that the set of characters, which condition the generalisation, are *not* the possible effect of more than one distinct set of fundamental properties. I do not know upon what ground we could establish a finite probability to this effect. The necessity for this seemingly arbitrary hypothesis strongly suggests that our conclusions should be in the form of inductive correlations, rather than of universal generalisations. Perhaps our generalisations should always run: 'It is probable that any given ϕ is f,' rather than, 'It is probable that all ϕ are f.' Certainly, what we commonly seem to hold with conviction is the belief that the sun will rise *to-morrow*, rather than the belief that the sun will *always* rise so long as the conditions explicitly known to us are fulfilled. This will be matter for further discussion in Part V., when Inductive Correlation is specifically dealt with.

10. There is a vagueness, it may be noticed, in the number of instances, which would be required on the above assumptions to establish a given numerical degree of probability, which corresponds to the vagueness in the degree of probability which we do actually attach to inductive conclusions. We assume that the necessary number of instances is finite, but we do not know what the number is. We know that the probability of a well-established induction is great, but, when we are asked to name its degree, we cannot. Common sense tells us that some inductive arguments are stronger than others, and that some are very strong. But how much stronger or how strong we cannot express. The probability of an induction is only numerically definite when we are able to make definite assumptions about the number of independent equiprobable influences at work. Otherwise, it is non-numerical, though bearing relations of greater and less to numerical probabilities according to the approximate limits within which our assumption as to the possible number of these causes lies.

11. Up to this point I have supposed, for the sake of simplicity, that it is necessary to make our assumptions as to the limitation of independent variety in an absolute form, to assume, that is to say, the finiteness of the system, to which the argument is applied, *for certain*. But we need not in fact go so far as this.

If our conclusion is C and our empirical evidence is E, then, in order to justify inductive methods, our premisses must include, in addition to E, a general hypothesis H such that C/H, the *à priori* probability of our conclusion, has a finite value. The effect of E is to increase the probability of C above its initial *à priori* value, C/HE being greater than C/H. But the method of strengthening C/H by the addition of evidence E is valid quite apart from the particular content of H. If, therefore, we have another general hypothesis H′ and other evidence E′, such that H/H′ has a finite value, we can, without being guilty of a circular argument, use evidence E′ by the same method as before to strengthen the probability H/H′. If we call H, namely, the absolute assertion of the finiteness of the system under consideration, the *inductive hypothesis*, and the process of strengthening C/H by the addition E the *inductive method*, it is not circular to use the inductive method to strengthen the inductive hypothesis itself, relative to some more primitive and less far-reaching assumption. If, therefore, we have any reason (H′) for attributing *à priori* a finite probability to the Inductive Hypothesis (H), then the actual conformity of experience *à posteriori* with expectations based on the assumption of H can be utilised by the inductive method to attribute an enhanced value to the probability of H. To this extent, therefore, we can support the Inductive Hypothesis by experience. In dealing with any particular question we can take the Inductive Hypothesis, not at its *à priori* value, but at the value to which experience in general has raised it. What we require *à priori*, therefore, is not the certainty of the Inductive Hypothesis, but a finite probability in its favour.[1]

Our assumption, in its most limited form, then, amounts to this, that we have a finite *à priori* probability in favour of the Inductive Hypothesis as to there being some limitation of independent variety (to express shortly what I have already explained in detail) in the objects of our generalisation. Our experience might have been such as to diminish this probability *à posteriori*. It has, in fact, been such as to increase it. It is because there has been so much repetition and uniformity in our experience that we place great confidence in it. To this extent the popular opinion that Induction depends upon experience for its validity is justified and does not involve a circular argument.

[1] I have implicitly assumed in the above argument that if H′ supports H, it strengthens an argument which H would strengthen. This is not *necessarily* the case for the reasons given on pp. 68 and 147. In these passages the necessary conditions for the above are elucidated. I am, therefore, assuming that in the case now in question these conditions actually are fulfilled.

12. I think that this assumption is adequate to its purpose and would justify our ordinary methods of procedure in inductive argument. It was suggested in the previous chapter that our theory of Analogy ought to be as applicable to mathematical as to material generalisations, if it is to justify common sense. The above assumptions of the limitation of independent variety sufficiently satisfy this condition. There is nothing in these assumptions which gives them a peculiar reference to material objects. We believe, in fact, that all the properties of numbers can be derived from a *limited* number of laws, and that the same set of laws governs all numbers. To apply empirical methods to such things as numbers renders it necessary, it is true, to make an assumption about the nature of numbers. But it is the same kind of assumption as we have to make about material objects, and has just about as much, or as little, plausibility. There is no new difficulty.

The assumption, also, that the system of Nature is finite is in accordance with the analysis of the underlying assumption of scientists, given at the close of the previous chapter. The hypothesis of atomic uniformity, as I have called it, while not formally equivalent to the hypothesis of the limitation of independent variety, amounts to very much the same thing. If the fundamental laws of connection changed altogether with variations, for instance, in the shape or size of bodies, or if the laws governing the behaviour of a complex had no relation whatever to the laws governing the behaviour of its parts when belonging to other complexes, there could hardly be a limitation of independent variety in the sense in which this has been defined. And, on the other hand, a limitation of independent variety seems necessarily to carry with it some degree of atomic uniformity. The underlying conception as to the character of the System of Nature is in each case the same.

13. We have now reached the last and most difficult stage of the discussion. The logical part of our inquiry is complete, and it has left us, as it is its business to leave us, with a question of epistemology. Such is the premiss or assumption which our logical processes need to work upon. What right have we to make it? It is no sufficient answer in philosophy to plead that the assumption is after all a very little one.

I do not believe that any conclusive or perfectly satisfactory answer to this question can be given, so long as our knowledge of the subject of epistemology is in so disordered and undeveloped a condition as it is in at present. No proper answer has yet been given to the inquiry—of what sorts of things are we *capable* of direct knowledge? The logician, therefore, is in a weak position, when he leaves his own subject and attempts to solve a particular instance of this general problem. He needs guidance as to what *kind* of reason we could have for such an assumption as the use of inductive argument appears to require.

On the one hand, the assumption may be absolutely *à priori* in the sense that it would be equally applicable to all possible objects. On the other hand, it may be seen to be applicable to some classes of objects only. In this case it can only arise out of some degree of particular knowledge as to the nature of the objects in question, and is to this extent dependent on experience. But if it is experience which in this sense enables us to know the assumption as true of certain amongst the objects of experience, it must enable us to know it in some manner which we may term direct and not as the result of an inference.

Now an assumption, that *all* systems of fact are finite (in the sense in which I have defined this term), cannot, it seems perfectly plain, be regarded as having absolute, universal validity in the sense that such an assumption is self-evidently applicable to every kind of object and to all possible experiences. It is not, therefore, in quite the same position as a self-evident *logical* axiom, and does not appeal to the mind in the same way. The most which can be maintained is that this assumption is true of *some* systems of fact, and, further, that there are some objects about which, as soon as we understand their nature, the mind is able to apprehend directly that the assumption in question *is* true.

In Chapter II. § 7, I wrote: "By some mental process of which it is difficult to give an account, we are able to pass from direct acquaintance with things to a knowledge of propositions about the things of which we have sensations or understand the meaning." Knowledge, so obtained, I termed direct knowledge. From a sensation of yellow and from an understanding of the meaning of 'yellow' and of 'colour,' we could, I suggested, have direct knowledge of the fact or proposition 'yellow is a colour;' we might also know that colour cannot exist without extension, or that two colours cannot be perceived at the same time in the same place. Other philosophers might use terms differently and express themselves otherwise; but the substance of what I was there trying to say is not very disputable. But when we come to the question as to what kinds of propositions we can come to know in this manner, we enter upon an unexplored field where no certain opinion is discoverable.

In the case of logical terms, it seems to be generally agreed that if we understand their meaning we can know directly propositions about them which go far beyond a mere expression of this meaning;—propositions of the kind which some philosophers have termed *synthetic*. In the case of non-logical or empirical entities, it seems sometimes to be assumed that our direct knowledge must be confined to what may be regarded as an expression or description of the meaning or sensation apprehended by us. If this view is correct the Inductive Hypothesis is not the kind of thing about which we can have direct knowledge as a result of our acquaintance with objects.

I suggest, however, that this view is incorrect, and that we are capable of direct knowledge about empirical entities which goes beyond a mere expression of our understanding or sensation of them. It may be useful to give the reader two examples, more familiar than the Inductive Hypothesis, where, as it appears to me, such knowledge is commonly assumed. The first is that of the causal irrelevance of mere position in time and space, commonly called the Uniformity of Nature. We do believe, and yet have no adequate inductive reason whatever for believing, that mere position in time and space cannot make any difference. This belief arises directly, I think, out of our acquaintance with the objects of experience and our understanding of the concepts of 'time' and 'space.' The second is that of the Law of Causation. We believe that every object in time has a 'necessary' connection[1] with some set of objects at a previous time. This belief also, I think, arises in the same way. It is to be noticed that neither of these beliefs clearly arises, in spite of the directness which may be claimed for them, out of any one single experience. In a way analogous to these, the validity of assuming the Inductive Hypothesis, as applied to a particular class of objects, appears to me to be justified.

Our justification for using inductive methods in an argument

[1] I do not propose to define the meaning of this.

about numbers arises out of our perceiving directly, when we understand the meaning of a number, that they are of the required character.[1] And when we perceive the nature of our phenomenal experiences, we have a direct assurance that in their case also the assumption is legitimate. We are capable, that is to say, of direct synthetic knowledge about the nature of the objects of our experience. On the other hand, there may be some kinds of objects, about which we have no such assurance and to which inductive methods are not reasonably applicable. It may be the case that some metaphysical questions are of this character and that those philosophers have been right who have refused to apply empirical methods to them.

14. I do not pretend that I have given any perfectly adequate reason for accepting the theory I have expounded, or any such theory. The Inductive Hypothesis stands in a peculiar position in that it seems to be neither a self-evident logical axiom nor an object of direct acquaintance; and yet it is just as difficult, as though the inductive hypothesis were either of these, to remove from the organon of thought the inductive method which can only be based on it or on something like it.

As long as the theory of knowledge is so imperfectly understood as now, and leaves us so uncertain about the grounds of many of our firmest convictions, it would be absurd to confess to a special scepticism about this one. I do not think that the foregoing argument has disclosed a reason for such scepticism. We need not lay aside the belief that this conviction gets its invincible certainty from some valid principle darkly present to our minds, even though it still eludes the peering eyes of philosophy.

[1] Since numbers are logical entities, it may be thought less unorthodox to make such an assumption in their case.

CHAPTER XXIII

SOME HISTORICAL NOTES ON INDUCTION

1. The number of books, which deal with inductive[1] theory, is extraordinarily small. It is usual to associate the subject with the names of Bacon, Hume, and Mill. In spite of the modern tendency to depreciate the first and the last of these, they are the principal names, I think, with which the history of induction ought to be associated. The next place is held by Laplace and Jevons. Amongst contemporary logicians there is an almost complete absence of constructive theory, and they content themselves for the most part with the easy task of criticising Mill, or with the more difficult one of following him.

That the inductive theories of Bacon and of Mill are full of errors and even of absurdities, is, of course, a commonplace of criticism. But when we ignore details, it becomes clear that they were really attempting to disentangle the essential issues. We depreciate them partly, perhaps, as a reaction from the view once held that they helped the progress of scientific discovery. For it is not plausible to suppose that Newton owed anything to Bacon, or Darwin to Mill. But with the logical problem their minds were truly occupied, and in the history of logical theory they should always be important.

It is true, nevertheless, that the advancement of science was the main object which Bacon himself, though not Mill, believed that his philosophy would promote. The *Great Instauration* was intended to promulgate an actual method of discovery entirely different from any which had been previously known.[2] It did

[1] See note at the end of this chapter on "The Use of the Term *Induction.*"

[2] He speaks of himself as being "in hac re plane protopirus, et vestigia nullius sequutus"; and in the *Praefatio Generalis* he compares his method to the mariner's compass, until the discovery of which no wide sea could be crossed (see Spedding and Ellis, vol. i. p. 24).

not do this, and against such pretensions Macaulay's well-known essay was not unjustly directed. Mill, however, expressly disclaimed in his preface any other object than to classify and generalise the practices "conformed to by accurate thinkers in their scientific inquiries." Whereas Bacon offered rules and demonstrations, hitherto unknown, with which any man could solve all the problems of science by taking pains, Mill admitted that "in the existing state of the cultivation of the sciences, there would be a very strong presumption against any one who should imagine that he had effected a revolution in the theory of the investigation of truth, or added any fundamentally new process to the practice of it."

2. The theories of both seem to me to have been injured, though in different degrees, by a failure to keep quite distinct the three objects: (1) of helping the scientist, (2) of explaining and analysing his practice, and (3) of justifying it. Bacon was really interested in the second as well as in the first, and was led to some of his methods by reflecting upon what distinguished good arguments from bad in actual investigations. To logicians his methods were as new as he claimed, but they had their origin, nevertheless, in the commonest inferences of science and daily life. But his main preoccupation was with the first, which did injury to his treatment of the third. He himself became aware as the work progressed that, in his anxiety to provide an infallible mode of discovery, he had put forth more than he would ever be able to justify.[1] His own mind grew doubtful, and the most critical parts of the description of the new method were never written. No one who has reflected much upon Induction need find it difficult to understand the progress and development of Bacon's thoughts. To the philosopher who first distinguished some of the complexities of empirical proof in a generalised, and not merely a particular, form, the prospects of systematising these methods must have seemed extraordinarily hopeful. The first investigator could not have anticipated that Induction, in spite of its apparent certainty, would prove so elusive to analysis.

Mill also was led, in a not dissimilar way, to attempt a too

[1] This view is taken in the edition of James Spedding and Leslie Ellis. Their introductions to Bacon's philosophical works seem to me to be very greatly superior to the accounts to be found elsewhere. They make intelligible, what seems, according to other commentaries, fanciful and without sense or reason.

simple treatment, and, in seeking for ease and certainty, to treat far too lightly the problem of justifying what he had claimed. Mill shirks, almost openly, the difficulties ; and scarcely attempts to disguise from himself or his readers that he grounds induction upon a circular argument.

3. Some of the most characteristic errors both of Bacon and of Mill arise, I think, out of a misapprehension, which it has been a principal object of this book to correct. Both believed, without hesitation it seems, that induction is capable of establishing a conclusion which is absolutely certain, and that an argument is invalid if the generalisation, which it supports, admits of exceptions in fact. "Absolute certainty," says Leslie Ellis,[1] "is one of the distinguishing characters of the Baconian induction." It was, in this respect, mainly that it improved upon the older induction *per enumerationem simplicem*. "The induction which the logicians speak of," Bacon argues in the *Advancement of Learning*, "is utterly vicious and incompetent. . . . For to conclude upon an enumeration of particulars, without instance contradictory, is no conclusion but a conjecture." The conclusions of the new method, unlike those of the old, are not liable to be upset by further experience. In the attempt to justify these claims and to obtain demonstrative methods, it was necessary to introduce assumptions for which there was no warrant.

Precisely similar claims were made by Mill, although there are passages in which he abates them,[2] for his own rules of procedure. An induction has no validity, according to him as according to Bacon, unless it is absolutely certain. The following passage[3] is significant of the spirit in which the subject was approached by him: "Let us compare a few cases of incorrect inductions with others which are acknowledged to be legitimate. Some, we know, which were believed for centuries to be correct, were nevertheless incorrect. *That all swans are white, cannot have been a good induction, since the conclusion has turned out erroneous.* The experience, however, on which the conclusion rested was genuine." Mill has not justly apprehended the relativity of all inductive arguments to the evidence, nor the element of uncertainty which is present, more

[1] *Op. cit.* vol. i. p. 23.
[2] When he deals with Plurality of Causes, for instance.
[3] Bk. iii. chap. iii. 3 (the italics are mine).

or less, in all the generalisations which they support.[1] Mill's methods would yield certainty, if they were correct, just as Bacon's would. It is the necessity, to which Mill had subjected himself, of obtaining certainty that occasions their want of reality. Bacon and Mill both assume that experiment can shape and analyse the evidence in a manner and to an extent which is not in fact possible. In the aims and expectations with which they attempt to solve the inductive problem, there is on fundamental points an unexpectedly close resemblance beween them.

4. Turning from these general criticisms to points of greater detail, we find that the line of thought pursued by Mill was essentially the same as that which had been pursued by Bacon, and, also, that the argument of the preceding chapters is, in spite of some real differences, a development of the same fundamental ideas which underlie, as it seems to me, the theories of Mill and Bacon alike.

We have seen that all empirical arguments require an initial probability derived from analogy, and that this initial probability may be raised towards certainty by means of pure induction or the multiplication of instances. In some arguments we depend mainly upon analogy, and the initial probability obtained by means of it (with the assistance, as a rule, of previous knowledge) is so large that numerous instances are not required. In other arguments pure induction predominates. As science advances and the body of pre-existing knowledge is increased, we depend increasingly upon analogy; and only at the earlier stages of our investigations is it necessary to rely, for the greater part of our support, upon the multiplication of instances. Bacon's great achievement, in the history of logical theory, lay in his being the first logician to recognise the importance of methodical analogy to scientific argument and the dependence upon it of most well-established conclusions. The *Novum Organum* is mainly concerned with explaining methodical ways of increasing what I have termed the Positive and Negative Analogies, and of avoiding false Analogies. The use of exclusions and rejections, to which

[1] This misapprehension may be connected with Mill's complete failure to grasp with any kind of thoroughness the nature and importance of the theory of probability. The treatment of this topic in the *System of Logic* is exceedingly bad. His understanding of the subject was, indeed, markedly inferior to the best thought of his own time.

Bacon attached supreme importance, and which he held to constitute the essential superiority of his method over those which preceded it, entirely consists in the determination of what characters (or natures as he would call them) belong to the positive and negative analogies respectively. The first two tables with which the investigation begins are, first, the table *essentiae et praesentiae*, which contains all known instances in which the given nature is present, and, second, the table *declinationis sive absentiae in proximo*, which contains instances corresponding in each case to those of the first table, but in which, notwithstanding this correspondence, the given nature is absent.[1] The doctrine of prerogative instances is concerned no less plainly with the methodical determination of Analogy. And the doctrine of idols is expounded for the avoidance of *false* analogies, standing, he says, in the same relation to the interpretation of Nature, as the doctrine of fallacies to ordinary logic.[2] Bacon's error lay in supposing that, because these methods were new to logic, they were therefore new to practice. He exaggerated also their precision and their certainty; and he underestimated the importance of pure induction. But there was, at bottom, nothing about his rules impracticable or fantastic, or indeed unusual.

5. Almost the whole of the preceding paragraph is equally applicable to Mill. He agreed with Bacon in depreciating the part played in scientific inquiry by pure induction, and in emphasising the importance of analogy to all systematic investigators. But he saw further than Bacon in allowing for the Plurality of Causes, and in admitting that an element of pure induction was therefore made necessary. "The Plurality of Causes," he says,[3] "is the only reason why mere number of instances is of any importance in inductive inquiry. The tendency of unscientific inquirers is to rely too much on number, without analysing the instances. . . . Most people hold their conclusions with a degree of assurance proportioned to the mere *mass* of the experience on which they appear to rest; not considering that by the addition of instances to instances, all of the same kind, that is, differing from one another only in points already recognised as immaterial, nothing whatever is added to the evidence of

[1] Ellis, vol. i. p. 33.
[2] Ellis, vol. i. p. 89.
[3] Book iv. chap. x. 2.

the conclusion. A single instance eliminating some antecedent which existed in all the other cases, is of more value than the greatest multitude of instances which are reckoned by their number alone." Mill did not see, however, that our knowledge of the instances is seldom complete, and that new instances, which are not known to differ from the former in material respects, may add, nevertheless, to the negative analogy, and that the multiplication of them may, for this reason, strengthen the evidence. It is easy to see that his methods of Agreement and Difference closely resemble Bacon's, and aim, like Bacon's, at the determination of the Positive and Negative Analogies. By allowing for Plurality of Causes Mill advanced beyond Bacon. But he was pursuing the same line of thought which alike led to Bacon's rules and has been developed in the chapters of this book. Like Bacon, however, he exaggerated the precision with which his canons of inquiry could be used in practice.

6. No more need be said respecting method and analysis. But in both writers the exposition of method is closely intermingled with attempts to justify it. There is nothing in Bacon which at all corresponds to Mill's appeals to Causation or to the Uniformity of Nature, and, when they seek for the ground of induction, there is much that is peculiar to each writer. It is my purpose, however, to consider in this place the details common to both, which seem to me to be important and which exemplify the only line of investigation which seems likely to be fruitful; and I shall pursue no further, therefore, their numerous points of difference.

The attempt, which I have made to justify the initial probability which Analogy seems to supply, primarily depends upon a certain limitation of independent variety and upon the derivation of all the properties of any given object from a limited number of primary characters. In the same way I have supposed that the number of primary characters which are capable of producing a given property is also limited. And I have argued that it is not easy to see how a finite probability is to be obtained unless we have in each case some such limitation in the number of the ultimate alternatives.

It was in a manner which bears fundamental resemblances to this that Bacon endeavoured to demonstrate the cogency of his method. He considers, he says, " the simple forms or difference of things which are few in number, and the degrees and co-ordinations whereof make all this variety." And in *Valerius Terminus* he argues "that every particular that worketh any effect is a thing compounded more or less of diverse single natures, more manifest and more obscure, and that it appeareth not to which of the natures the effect is to be ascribed."[1] It is indeed essential to the method of exclusions that the matter to which it is applied should be somehow resolvable into a finite number of elements. But this assumption is not peculiar, I think, to Bacon's method, and is involved, in some form or other, in every argument from Analogy. In making it Bacon was initiating, perhaps obscurely, the modern conception of a finite number of laws of nature out of the combinations of which the almost boundless variety of experience ultimately arises. Bacon's error was double and lay in supposing, first, that these distinct elements lie upon the surface and consist in visible characters, and second, that their natures are, or easily can be, known to us, although the part of the *Instauration*, in which the manner of conceiving simple natures was to be explained, he never wrote. These beliefs falsely simplified the problem as he saw it, and led him to exaggerate the ease, certainty, and fruitfulness of the new method. But the view that it is possible to reduce all the phenomena of the universe to combinations of a limited number of simple elements—which is, according to Ellis,[2] the central point of Bacon's whole system—was a real contribution to philosophy.

7. The assumption that every event can be analysed into a limited number of ultimate elements, is never, so far as I am aware, explicitly avowed by Mill. But he makes it in almost every chapter, and it underlies, throughout, his mode of procedure. His methods and arguments would fail immediately, if we were to suppose that phenomena of infinite complexity, due to an infinite number of independent elements, were in question, or if an infinite plurality of causes had to be allowed for.

In distinguishing, therefore, analogy from pure induction, and in justifying it by the assumption of a *limited* complexity in the problems which we investigate, I am, I think, pursuing, with numerous differences, the line of thought which Bacon first

[1] Quoted by Ellis, vol. i. p. 41.
[2] Vol. i. p. 28.

pursued and which Mill popularised. The method of treatment is dissimilar, but the subject-matter and the underlying beliefs are, in each case, the same.

8. Between Bacon and Mill came Hume. Hume's sceptical criticisms are usually associated with causality; but argument by induction—inference from past particulars to future generalisations—was the real object of his attack. Hume showed, not that inductive methods were false, but that their validity had never been established and that all possible lines of proof seemed equally unpromising. The *full* force of Hume's attack and the nature of the difficulties which it brought to light were never appreciated by Mill, and he makes no adequate attempt to deal with them. Hume's statement of the case against induction has never been improved upon; and the successive attempts of philosophers, led by Kant, to discover a transcendental solution have prevented them from meeting the hostile arguments on their own ground and from finding a solution along lines which might, conceivably, have satisfied Hume himself.

9. It would not be just here to pass by entirely the name of the great Leibniz, who, wiser in correspondence and fragmentary projects than in completed discourses, has left to us sufficient indications that his private reflections on this subject were much in advance of his contemporaries'. He distinguished three degrees of conviction amongst opinions, logical certainty (or, as we should say, propositions known to be formally true), physical certainty which is only logical probability, of which a well-established induction, as that man is a biped, is the type, and physical probability (or, as we should say, an inductive correlation), as for example that the south is a rainy quarter.[1] He condemned generalisations based on mere repetition of instances, which he declared to be without logical value, and he insisted on the importance of *Analogy* as the basis of a valid induction.[2] He regarded a hypothesis as more probable in proportion to its *simplicity* and its *power*, that is to say, to the number of the phenomena it would explain and the fewness of the assumptions it involved. In particular a power of accurate prediction and of explaining phenomena or experiments pre-

[1] Couturat, *Opuscules et fragments inédits de Leibniz*, p. 232.

[2] Couturat, *La Logique de Leibniz d'après des documents inédits*, pp. 262, 267.

viously untried is a just ground of secure confidence, of which he cites as a nearly perfect example the key to a cryptogram.[1]

10. Whewell and Jevons furnished logicians with a storehouse of examples derived from the practice of scientists. Jevons, partly anticipated by Laplace, made an important advance when he emphasised the close relation between Induction and Probability. Combining insight and error, he spoilt brilliant suggestions by erratic and atrocious arguments. His application of Inverse Probability to the inductive problem is crude and fallacious, but the idea which underlies it is substantially good. He, too, made explicit the element of Analogy, which Mill, though he constantly employed it, had seldom called by its right name. There are few books, so superficial in argument yet suggesting so much truth, as Jevons's *Principles of Science*.

11. Modern text-books on Logic all contain their chapters on Induction, but contribute little to the subject. Their recognition of Mill's inadequacy renders their exposition, which, in spite of criticisms, is generally along his lines, nerveless and confused. Where Mill is clear and offers a solution, they, confusedly criticising, must withhold one. The best of them, Sigwart and Venn, contain criticism and discussion which is interesting, but constructive theory is lacking. Hitherto Hume has been master, only to be refuted in the manner of Diogenes or Dr. Johnson.

[1] Letter to Conring, 19th March 1678.

NOTES ON PART III

(i.) On the Use of the Term *Induction*

1. Induction is in origin a translation of the Aristotelian ἐπαγωγή. This term was used by Aristotle in two quite distinct senses—first, and principally, for the process by which the observation of particular instances, in which an abstract notion is exemplified, enables us to realise and comprehend the abstraction itself; secondly, for the type of argument in which we generalise after the complete enumeration and assertion of *all* the particulars which the generalisation embraces. From this second sense it was sometimes extended to cases in which we generalise after an *incomplete* enumeration. In post-Aristotelian writers the induction *per enumerationem simplicem* approximates to induction in Aristotle's second sense, as the number of instances is increased. To Bacon, therefore, "the induction of which the logicians speak" meant a method of argument by multiplication of instances. He himself deliberately extended the use of the term so as to cover all the systematic processes of empirical generalisation. But he also used it, in a manner closely corresponding to Aristotle's *first* use, for the process of forming scientific conceptions and correct notions of "simple natures." [1]

2. The modern use of the term is derived from Bacon's. Mill defines it as "the operation of discovering and proving general propositions." His philosophical system required that he should define it as widely as this; but the term has really been used, both by him and by other logicians, in a narrower sense, so as to cover those methods of proving general propositions, which we call empirical, and so as to exclude generalisations, such as those of mathematics, which have been proved formally. Jevons was led, partly by the linguistic resemblance, partly because in the one case we proceed from the particular to the general and in the other from the general to the particular, to define Induction as the inverse process of Deduction. In contemporary logic Mill's use prevails; but there

[1] See Ellis's edition of Bacon's *Works*, vol. i. p. 37. On the first occasion on which Induction is mentioned in the *Novum Organum*, it is used in this secondary sense.

is, at the same time, a suggestion—arising from earlier usage, and because Bacon and Mill never quite freed themselves from it—of argument by mere multiplication of instances. I have thought it best, therefore, to use the term *pure induction* to describe arguments which are based upon the *number* of instances, and to use *induction* itself for all those types of arguments which combine, in one form or another, pure induction with analogy.

(ii.) On the Use of the Term *Cause*

1. Throughout the preceding argument, as well as in Part II., I have been able to avoid the metaphysical difficulties which surround the true meaning of *cause.* It was not necessary that I should inquire whether I meant by *causal* connection an invariable connection in fact merely, or whether some more intimate relation was involved. It has also been convenient to speak of causal relations between objects which do not strictly stand in the position of cause and effect, and even to speak of *a probable cause,* where there is no implication of necessity and where the antecedents will sometimes lead to particular consequents and sometimes will not. In making this use of the term, I have followed a practice not uncommon amongst writers on probability, who constantly use the term *cause,* where *hypothesis* might seem more appropriate.[1]

One is led, almost inevitably, to use 'cause' more widely than 'sufficient cause' or than 'necessary cause,' because, the necessary causation of particulars by particulars being rarely apparent to us, the strict sense of the term has little utility. Those antecedent circumstances, which we are usually content to accept as causes, are only so in strictness under a favourable conjunction of innumerable other influences.

2. As our knowledge is partial, there is constantly, in our use of the term *cause,* some reference implied or expressed to a limited body of knowledge. It is clear that, whether or not, as Cournot[2] maintains, there are such things as independent series in the order of causation, there is often a sense in which we may hold that there is a closer intimacy between some series than between others. This intimacy is relative, I think, to particular information, which is actually known to us, or which is within our reach. It will be useful, therefore, to give precise definitions of these wider senses in which it is often convenient to use the expression *cause.*

---

[1] Cf. Czuber, *Wahrscheinlichkeitsrechnung,* p. 139. In dealing with Inverse Probability Czuber explains that he means by *possible cause* the various *Bedingungskomplexe* from which the cause *can* result.

[2] See Chapter XXIV. § 3.

We must first distinguish between assertions of law and assertions of fact, or, in the terminology of Von Kries,[1] between nomologic and ontologic knowledge. It may be convenient in dealing with some questions to frame this distinction with reference to the special circumstances. But the distinction generally applicable is between propositions which contain no reference to *particular* moments of time, and existential propositions which cannot be stated without reference to specific points in the time series. The Principle of the Uniformity of Nature amounts to the assertion that natural laws are all, in this sense, timeless. We may, therefore, divide our *data* into two portions k and l, such that k denotes our formal and nomologic evidence, consisting of propositions whose predication does not involve a particular time reference, and l denotes the existential or ontologic propositions.

3. Let us now suppose that we are investigating two existential propositions a and b, which refer two events A and B to particular moments of time, and that A is referred to moments which are all prior to those at which B occurred. What various meanings can we give to the assertion that A and B are *causally connected*?

(i.) If $b/ak = 1$, A is a sufficient cause of B. In this case A is a cause of B in the strictest sense. b can be *inferred* from a, and no additional knowledge consistent with k can invalidate this.

(ii.) If $b/\bar{a}k = 0$, A is a necessary cause of B.

(iii.) If k includes all the laws of the existent universe, then A is *not* a sufficient cause of B unless $b/ak = 1$. The Law of Causation, therefore, which states that every existent has to some other previous existent the relation of effect to sufficient cause, is equivalent to the proposition that, if k is the body of natural law, then, if b is true, there is always another true proposition a, which asserts existences prior to B, such that $b/ak = 1$. No use has been made so far of our existential knowledge l, which is irrelevant to the definitions preceding.

(iv.) If $b/akl = 1$ and $b/kl \neq 1$, A is a sufficient cause of B under conditions l.

(v.) If $b/\bar{a}kl = 0$ and $b/kl \neq 0$, A is a necessary cause of B under conditions l.

(vi.) If there is any existential proposition h such that $b/ahk = 1$ and $b/hk \neq 1$, A is, relative to k, a possible sufficient cause of B.

(vii.) If there is an existential proposition h such that $b/\bar{a}hk = 0$ and $b/hk \neq 0$, A is, relative to k, a possible necessary cause of B.

(viii.) If $b/ahkl = 1$, $b/hk \neq 1$, and $h/akl \neq 0$, A is, relative to k, a possible sufficient cause of B under conditions l.

(ix.) If $b/\bar{a}hkl = 0$, $b/hkl \neq 0$, $h/\bar{a}kl \neq 0$, and $h/akl \neq 0$, A is, relative to k, a possible necessary cause of B under conditions l.

---

[1] *Die Principien der Wahrscheinlichkeitsrechnung*, p. 86.

Thus an event is a possible necessary cause of another, relative to given nomologic data, if circumstances can arise, not inconsistent with our existential data, in which the first event will be indispensable if the second is to occur.

(x.) Two events are *causally independent* if no part of either is, relative to our nomologic data, a possible cause of any part of the other under the conditions of our existential knowledge. The greater the scope of our existential knowledge, the greater is the likelihood of our being able to pronounce events causally dependent or independent.

4. These definitions preserve the distinction between 'causally independent' and 'independent for probability,'—the distinction between *causa essendi* and *causa cognoscendi*. If $b/ahkl \neq b/\bar{a}hkl$, where a and b may be any propositions whatever and are not limited as they were in the causal definitions, we have 'dependence for probability,' and a is a *causa cognoscendi* for b, relative to data kl. If a and b are causally dependent, according to definition (x.), b is a possible *causa essendi*, relative to data kl.

But, after all, the essential relation is that of 'independence for probability.' We wish to know whether knowledge of one fact throws light *of any kind* upon the likelihood of another. The theory of causality is only important because it is thought that by means of its assumptions light *can* be thrown by the experience of one phenomenon upon the expectation of another.

PART IV

SOME PHILOSOPHICAL APPLICATIONS OF PROBABILITY

CHAPTER XXIV

THE MEANINGS OF OBJECTIVE CHANCE, AND OF RANDOMNESS

1. MANY important differences of opinion in the treatment of Probability have been due to confusion or vagueness as to what is meant by Randomness and by Objective Chance, as distinguished from what, for the purposes of this chapter, may be termed Subjective Probability. It is agreed that there is a sort of Probability which depends upon knowledge and ignorance, and is relative, in some manner, to the mind of the subject; but it is supposed that there is also a more objective Probability which is not thus dependent, or less completely so, though precisely what this conception stands for is not plain. The relation of Randomness to the other concepts is also obscure. The problem of clearing up these distinctions is of importance if we are to criticise certain schools of opinion intelligently, as well as to the treatment of the foundations of Statistical Inference which is to be attempted in Part V.

There are at least three distinct issues to be kept apart. There is the antithesis between knowledge and ignorance, between events, that is to say, which we have some reason to expect, and events which we have no reason to expect, which gives rise to the theory of subjective probability and subjective chance; and, connected with this, the distinction between 'random' selection and 'biassed' selection. There are next objective probability and objective chance, which are as yet obscure, but which are commonly held to arise out of the antithesis between 'cause' and 'chance,' between events, that is to say, which are causally connected and events which are not causally connected. And there is, lastly, the antithesis between chance and design, between 'blind causes' and 'final causes,' where we oppose a 'chance' event to one, part of whose cause is a volition following on a conscious desire for the event.[1]

2. The method of this treatise has been to regard subjective probability as fundamental and to treat all other relevant conceptions as derivative from this. That there is such a thing as probability in this sense has been admitted by all sensible philosophers since the middle of the eighteenth century at least.[2] But there is also, many writers have supposed, something else which may be fitly described as objective probability; and there is, besides, a long tradition in favour of the view that it is this (whatever it may be) which is logically and philosophically important, subjective probability being a vague and mainly psychological conception about which there is very little to be said.

The distinction exists already in Hume: "Probability is of two kinds, either when the object is really in itself uncertain, and to be determined by chance; or when, though the object be already certain, yet 'tis uncertain to our judgment, which finds a number of proofs on each side of the question."[3] But the distinction is not elucidated, and one can only infer from other passages that Hume did not intend to imply in this passage the existence of objective chance in a sense contradictory to a determinist theory of the Universe. In Condorcet all is confused; and in Laplace nearly all. In the nineteenth century the distinction begins to grow explicit in the writings of Cournot. "Les explications que j'ai données . . . ," he writes in the preface to his *Exposition*, "sur le double sens du mot de probabilité, qui tantôt se rapporte à une certaine mesure de nos connaissances, et tantôt à une mesure de la possibilité des choses, indépendamment de la connaissance que nous en avons: ces explications, dis-je, me semblent propres à resoudre les difficultés qui ont rendu jusqu'ici suspecte à de bons esprits toute la théorie de la probabilité mathématique." It will be worth while to pause for a moment to consider the ideas of Cournot.

[1] This is discussed in Chapter XXV. § 4.

[2] D'Alembert, collecting (largely from Hume, many passages being translated almost *verbatim*) in the *Encyclopédie méthodique* the most up-to-date commonplaces of the subject, found it natural to write: "Il n'y a point de hasard à proprement parler; mais il y a son équivalent: l'ignorance, où nous sommes des vraies causes des événemens, a sur notre esprit l'influence qu'on suppose au hasard." Compare also the sentences from Spinoza quoted on p. 117 above.

[3] *A Treatise of Human Nature*, Book ii. part iii. section ix.

3. Cournot, while admitting that there is such a thing as subjective chance, was concerned to dispute the opinion that chance is *merely* the offspring of ignorance, saying that in this case "le calcul des chances" is merely "un calcul des illusions." The chance, upon which "le calcul des chances" is based, is something different, and depends, according to him, on the combination or convergence of phenomena belonging to *independent* series. By "independent series" he means series of phenomena which develop as parallel or successive series without any causal interdependence or link of solidarity whatever.[1] No one, he says by way of example, seriously believes that in striking the ground with his foot he puts out the navigator in the Antipodes, or disturbs the system of Jupiter's satellites. Separate trains of events, that is to say, have been set going by distinct initial acts of creation, so to speak.[2] Every event is causally connected with previous events belonging to its own series, but it cannot be modified by contact with events belonging to a different series. A 'chance' event is a complex due to the concurrence in time or place of events belonging to causally independent series.

This theory, as it stands, is evidently unsatisfactory. Even if there are series of phenomena which are independent in Cournot's sense, it is not clear how we can know which they are, or how we can set up a calculus which presumes an acquaintance with them. Just as it is likely that we are all cousins if we go back far enough, so there may be, after all, remote relationships between ourselves and Jupiter. A remote connection or a reaction quantitatively small is a matter of degree and not by any means the same thing as absolute independence. Nevertheless Cournot has contributed something, I think, to the stock of our ideas. He has

[1] "Le mot hasard," Cournot writes in his *Essai sur les fondements de nos connaissances*, "n'indique pas une cause substantielle, mais une idée : cette idée est celle de la combinaison entre plusieurs séries de causes ou de faits qui se développent chacun dans sa série propre, indépendamment les uns des autres." This is very like the definition given by Jean de la Placette in his *Traité des jeux de hasard*, to which Cournot refers : "Pour moi, je suis persuadé que le hasard renferme quelque chose de réal et de positif, savoir un concours de deux ou plusieurs événements contingents, chacun desquels a ses causes, mais en sorte que leur concours n'en a aucune que l'on connaisse."

[2] *Essai sur les fondements de nos connaissances*, i. 134 : "La nature ne se gouverne pas par une loi unique . . . ses lois ne sont pas toutes dérivées les unes des autres, ou dérivées toutes d'une loi supérieure par une nécessité purement logique . . . nous devons les concevoir au-contraire comme ayant pu être décrétées séparément d'une infinité de manières."

hinted at, even if he has not disentangled, one of the elements in a common conception of chance; and of the notion, which he seems to have in his mind, we must in due course take account.[1]

4. In the writings of Condorcet, I have said above, all is confused. But in Bertrand's criticism of him a relevant distinction, though not elucidated, is brought before the mind. "The motives for believing," wrote Condorcet, "that, from ten million white balls mixed with one black, it will not be the black ball which I shall draw at the first attempt is *of the same kind* as the motive for believing that the sun will not fail to rise to-morrow." "The assimilation of the two cases," Bertrand writes in criticism of the above,[2] "is not legitimate: one of the probabilities is objective, the other subjective. The probability of drawing the black ball at the first attempt is $\frac{1}{10,000,000}$, neither more nor less. Whoever evaluates it otherwise makes a mistake. The probability that the sun will rise varies from one mind to another. A scientist might hold on the basis of a false theory, without being utterly irrational, that the sun will soon be extinguished; he would be within his rights, just as Condorcet is within his; both would exceed their rights in accusing of error those who think differently." Before commenting on this distinction, let us have before us also some interesting passages by Poincaré.

5. We certainly do not use the term 'chance,' Poincaré points out, as the ancients used it, in opposition to determinism. For us therefore the natural interpretation of 'chance' is subjective, —"Chance is only the measure of our ignorance. Fortuitous phenomena are, by definition, those, of the laws of which we are

[1] Cournot's work on Probability has been highly praised by authorities as diverse and distinguished as Boole and Von Kries, and has been made the foundation of a school by some recent French philosophers (see the special number of the *Revue de métaphysique et de morale*, devoted to Cournot and published in 1905, and the bibliography at the end of the present volume *passim*). The best account with which I am acquainted, of Cournot's theory of probability, is to be found in A. Darbon's *Le Concept du hasard*. Cournot's philosophy of the subject is developed, not so much in his *Exposition de la théorie des chances*, as in later works, especially in his *Essai sur les fondements de nos connaissances*. Cournot never touched any subject without contributing something to it, but, on the whole, his work on Probability is, in my opinion, disappointing. No doubt his *Exposition* is superior to other French text-books of the period, of which there is so large a variety, and his work, both here and elsewhere, is not without illuminating ideas: but the philosophical treatment is so confused and indefinite that it is difficult to make much of it beyond the one specific point treated above.

[2] *Calcul des probabilités*, p. xix.

ignorant." But Poincaré immediately adds: "Is this definition very satisfactory? When the first Chaldaean shepherds followed with their eyes the movements of the stars, they did not yet know the laws of astronomy, but would they have dreamed of saying that the stars move by chance? If a modern physicist is studying a new phenomenon, and if he discovers its law on Tuesday, would he have said on Monday that the phenomenon was fortuitous?"[1]

There is also another type of case in which "chance must be something more than the name we give to our ignorance." Among the phenomena, of the causes of which we are ignorant, there are some, such as those dealt with by the manager of a life insurance company, about which the calculus of probabilities can give real information. Surely it cannot be thanks to our ignorance, Poincaré urges, that we are able to arrive at valuable conclusions. If it were, it would be necessary to answer an inquirer thus: "You ask me to predict the phenomena that will be produced. If I had the misfortune to know the laws of these phenomena, I could not succeed except by inextricable calculations, and I should have to give up the attempt to answer you; but since I am fortunate enough to be ignorant of them, I will give you an answer at once. And, what is more extraordinary still, my answer will be right." The ignorance of the manager of the life insurance company as to the prospects of life of his individual policy-holders does not prevent his being able to pay dividends to his shareholders.

Both these distinctions seem to be real ones, and Poincaré proceeds to examine further instances in which we seem to distinguish *objectively* between events according as they are or are not due to 'chance.' He takes the case of a cone balanced upon its tip; we know for certain that it will fall, but not on which side—chance will determine. "A very small cause which escapes our notice determines a considerable effect that we cannot fail to see, and then we say that that effect is due to chance." The weather, and the distribution of the minor planets on the Zodiac, are analogous instances. And what we term 'games of chance' afford, it has always been recognised, an almost perfect

[1] *Calcul des probabilités* (2nd edition), p. 2. This passage also appears in an article in the *Revue du mois* for 1907 and in the author's *Science et méthode*, of the English translation of which I have made use above,—at the cost of doing incomplete justice to Poincaré's most admirable style.

example. "It may happen that small differences in the initial conditions produce very great ones in the final phenomena. A small error in the former will produce an enormous error in the latter. Prediction becomes impossible, and we have the fortuitous phenomenon:" "The greatest chance is the birth of a great man. It is only by chance that the meeting occurs of two genital cells of different sex that contain precisely, each on its side, the mysterious elements, the mutual reaction of which is destined to produce genius. . . . How little it would have taken to make the spermatozoid which carried them deviate from its course. It would have been enough to deflect it a hundredth part of an inch, and Napoleon would not have been born and the destinies of a continent changed. No example can give a better comprehension of the true character of chance."

Poincaré calls attention next to another class of events, which we commonly assign to 'chance,' the distinguishing characteristic of which seems to be that their causes are very numerous and complex,—the motions of molecules of gas, the distribution of drops of rain, the shuffling of a pack of cards, or the errors of observation. Thirdly there is the type, usually connected with one of the first two, and specially emphasised, as we have seen above, by Cournot, in which something comes about through the concurrence of events which we regard as belonging to distinct causal trains,—a man is walking along the street and is killed by the fall of a tile.

6. When we attribute such events, as those illustrated by Poincaré, to *chance*, we certainly do not mean merely to assert that we do not know how they arose or that we had no special reason for anticipating them *à priori*. So far from this being the case, we mean to make a definite assertion as to the kind of way in which they arose;—though exactly what we mean to assert about them it is extremely difficult to say.

Now a careful examination of all the cases in which various writers claim to detect the presence of 'objective chance' confirms the view that 'subjective chance,' which is concerned with knowledge and ignorance, is fundamental, and that so-called 'objective chance,' however important it may turn out to be from the practical or scientific point of view, is really a special kind of 'subjective chance' and a derivative type of the latter. For none of the adherents of 'objective chance' wish to question the determinist character of natural order; and the possibility of this objective chance of theirs seems always to depend on the possibility that a particular kind of knowledge either is ours or is within our powers and capacity. Let me try to distinguish as exactly as I can the criterion of objective chance.

7. When we say that an event has happened by chance, we do not mean that previous to its occurrence the event was, on the available evidence, very improbable; this may or may not have been the case. We say, for example, that if a coin falls heads it is 'by chance,' whereas its falling heads is not at all improbable. The term 'by chance' has reference rather to the state of our information about the concurrence of the event considered and the event premised. The fall of the coin is a chance event if our knowledge of the circumstances of the throw is *irrelevant* to our expectation of the possible alternative results. If the number of alternatives is very large, then the occurrence of the event is not only subject to chance but is also very improbable. In general two events may be said to have a chance connection, in the subjective sense, when knowledge of the first is irrelevant to our expectation of the second, and produces no additional presumption for or against it; when, that is to say, the probabilities of the propositions asserting them are *independent* in the sense defined in Chapter XII. § 8.

The above definition deals with chance in the widest sense. What is the *differentia* of the narrower group of cases to which it is desired to apply the term '*objective* chance'? The occurrence of an event may be said to be subject to objective chance, I think, when it is not only a chance event in the above sense, but when we also have good reason to suppose that the addition of further knowledge of a given kind, if it were procurable, would not affect its chance character. We must consider, that is to say, the probability which is relative not to *actual* knowledge but to the *whole* of a *certain kind* of knowledge. We may be able to infer from our evidence that, even with certain kinds of additions to our knowledge, the connections between the events would still be subject to chance in the sense just defined, and we may be able to infer this without actually having the additional information in question. If, however complete our knowledge of certain kinds of things might be, there would still exist independence between the propositions, the conjunction of which we are investigating, then we may say there is an *objective* sense in which the actual conjunction of these propositions is due to chance.

8. This is, I think, the right line of inquiry. It remains to decide, *what* kinds of information must be irrelevant to the connection, in order that the presence of objective chance may be established.

When we attribute a coincidence to objective chance, we mean not only that we do not actually know a law of connection, but, speaking roughly, that there is no law of connection to be known. And when we say that the occurrence of one alternative rather than another is due to chance, we mean not only that we know no principle by which to choose between the alternatives, but also that no such principle is knowable. This use of the term closely corresponds to what Venn means by the term 'casual': "We call a coincidence casual, I apprehend, when we mean to imply that no knowledge of one of the two elements, which we can suppose to be practically attainable, would enable us to expect the other."[1]

To make this more precise, we must revive our distinction,[2] between nomologic knowledge and ontologic knowledge, between knowledge of laws and knowledge of facts or existence. Given certain facts $f(a)$ about a and certain laws of connection, L, we can infer certainly or probably other facts $\phi(a)$ about a. If a *complete* knowledge of laws of connection together with $f(a)$ yields no appreciable probability for preferring $\phi(a)$ to other alternatives, then I suggest that an actual connection between ϕ and f in a particular instance may be said to be due to chance in a sense which usage justifies us in calling objective. We do not, in fact, when we speak of objective chance, always use it in so strict a sense as this, but this is, I think, the underlying conception to which current usage approximates. Current usage diverges from this sense mainly for two reasons. We speak of objective chance if in the above conditions our grounds for preference, though appreciable, are very small; and we are not insistent to assert the rule of chance if a comparatively *slight* addition to our ontologic knowledge would render the probability or the grounds for preference appreciable.

[1] *Logic of Chance*, p. 245.

[2] See Part III. Note (ii.) § 2, p. 275.

To sum up the above, an event is due to objective chance if in order to predict it, or to prefer it to alternatives, at present equi-probable, with any high degree of probability, it would be necessary to know a great many more facts of existence about it than we actually do know, and if the addition of a wide knowledge of general principles would be little use.

It must be added that we make a distinction between facts of existence which are highly variable from case to case and those which are constant or nearly constant over a certain field of observation or experience. Within the limits of this field we regard the *permanent* facts of existence as being, from the standpoint of chance, in nearly the same position as laws. A connection is not due to chance, therefore, if a knowledge of the permanent facts of existence could lead to their prediction.

To sum up again therefore,—if within a given field of observation or experience a knowledge of those facts of existence which are permanent or invariable within that field, together with a knowledge of all the relevant fundamental causal laws or general principles, and of a *few* other facts of existence, would not permit us, given $f(a)$, to attribute an appreciable probability to $\phi(a)$ (or an appreciable probability to the alternative $\phi_1(a)$ rather than $\phi_2(a)$); then the conjunction of $\phi(a)$ (or of $\phi_1(a)$ rather than $\phi_2(a)$ with $f(a)$) is due to objective chance.

9. If we return to the examples of Poincaré, the above definition appears to conform satisfactorily with the usages of common sense. It is when an *exact* knowledge of fact, as distinguished from principle, is required for even approximate prediction that the expression 'objective chance' seems applicable. But neither our definition nor usage is precise as to the *amount* of knowledge of fact which must be required for prediction, in order that, in the absence of it, the event may be regarded as subject to objective chance.

It may be added that the expression 'chance' can be used with reference to general statements as well as to particular facts. We say, for example, that it is a matter of chance if a man dies on his birthday, meaning that, as a general principle and in the absence of special information bearing on a particular case, there is no presumption whatever in favour of his dying on his birthday rather than on any other day. If as a general rule there were celebrations on such a day such as would be not unlikely to accelerate death, we should say that a man's dying on his birthday was not altogether a matter of chance. If we knew no such general rule but did not know enough about birthdays to be assured that there was no such rule, we could not call the chance 'objective'; we could only speak of it thus, if on the evidence before us there was a strong presumption against the existence of any such general rule.

10. The philosophical and scientific importance of objective chance as defined above cannot be made plain, until Part V., on the Foundations of Statistical Inference, has been reached. There it will appear in more than one connection, but chiefly in connection with the application of Bernoulli's formula. In cases where the use of this formula is valid, important inferences can be drawn; and it will be shown that, when the conditions for objective chance are approximately satisfied, it is probable that the conditions for the application of Bernoulli's formula will be approximately satisfied also.

11. The term *random* has been used, it is well recognised, in several distinct senses. Venn [1] and other adherents of the 'frequency' theory have given to it a precise meaning, but one which has avowedly very little relation to popular usage. A random sample, says Peirce,[2] is one "taken according to a precept or method, which, being applied over and over again indefinitely, would in the long run result in the drawing of any one set of instances as often as any other set of the same number." The same fundamental idea has been expressed with greater precision by Professor Edgeworth in connection with his investigations into the law of error.[3] It is a fatal objection, in my opinion, to this mode of defining randomness, that in general we can only know whether or not we have a random sample when our knowledge is nearly complete. Its divergence from ordinary usage is well illustrated by the fact that there would be perfect randomness in the distribution of stars in the heavens, as Venn explicitly points out, if they were disposed in an exact and symmetrical pattern.[4]

[1] *Logic of Chance*, chap. v., "The Conception *Randomness* and its Scientific Treatment."

[2] "A Theory of Probable Inference" (published in Johns Hopkins *Studies in Logic*), p. 152.

[3] "Law of Error," *Camb. Phil. Trans.*, 1904, p. 128.

[4] But it may be added that this seems inconsistent with Venn's conception of randomness as that of aggregate order and *individual irregularity*; nor is it concordant with Venn's typically random diagram (p. 118). His usage, therefore, is sometimes nearer than his definition to the popular usage.

I do not believe, therefore, that this kind of definition is a useful one. The term must be defined with reference to probability, not to what will happen "in the long run"; though there may be two senses of it, corresponding to subjective and objective probability respectively.

The most important phrase in which the term is used is that of 'a random selection' or 'taken at random.' When we apply this term to a particular member of a series or collection of objects, we may mean one of two things. We may mean that our knowledge of the method of choosing the particular member is such that *à priori* the member chosen is as likely to be any one member of the series as any other. We may also mean, not that we have *no* knowledge as to which particular member is in question, but that such knowledge as we have respecting the particular member, as distinguished from other members of the series, is irrelevant to the question as to whether or not this member has the characteristic under examination. In the first case the particular member is a random member of the series for *all* characteristics; in the second case it is a random member for some only. As the second case is the more general, we had better take that for the purpose of defining 'random selection.'

The point will be brought out further if we discuss the more difficult use of the term. What exactly do we mean by the statement: "Any number, taken at random, is equally likely to be odd or even"? According to the frequency theory, this simply means that there are as many odd numbers as there are even. Taking it in a sense corresponding to subjective chance (and to the explanations given above), I propose as a definition the following: a is taken at random from the class S for the purposes of the propositional function $S(x) . \phi(x)$, relative to evidence h, if 'x is a' is irrelevant to the probability $\phi(x)/S(x) . h$. Thus 'the number of the inhabitants of France is odd' is, relative to my knowledge, a random instance of the propositional function 'x is an odd number,' since 'a is the number of the inhabitants of France' is irrelevant to the probability of 'a is odd.'[1] Thus to say that a number taken at random is as likely to be odd as even, means that there is a

[1] In the above $S(x)$ stands for 'x is a number,' $\phi(x)$ stands for 'x is odd,' a stands for 'the number of inhabitants of France.'

probability $\frac{1}{2}$ that any instance taken at random of the generalisation 'all numbers are odd' (or of the corresponding generalisation 'all numbers are even') is true; an instance being taken at random in respect of evenness or oddness, if our knowledge about it satisfies the conditions defined above. Whether or not a given instance is taken at random, depends, therefore, upon what generalisation is in question.

12. We may or may not have reason to believe that, if we take a series of random selections, the proportionate number of occurrences of one particular type of result will very probably lie within certain limits. For reasons to be explained in Chapter XXIX., random selection relative to such information may conveniently be termed 'random selection under Bernoullian conditions.' It is this kind of random selection which is scientifically and statistically important. But, as this corresponds to 'objective chance,' it is convenient to have a wider definition of 'random selection' unqualified, corresponding to 'subjective chance'; and it is this wider definition which is given above.

The term opposite to 'random selection' in ordinary usage is 'biassed selection.' When I use this phrase without qualification I shall use it as the opposite of 'random selection' in the wider unqualified sense.

CHAPTER XXV

SOME PROBLEMS ARISING OUT OF THE DISCUSSION OF CHANCE

1. There are two classical problems in which attempts have been made to attribute certain astronomical phenomena to a specific cause, rather than to objective chance in some such sense as has been defined in the preceding chapter.

The first of these is concerned with the inclinations to the ecliptic of the orbits of the planets of the solar system. This problem has a long history, but it will be sufficient to take De Morgan's statement of it.[1] If we suppose that each of the orbits might have *any* inclination, we obtain a vast number of combinations of which only a small number are such that their sum is as small or smaller than the sum of those of the actual system. But the very existence of ourselves and our world can be shown to imply that one of this small number has been selected, and De Morgan derives from this an enormous presumption that "there was a necessary cause in the formation of the solar system for the inclinations being what they are."

The answer to this was pointed out by D'Alembert [2] in criticis-

[1] Article on *Probabilities* in *Encyclopaedia Metropolitana*, p. 412, § 46. De Morgan takes this without acknowledgment from Laplace, *Théorie analytique des probabilités* (1st edition), pp. 257, 258. Laplace also allows for the fact that all the planets move in the same sense as the earth. He concludes: "On verra que l'existence d'une cause commune qui a dirigé tous ces mouvemens dans le sens de la rotation du soleil, et sur des plans peu inclinés à celui de son équateur, est indiquée avec une probabilité bien supérieure à celle du plus grand nombre des faits historiques sur lesquels on ne se permet aucun doute." Laplace had in his turn borrowed the example, also without acknowledgment, from Daniel Bernoulli. See also D'Alembert, *Opuscules mathématiques*, vol. iv., 1768, pp. 89 and 292.

[2] *Op. cit.* p. 292. "Il y a certainement d'infini contre un à parier que les Planètes ne devraient pas se trouver dans le même plan; ce n'est pas une raison pour en conclure que cette disposition, si elle avoit lieu, auroit nécessairement d'autre cause que le hasard; car il y auroit de même l'infini contre un à parier

ing Daniel Bernoulli. De Morgan could have reached a similar result *whatever* the configuration might have happened to be. *Any* arbitrary disposition over the celestial sphere is vastly improbable *à priori*, that is to say in the absence of known laws tending to favour particular arrangements. It does not follow from this, as De Morgan argues, that any actual disposition possesses *à posteriori* a peculiar significance.

2. The second of these problems is known as Michell's problem of binary stars. Michell's Memoir was published in the *Philosophical Transactions* for 1767.[1] It deals with the question as to whether stars which are optically double, *i.e.* which are so situated as to appear close together to an observer on the earth—are also physically so "either by an original act of the Creator, or in consequence of some general law, such perhaps as gravity." He argues that if the stars "were scattered by mere chance as it might happen . . . it is manifest . . . that every star being as likely to be in any one situation as another, the probability that any one particular star should happen to be within a certain distance (as, for example, one degree) of any other given star would be represented . . . by a fraction whose numerator would be to its denominator as a circle of one degree radius to a circle whose radius is the diameter of a great circle . . . that is, about 1 in 13131." From this beginning he derives an immense presumption against the scattering of the several contiguous stars that may be observed "by mere chance as it might happen." And he goes on to argue that, if there are causal laws directly tending to produce the observed proximities, we may reasonably suppose that the proximities are actual, and not merely optical and apparent. The fact that Michell's induction was confirmed by the later investigations of Herschell adds interest to the speculation. But apart from this the argument is evidently

---

que les Planètes pourroient n'avoir pas une certaine disposition déterminée à volonté. . . ."

D'Alembert is employing the instance for his own purposes, in order to build up an *ad hominem* argument in favour of his theory concerning 'runs' against D. Bernoulli (see also p. 317).

[1] See also Todhunter's *History*, pp. 332-4; Venn, *Logic of Chance*, p. 260; Forbes, "On the Alleged Evidence for a Physical Connexion between Stars forming Binary or Multiple Groups, deduced from the Doctrine of Chances," *Phil. Mag.*, 1850, and Boole, "On the Theory of Probabilities and in particular on Michell's Problem of the Distribution of the Fixed Stars," *Phil. Mag.*, 1851.

subtler than in the first example. Michell argues that there are more stars optically contiguous, than would be likely if there were no special cause acting towards this end, and further that, if such a cause is in operation, it must be *real*, and not merely optical, contiguity that results from it.

Let us analyse the argument more closely. By "mere chance as it might happen" Michell cannot be supposed to mean "uncaused." He is thinking of objective chance in the sense in which I have defined this in the preceding chapter. We speak of a chance occurrence when it is brought about by the coincidence of forces and circumstances so numerous and complex that knowledge sufficient for its prediction is of a kind altogether out of our reach. Michell uses the term vaguely but means, I think, something of this kind: An event is due to *mere chance* when it can only occur if a large number of independent[1] conditions are fulfilled simultaneously. The alternatives which Michell is discussing are therefore these: Are binary stars merely due to the interaction of a vast variety of stellar laws and positions or are they the result of a few fundamental tendencies, which might be the subject of knowledge and which would lead us to expect such stars in relative profusion?

The existence of numerous binary stars may give a real inductive argument in favour of their arising out of the interaction of a relatively small number of independent causes. But it is not possible to arrive at such precise results as Michell's. If there is some finite probability *à priori* that binary stars, when they arise, do arise in this way, then, since the frequent coincidence of a given set of independent causes relatively few in number is more likely than that of a set relatively numerous, the observation of binary stars will raise this probability *à posteriori* to an extent which depends upon the relative profusion in which such stars appear. If, in short, the first of the two alternatives proposed above is assumed, there is no greater presumption for a distribution, covering a part of the heavens, in which binary stars appear, than for any other distribution; if the second is assumed, there *is* a greater presumption. The observation of numerous distributions in which binary stars appear *increases*, therefore, by the inverse principle, any *à priori* probability which may exist in favour of the second hypothesis.

[1] See § 3 of Note (ii.) to Part III.

But more than this the argument cannot justify. That Michell's argument is, as it stands, no more valid than De Morgan's, becomes plain when we notice that he would still have a high probability for his conclusion even if *only one* binary star had been observed. The valuable part of the argument must clearly turn upon the observation of *numerous* binary stars.

Let us now turn to Michell's second step. He argues that, if binary stars arise out of the interaction of a small number of independent forces, they must be physically and not merely optically double. The force of this argument seems to depend upon our possessing previous knowledge as to the nature of the principal natural laws, and upon an assumption, arising out of this, that there are not likely to be forces tending to arrange stars, in reality at great distances from one another, so as to *appear* double from this particular planet. But Michell, in arguing thus, was neglecting the possibility that the optical connection between the stars might be due to the observer and his means of observation. It was not impossible that there should be a law, connected with the transmission of light for example, which would cause stars to appear to an observer to be much nearer together than they really are.

While, therefore, a relative profusion of binary stars constitutes evidence favourably relevant to Michell's conclusion, the argument is more complex and much less conclusive than he seems to have supposed. This is a criticism which is applicable to many such arguments. The simplicity of the evidence, which arises out of the lack of much relevant information, is liable, unless we are careful, to lead us into deceptive calculations and into assertions of high numerical probabilities, upon which we should never venture in cases where the evidence is full and complicated, but where, in fact, the conclusion is established far more strongly. The enormously high probability in favour of his conclusion, to which Michell's calculations led him, should itself have caused him to suspect the accuracy of the reasoning by which he reached it.

3. Some more recent problems of this type seem, however, so far as I am acquainted with them, to follow safer lines of argument. The most important are concerned with the existence of star drifts. It seems to me not at all impossible to possess *data* on which a valid argument can be constructed from the

observation of optically apparent star drifts to the probability of a real uniformity of motion amongst certain sets of stars relatively to others.

Another problem, somewhat analogous to the preceding, has been recently discussed by Professor Karl Pearson.[1] The title might prove a little misleading, perhaps, until the explanation has been reached of the sense in which the term 'random' is used in it. But Professor Pearson uses the term in a perfectly precise sense. He defines a random distribution as one in which spherical shells of equal volume about the sun as centre contain the same number of stars.[2] He argues that the observed facts render probable the following disjunction: Either the distribution of stars is not random in the sense defined above, *or* there is a correlation between their distance and their brilliancy, such as might be produced, for example, by the absorption of light in its transmission through space, *or* the space within which they all lie is limited in volume and not spherical in form.[3] But it is useless to employ the term *random* in this sense in such inquiries as Michell's. For there is no reason to suppose that a non-random distribution is more likely than a random distribution to depend upon the interaction of a small number of independent forces, and there might even exist a presumption the other way. This arbitrary interpretation of randomness does not help us to the solution of any interesting problem.

4. The discussion of *final* causes and of the argument from design has suffered confusion from its supposed connection with theology. But the logical problem is plain and can be determined upon formal and abstract considerations. The argument is in all cases simply this—an event has occurred and has been observed which would be very improbable *à priori* if we did not know that it had actually happened; on the other hand, the event is of such a character that it might have been not unreasonably predicted if we had assumed the existence of a conscious agent whose motives are of a certain kind and whose powers are sufficient.

[1] "On the Improbability of a Random Distribution of the Stars in Space," *Proceedings of Royal Society*, series A, vol. 84, pp. 47-70, 1910.

[2] It is, therefore, independent of direction, and the distribution is random even if the stars are massed in particular quarters of the heavens. The definition is, therefore, exceedingly arbitrary.

[3] This should run more correctly, I think, "not a sphere *with the sun as centre*."

Symbolically: Let h be our original *data*, a the occurrence of the event, b the existence of the supposed conscious agent. Then a/h is assumed very small in comparison with a/bh; and we require b/ah, the probability, that is to say, of b after a is known. The inverse principle of probability already demonstrated shows that $b/ah = a/bh . \frac{b/h}{a/h}$, and b/ah is therefore not determinate in terms of a/bh and a/h alone. Thus we cannot measure the probability of the conscious agent's existence *after* the event, unless we can measure its probability *before* the event. And it is our ignorance of this, as a rule, that we are endeavouring to remedy. The argument tells us that the existence of the hypothetical agent is more likely after the event than before it; but, as in the case of the general inductive problem dealt with in Part III., unless there is an *appreciable* probability first, there cannot be an appreciable probability afterwards. No conclusion, therefore, which is worth having, can be based on the argument from design *alone*; like induction, this type of argument can only strengthen the probability of conclusions, for which there is something to be said on *other* grounds. We cannot say, for example, that the human eye is due to design more probably than not, unless we have some reason, apart from the nature of its construction, for suspecting conscious workmanship. But the necessary *à priori* probability, derived from some other source, may sometimes be forthcoming. The man who upon a desert island picks up a watch, or who sees the symbol *John Smith* traced upon the sand, can use with reason the argument from design. For he has *other* grounds for supposing that beings, capable of designing such objects, do exist, and that their presence on the island, now or formerly, is appreciably possible.

5. The most important problems at the present day, in which arguments of this kind are employed, are those which arise in connection with psychical research.[1] The analysis of the 'cross-

[1] The probability that a remarkable success in naming playing cards is due to psychic agency, was discussed by Professor Edgeworth in *Metretike*. This was, I think, the first application of probabilities to these questions. See also *Proceedings of the Society for Psychical Research*, Parts VIII. and X.; Professor Edgeworth's article on *Psychical Research and Statistical Method*, Stat. Journ. vol. lxxxii. (1919) p. 222; and *Experiments in Psychical Research at Leland Stanford Junior University*, by J. Coover.

correspondences,' which have played so large a part in recent discussions, presents many points of difficulty which are not dissimilar to those which arise in other scientific inquiries of great complexity in which our initial knowledge is small. An important part of the *logical* problem, therefore, is to distinguish the *peculiarity* of psychical problems and to discover what special evidence they demand beyond what is required when we deal with other questions. There is a certain tendency, I think, arising out of the belief that psychical problems are in some way peculiar, to raise sceptical doubts against them, which are equally valid against *all* scientific proofs. Without entering into any questions of detail, let us endeavour to separate those difficulties which seem peculiar to psychical research from those which, however great, are not different from the difficulties which confront students of heredity, for instance, and which are not less likely than these to yield ultimately to the patience and the insight of investigators.

For this purpose it is necessary to recur, briefly, to the analysis of Part III. It was argued there that the methods of empirical proof, by which we strengthen the probability of our conclusions, are not at all dissimilar, when we apply them to the discovery of formal truth, and when we apply them to the discovery of the laws which relate material objects, and that they may possibly prove useful even in the case of metaphysics; but that the *initial* probability which we strengthen by these means is differently obtained in each class of problem. In logic it arises out of the postulate that apparent self-evidence invests what seems self-evident with *some* degree of probability; and in physical science, out of the postulate that there is a limitation to the amount of *independent* variety amongst the qualities of material objects. But both in logic and in physical science we may wish to consider hypotheses which it is not possible to invest with any *à priori* probability and which we entertain solely on account of the known truth of many of their consequences. An axiom which has no self-evidence, but which it seems necessary to combine with other axioms which are self-evident in order to deduce the generally accepted body of formal truth, stands in this category. A scientific entity, such as the ether or the electron, whose qualities have *never* been observed but whose existence we postulate for purposes of explanation, stands in it also. If the analysis of Part III. is correct, we can never attribute a finite probability[1] to the truth of such axioms or to the existence of such scientific entities, however many of their consequences we find to be true. They may be *convenient* hypotheses, because, if we confine ourselves to *certain classes* of their consequences, we are not likely to be led into error; but they stand, nevertheless, in a position altogether different from that of such generalisations as we have reason to invest with an initial probability.

Let us now apply these distinctions to the problems of psychical research. In the case of some of them we can obtain the initial probability, I think, by the same kind of postulates as in physical science, and our conclusions need not be open to a greater degree of doubt than these. In the case of others we cannot; and these must remain, unless some method is open to us peculiar to psychical research, as tentative unproved hypotheses in the same category as the ether.

The best example of the first class is afforded by telepathy. We know that the consciousnesses which, if our hypothesis is correct, act upon one another, do exist; and I see no *logical* difference between the problem of establishing a law of telepathy and that of establishing the law of gravitation. There is at present a *practical* difference on account of the much narrower scope of our knowledge, in the case of telepathy, of cognate matters. We can, therefore, be much less certain; but there seems no reason why we should necessarily remain less certain after more evidence has been accumulated. It is important to remember that, in the case of telepathy, we are merely discovering a relation between objects *which we already know to exist.*

The best example of the other class is afforded by attempts to attribute psychic phenomena to the agency of 'spirits' other than human beings. Such arguments are weakened at present by the fact that no phenomena are known, so far as I am aware, which cannot be explained, though improbably in some cases, in other ways. But even if phenomena were to be observed of

[1] I am *assuming* that there is *no* argument, arising either from self-evidence or analogy, in addition to the argument arising from the truth of their consequences, in favour of the truth of such axioms or the existence of such objects; but I daresay that this may not certainly be the case. The reader may be reminded also that, when I deny a finite probability this is not the same thing as to affirm that the probability is infinitely small. I mean simply that it is not greater than some numerically measurable probability.

which no known agency could afford even an improbable explanation, the hypothesis of 'spirits' would still lie in the same logical limbo as the hypothesis of the 'ether,' in which they might be supposed not inappropriately to move.

Such an hypothesis as the existence of 'spirits' could only become substantial if some peculiar method of knowledge were within our power which would yield us the initial probability which is demanded. That such a method exists, it is not infrequently claimed. If we can directly perceive these 'spirits,' as many of those who are described in James's *Varieties of Religious Experience* think they can, the problem is, logically, altogether changed. We have, in fact, very much the same kind of reason, though it may be with less probability, that we have for believing in the existence of other people. The preceding paragraph applies only to attempts at proving the existence of 'spirits' from such evidence as is discussed by the Society for Psychical Research.

In between these two extremes comes a class of cases, with regard to which it is extremely difficult to come to a decision—that of attempts to attribute psychic phenomena to the conscious agency of the dead. I wish to discuss here, not the nature of the existing evidence, but the question whether it is possible for *any* evidence to be convincing. In this case the object whose existence we are endeavouring to demonstrate resembles *in many respects* objects which we know to exist. The question of epistemology, which is before us, is this: Is it necessary, in order that we may have an initial probability, that the object of our hypothesis should resemble in every relevant particular some *one* object which we know to exist, or is it sufficient that we should know instances of all its supposed qualities, though never in combination? It is clear that *some* qualities may be irrelevant—position in time and space, for example—and that 'every relevant particular' need not include these. But can the initial probability exist if our hypothesis assumes qualities, which have plainly some degree of relevance, in *new* combinations? If we have no knowledge of consciousness existing apart from a living body, can *indirect* evidence of whatever character afford us any probability of such a thing? Could any evidence, for example, persuade us that a tree felt the emotion of amusement, even if it laughed repeatedly when we made jokes? Yet the analogy which we demand seems to be a matter of degree ; for it does not seem unreasonable to attribute consciousness to dogs, although this constitutes a combination of qualities unlike in many respects to any which we *know* to exist.

This discussion, however, is wandering from the subject of probability to that of epistemology, and it will not be solved until we possess a more comprehensive account of this latter subject than we have at present. I wish only to distinguish between those cases in which we obtain the initial probability in the same manner as in physical science from those in which we must get it, if at all, in some other way. The distinctions I have made are sufficiently summarised by a recapitulation of the following comparisons : We compared the proof of telepathy to the proof of gravitation, the proof of non-human 'spirits' to the proof of the ether, and, much less closely, the proof of the consciousness of the dead to the proof of the consciousness of trees, or, perhaps, of dogs.

Before passing to the next of the rather miscellaneous topics of this chapter, it may be worth while to add that we should be very chary of applying to problems of psychical research the *calculus* of probabilities. The alternatives seldom satisfy the conditions for the application of the Principle of Indifference, and the initial probabilities are not capable of being measured numerically. If, therefore, we endeavour to *calculate* the probability that some phenomenon is due to 'abnormal' causes, our mathematics will be apt to lead us into unjustifiable conclusions.

6. Uninstructed common sense seems to be specially unreliable in dealing with what are termed 'remarkable occurrences.' Unless a 'remarkable occurrence' is simply one which produces on us a particular psychological effect, that of surprise, we can only define it as an event which *before* its occurrence is very improbable on the available evidence. But it will often occur—whenever, in fact, our *data* leave open the possibility of a large number of alternatives and show no preference for any of them —that *every* possibility is exceedingly improbable *à priori*. It follows, therefore, that what actually occurs does not derive any peculiar significance merely from the fact of its being 'remarkable' in the above sense. Something further is required before we can build with success. Yet Michell's argument and the argu-

ment from design derive a good deal of their plausibility, I think, from the 'remarkable' character of the actual constitution whether of the heavens or of the universe, in forgetfulness of the fact that it is impossible to propound *any* constitution which would if it existed be other than 'remarkable.' It is supposed that a remarkable occurrence is specially in need of an explanation, and that any *sufficient* explanation has a high probability in its favour. That an explanation is particularly required, possesses a measure of truth; for it is likely that our original *data* were much lacking in completeness, and the occurrence of the extraordinary event brings to light this deficiency. But that we are not justified in adopting with confidence any sufficient explanation, has been shown already.

Such arguments, however, get a part of their plausibility from a quite different source. There is a general supposition that some kinds of occurrences are more likely than others to be *susceptible* of an explanation *by us*; and, therefore, any explanation which deals with such cases falls in prepared soil. Results which, judging from ourselves, conscious agents would be likely to produce fall into this category. Results which would be probable, supposing a direct and predominant causal dependence between the elements whose concomitance is remarked, belong to it also. There is, in fact, a sort of argument from analogy as to whether certain sorts of phenomena are or are not likely to be due to 'chance.' This may explain, for example, why the particular concurrence of atoms that go to compose the human eye, why a series of correct guesses in naming playing cards, why special symmetry or special asymmetry amongst the stars, seem to require explanation in no ordinary degree. *Prior* to an explanation these particular concurrences or series or distributions are no more improbable than any other. But the causes of such conjunctions as these are more likely to be discoverable by the human mind than are the causes of others, and the attempt to explain them deserves, therefore, to be more carefully considered. This supposition, derived by analogy or induction from those cases in which we believe the causes to be known to us, has, perhaps, some weight. But the direct application of the Calculus of Probabilities can do no more in these cases than suggest matter for investigation. The fact that a man has made a long series of correct guesses in cases where he is cut off from the ordinary channels of communication, is a fact worthy of investigation, because it is more likely to be susceptible of a *simple* causal explanation, which may have many applications, than a case in which false and true guesses follow one another with no apparent regularity.

7. In the case of empirical laws, such as Bode's law, which have no more than a very slight connection with the general body of scientific knowledge, it is sometimes thought that the law is more probable if it is proposed *before* the examination of some or all of the available instances than if it is proposed after their examination. Supposing, for example, that Bode's law is accurately true for seven planets, it is held that the law would be more probable if it was suggested after the examination of six and was confirmed by the subsequent discovery of the seventh, than it would be if it had not been propounded until after all seven had been observed. The arguments in favour of such a conclusion are well put by Peirce:[1] "All the qualities of objects may be conceived to result from variations of a number of continuous variables; hence any lot of objects possesses some character in common, not possessed by any other." Hence if the common character is not predesignate we can conclude nothing. Cases must not be used to prove a generalisation which has only been suggested by the cases themselves. He takes the first five poets from a biographical dictionary with their ages at death:

Aagard . . .	48	Abunowas . .	48
Abeille . . .	76	Accords . .	45
Abulola . . .	84		

"These five ages have the following characters in common:

"1. The difference of the two digits composing the number, divided by three, leaves a remainder of *one*.

"2. The first digit raised to the power indicated by the second, and then divided by three, leaves a remainder of *one*.

"3. The sum of the prime factors of each age, including *one* as a prime factor, is divisible by *three*."

He compares a generalisation regarding the ages of poets based

[1] C. S. Peirce, *A Theory of Probable Inference*, pp. 162-167; published in Johns Hopkins *Studies in Logic*, 1883.

on this evidence to Dr. Lyon Playfair's argument about the specific gravities of the three allotropic forms of carbon:

Diamond	$3{\cdot}48 = \sqrt[2]{12}$
Graphite	$2{\cdot}29 = \sqrt[3]{12}$
Charcoal	$1{\cdot}88 = \sqrt[4]{12}$

approximately, the atomic weight of carbon being 12. Dr. Playfair thinks that the above renders it probable that the specific gravities of the allotropic forms of other elements would, if we knew them, be found to equal the different roots of their atomic weight.

The weakness of these arguments, however, has a different explanation. These inductions are very improbable, because they are out of relation to the rest of our knowledge and are based on a very small number of instances. The apparent absurdity, moreover, of the inductive law of Poets' Ages is increased by the fact that we take account of the knowledge we actually possess that the ages of poets are not in fact connected by any such law. If we knew nothing whatever about poets' ages except what is stated above, the induction would be as valid as any other which is based on a very weak analogy and a very small number of instances and is unsupported by indirect evidence.

The peculiar virtue of prediction or predesignation is altogether imaginary. The number of instances examined and the analogy between them are the essential points, and the question as to whether a particular hypothesis happens to be propounded before or after their examination is quite irrelevant. If all our inductions had to be thought of before we examined the cases to which we apply them, we should, doubtless, make fewer inductions; but there is no reason to think that the few we should make would be any better than the many from which we should be precluded. The plausibility of the argument is derived from a different source. If an hypothesis is proposed *à priori*, this commonly means that there is some ground for it, arising out of our previous knowledge, *apart from* the purely inductive ground, and if such is the case the hypothesis is clearly stronger than one which reposes on inductive grounds only. But if it is a mere guess, the lucky fact of its preceding some or all of the cases which verify it adds nothing whatever to its value. It is the union of prior knowledge, with the inductive grounds which arise out of the immediate instances, that lends weight to an hypothesis, and not the occasion on which the hypothesis is first proposed. It is sometimes said, to give another example, that the daily fulfilment of the predictions of the *Nautical Almanack* constitutes the most cogent proof of the laws of dynamics. But here the essence of the verification lies in the variety of cases which can be brought accurately under our notice by means of the *Almanack*, and in the fact that they have all been obtained on a uniform principle, *not* in the fact that the verification is preceded by a prediction.

The same point arises not uncommonly in statistical inquiries. If a theory is first proposed and is then confirmed by the examination of statistics, we are inclined to attach more weight to it than to a theory which is constructed in order to suit the statistics. But the fact that the theory which precedes the statistics is more likely than the other to be supported by general considerations —for it has not, presumably, been adopted for no reason at all— constitutes the only valid ground for this preference. If it does *not* receive more support than the other from general considerations, then the circumstances of its origin are no argument in its favour. The opposite view, which the unreliability of some statisticians has brought into existence,—that it is a positive advantage to approach statistical evidence *without* preconceptions based on general grounds, because the temptation to 'cook' the evidence will prove otherwise to be irresistible,—has no *logical* basis and need only be considered when the impartiality of an investigator is in doubt.

CHAPTER XXVI

THE APPLICATION OF PROBABILITY TO CONDUCT

1. Given as our basis what knowledge we actually have, the probable, I have said, is that which it is rational for us to believe. This is not a definition. For it is not rational for us to believe that the probable is true; it is only rational to have a probable belief in it or to believe it in preference to alternative beliefs. To believe one thing *in preference* to another, as distinct from believing the first true or more probable and the second false or less probable, must have reference to action and must be a loose way of expressing the propriety of *acting* on one hypothesis rather than on another. We might put it, therefore, that the probable is the hypothesis on which it is rational for us to act. It is, however, not so simple as this, for the obvious reason that of two hypotheses it may be rational to act on the less probable if it leads to the greater good. We cannot say more at present than that the probability of a hypothesis is one of the things to be determined and taken account of before acting on it.

2. I do not know of passages in the ancient philosophers which explicitly point out the dependence of the duty of pursuing goods on the reasonable or probable expectation of attaining them relative to the agent's knowledge. This means only that analysis had not disentangled the various elements in rational action, not that common sense neglected them. Herodotus puts the point quite plainly. "There is nothing more profitable for a man," he says, "than to take good counsel with himself; for even if the event turns out contrary to one's hope, still one's decision was right, even though fortune has made it of no effect: whereas if a man acts contrary to good counsel, although by luck he gets what he had no right to expect, his decision was not any the less foolish." [1]

[1] Herod. vii. 10.

3. The first contact of theories of probability with modern ethics appears in the Jesuit doctrine of probabilism. According to this doctrine one is justified in doing an action for which there is *any* probability, however small, of its results being the best possible. Thus, if any priest is willing to permit an action, that fact affords some probability in its favour, and one will not be damned for performing it, however many other priests denounce it.[1] It may be suspected, however, that the object of this doctrine was not so much duty as safety. The priest who permitted you so to act assumed thereby the responsibility. The correct application of probability to conduct naturally escaped the authors of a juridical ethics, which was more interested in the fixing of responsibility for definite acts, and in the various specified means by which responsibility might be disposed of, than in the greatest possible sum-total of resultant good.

A more correct doctrine was brought to light by the efforts of the philosophers of the Port Royal to expose the fallacies of probabilism. "In order to judge," they say, "of what we ought to do in order to obtain a good and to avoid an evil, it is necessary to consider not only the good and evil in themselves, but also the probability of their happening and not happening, and to regard geometrically the proportion which all these things have, taken together."[2] Locke perceived the same point, although not so clearly.[3] By Leibniz this theory is advanced more explicitly; in such judgments, he says, "as in other estimates disparate and heterogeneous and, so to speak, of more than one dimension, the greatness of that which is discussed is in reason composed of both estimates (*i.e.* of goodness and of probability), and is like a rectangle, in which there are two considerations, viz. that of length and that of breadth. . . . Thus we should

[1] Compare with this doctrine the following curious passage from Jeremy Taylor:—"We being the persons that are to be persuaded, we must see that we be persuaded reasonably. And it is unreasonable to assent to a lesser evidence when a greater and clearer is propounded: but of that every man for himself is to take cognisance, if he be able to judge; if he be not, he is not bound under the tie of necessity to know anything of it. That that is necessary shall be certainly conveyed to him: God, that best can, will certainly take care for that; for if he does not, it becomes to be not necessary; or if it should still remain necessary, and he be damned for not knowing it, and yet to know it be not in his power, then who can help it! There can be no further care in this business."

[2] *The Port Royal Logic* (1662), Eng. Trans. p. 367.

[3] *Essay concerning Human Understanding*, book ii. chap. xxi. § 66.

still need the art of thinking and that of estimating probabilities, besides the knowledge of the value of goods and evils, in order properly to employ the art of consequences."[1]

In his preface to the *Analogy* Butler insists on "the absolute and formal obligation" under which even a low probability, if it is the greatest, may lay us: "To us probability is the very guide of life."

4. With the development of a utilitarian ethics largely concerned with the summing up of consequences, the place of probability in ethical theory has become much more explicit. But although the general outlines of the problem are now clear, there are some elements of confusion not yet dispersed. I will deal with some of them.

In his *Principia Ethica* (p. 152) Dr. Moore argues that "the first difficulty in the way of establishing a probability that one course of action will give a better total result than another, lies in the fact that we have to take account of the effects of both throughout an infinite future. . . . We can certainly only pretend to calculate the effects of actions within what may be called an 'immediate future.' . . . We must, therefore, certainly have some reason to believe that no consequences of our action in a further future will generally be such as to reverse the balance of good that is probable in the future which we can foresee. This large postulate must be made, if we are ever to assert that the results of one action will be even probably better than those of another. Our utter ignorance of the far future gives us no justification for saying that it is even probably right to choose the greater good within the region over which a probable forecast may extend."

This argument seems to me to be invalid and to depend on a wrong philosophical interpretation of probability. Mr. Moore's reasoning endeavours to show that there is not even a *probability* by showing that there is not a *certainty*. We must not, of course, have reason to believe that remote consequences will *generally* be such as to reverse the balance of immediate good. But we need not be certain that the opposite is the case. If good is additive, if we have reason to think that of two actions one produces more good than the other in the near future, and if we have no means of discriminating between their results in the distant

[1] *Nouveaux Essais*, book ii. chap. xxi.

future, then by what seems a legitimate application of the Principle of Indifference we may suppose that there is a probability in favour of the former action. Mr. Moore's argument must be derived from the empirical or frequency theory of probability, according to which we must know for certain what will happen *generally* (whatever that may mean) before we can assert a probability.

The results of our endeavours are very uncertain, but we have a genuine probability, even when the evidence upon which it is founded is slight. The matter is truly stated by Bishop Butler: "From our short views it is greatly uncertain whether this endeavour will, in particular instances, produce an overbalance of happiness upon the whole; since so many and distant things must come into the account. And that which makes it our duty is that there is some appearance that it will, and no positive appearance to balance this, on the contrary side. . . ."[1]

The difficulties which exist are not chiefly due, I think, to our ignorance of the remote future. The possibility of our knowing that one thing rather than another is our duty depends upon the assumption that a greater goodness in any part makes, in the absence of evidence to the contrary, a greater goodness in the whole more probable than would the lesser goodness of the part. We assume that the goodness of a part is *favourably* relevant to the goodness of the whole. Without this assumption we have no reason, not even a probable one, for preferring one action to any other on the whole. If we suppose that goodness is always *organic*, whether the whole is composed of simultaneous or successive parts, such an assumption is not easily justified. The case is parallel to the question, whether physical law is organic or atomic, discussed in Chapter XXI. § 6.

Nevertheless we can admit that goodness is partly organic and still allow ourselves to draw probable conclusions. For the alternatives, that *either* the goodness of the whole universe throughout time is organic *or* the goodness of the universe is the arithmetic sum of the goodnesses of infinitely numerous and infinitely divided parts, are not exhaustive. We may suppose that the goodness of conscious persons is organic for each distinct

[1] This passage is from the *Analogy*. The Bishop adds: " . . . and also that such benevolent endeavour is a cultivation of that most excellent of all virtuous principles, the active principle of benevolence."

and individual personality. Or we may suppose that, when conscious units are in conscious relationship, then the whole which we must treat as organic includes both units. These are only examples. We must suppose, in general, that the units whose goodness we must regard as organic and indivisible are not always larger than those the goodness of which we can perceive and judge directly.

5. The difficulties, however, which are most fundamental from the standpoint of the student of probability, are of a different kind. Normal ethical theory at the present day, if there can be said to be any such, makes two assumptions: first, that degrees of goodness are numerically measurable and arithmetically additive, and second, that degrees of probability also are numerically measurable. This theory goes on to maintain that what we ought to add together, when, in order to decide between two courses of action, we sum up the results of each, are the 'mathematical expectations' of the several results. 'Mathematical expectation' is a technical expression originally derived from the scientific study of gambling and games of chance, and stands for the product of the possible gain with the probability of attaining it.[1] In order to obtain, therefore, a measure of what ought to be our preference in regard to various alternative courses of action, we must sum for each course of action a series of terms made up of the amounts of good which may attach to each of its possible consequences, each multiplied by its appropriate probability.

The first assumption, that quantities of goodness are duly subject to the laws of arithmetic, appears to me to be open to a certain amount of doubt. But it would take me too far from my proper subject to discuss it here, and I shall allow, for the purposes of further argument, that in some sense and to some extent this assumption can be justified. The second assumption, however, that degrees of probability are wholly subject to the laws of arithmetic, runs directly counter to the view which has

[1] Priority in the conception of mathematical expectation can, I think, be claimed by Leibniz, *De incerti aestimatione*, 1678 (Couturat, *Logique de Leibniz*, p. 248). In a letter to Placcius, 1687 (Dutens, vi. i. 36 and Couturat, *op. cit.* p. 246) Leibniz proposed an application of the same principle to jurisprudence, by virtue of which, if two litigants lay claim to a sum of money, and if the claim of the one is twice as probable as that of the other, the sum should be divided between them in that proportion. The doctrine, seems sensible, but I am not aware that it has ever been acted on.

been advocated in Part I. of this treatise. Lastly, if both these points be waived, the doctrine that the 'mathematical expectations' of alternative courses of action are the proper measures of our degrees of preference is open to doubt on two grounds—first, because it ignores what I have termed in Part I. the 'weights' of the arguments, namely, the amount of evidence upon which each probability is founded; and second, because it ignores the element of 'risk' and assumes that an even chance of heaven or hell is precisely as much to be desired as the certain attainment of a state of mediocrity. Putting on one side the first of these grounds of doubt, I will treat each of the others in turn.

6. In Chapter III. of Part I. I have argued that only in a strictly limited class of cases are degrees of probability numerically measurable. It follows from this that the 'mathematical expectations' of goods or advantages are not always numerically measurable; and hence, that even if a meaning can be given to the sum of a series of non-numerical 'mathematical expectations,' not every pair of such sums are numerically comparable in respect of more and less. Thus even if we know the degree of advantage which might be obtained from each of a series of alternative courses of actions and know also the probability in each case of obtaining the advantage in question, it is not always possible by a mere process of arithmetic to determine which of the alternatives ought to be chosen. If, therefore, the question of right action is under all circumstances a determinate problem, it must be in virtue of an intuitive judgment directed to the situation as a whole, and not in virtue of an arithmetical deduction derived from a series of separate judgments directed to the individual alternatives each treated in isolation.

We must accept the conclusion that, if one good is greater than another, but the probability of attaining the first less than that of attaining the second, the question of which it is our duty to pursue may be indeterminate, unless we suppose it to be within our power to make direct quantitative judgments of probability and goodness jointly. It may be remarked, further, that the difficulty exists, whether the numerical indeterminateness of the probability is intrinsic or whether its numerical value is, as it is according to the Frequency Theory and most other theories, simply unknown.

7. The second difficulty, to which attention is called above,

is the neglect of the 'weights' of arguments in the conception of 'mathematical expectation.' In Chapter VI. of Part I. the significance of 'weight' has been discussed. In the present connection the question comes to this—if two probabilities are equal in degree, ought we, in choosing our course of action, to prefer that one which is based on a greater body of knowledge?

The question appears to me to be highly perplexing, and it is difficult to say much that is useful about it. But the degree of completeness of the information upon which a probability is based does seem to be relevant, as well as the actual magnitude of the probability, in making practical decisions. Bernoulli's maxim,[1] that in reckoning a probability we must take into account all the information which we have, even when reinforced by Locke's maxim that we must get all the information we can,[2] does not seem completely to meet the case. If, for one alternative, the available information is necessarily small, that does not seem to be a consideration which ought to be left out of account altogether.

8. The last difficulty concerns the question whether, the former difficulties being waived, the 'mathematical expectation' of different courses of action accurately measures what our preferences ought to be—whether, that is to say, the undesirability of a given course of action increases in direct proportion to any increase in the uncertainty of its attaining its object, or whether some allowance ought to be made for 'risk,' its undesirability increasing more than in proportion to its uncertainty.

In fact the meaning of the judgment, that we ought to act in such a way as to produce most probably the greatest sum of goodness, is not perfectly plain. Does this mean that we ought so to act as to make the sum of the goodnesses of each of the possible consequences of our action multiplied by its probability a maximum? Those who rely on the conception of 'mathematical expectation' must hold that this is an indisputable proposition. The justifications for this view most commonly advanced resemble that given by Condorcet in his "Réflexions

[1] *Ars Conjectandi*, p. 215: "Non sufficit expendere unum alterumve argumentum, sed conquirenda sunt omnia, quae in cognitionem nostram venire possunt, atque ullo modo ad probationem rei facere videntur."

[2] *Essay concerning Human Understanding*, book ii. chap. xxi. § 67: "He that judges without informing himself to the utmost that he is capable, cannot acquit himself of *judging amiss*."

sur la règle générale, qui prescrit de prendre pour valeur d'un événement incertain, la probabilité de cet événement multipliée par la valeur de l'événement en lui-même," [1] where he argues from Bernoulli's theorem that such a rule will lead to satisfactory results if a very large number of trials be made. As, however, it will be shown in Chapter XXIX. of Part V. that Bernoulli's theorem is not applicable in by any means every case, this argument is inadequate as a general justification.

In the history of the subject, nevertheless, the theory of 'mathematical expectation' has been very seldom disputed. As D'Alembert has been almost alone in casting serious doubts upon it (though he only brought himself into disrepute by doing so), it will be worth while to quote the main passage in which he declares his scepticism : " Il me sembloit " (in reading Bernoulli's *Ars Conjectandi*) " que cette matière avoit besoin d'être traitée d'une manière plus claire ; je voyois bien que l'espérance étoit plus grande, 1° que la somme espérée étoit plus grande, 2° que la probabilité de gagner l'étoit aussi. Mais je ne voyois pas avec la même évidence, et je ne le vois pas encore, 1° que la probabilité soit estimée exactement par les méthodes usitées ; 2° que quand elle le seroit, l'espérance doive être proportionnelle à cette probabilité simple, plutôt qu'à une puissance ou même à une fonction de cette probabilité ; 3° que quand il y a plusieurs combinaisons qui donnent différens avantages ou différens risques (qu'on regarde comme des avantages négatifs) il faille se contenter d'*ajouter* simplement ensemble toutes les espérances pour avoir l'espérance totale." [2]

In extreme cases it seems difficult to deny some force to D'Alembert's objection ; and it was with reference to extreme cases that he himself raised it. Is it certain that a larger good, which is extremely improbable, is precisely equivalent ethically to a smaller good which is proportionately more probable ? We may doubt whether the moral value of speculative and cautious action respectively can be weighed against one another in a simple arithmetical way, just as we have already doubted whether a good whose probability can only be determined on a slight basis of evidence can be compared by means merely of the

[1] *Hist. de l'Acad.*, Paris, 1781.

[2] *Opuscules mathématiques*, vol. iv., 1768 (extraits de lettres), pp. 284, 285. See also p. 88 of the same volume.

magnitude of this probability with another good whose likelihood is based on completer knowledge.

There seems, at any rate, a good deal to be said for the conclusion that, other things being equal, that course of action is preferable which involves least risk, and about the results of which we have the most complete knowledge. In marginal cases, therefore, the coefficients of weight and risk as well as that of probability are relevant to our conclusion. It seems natural to suppose that they should exert some influence in other cases also, the only difficulty in this being the lack of any principle for the calculation of the degree of their influence. A high weight and the absence of risk increase *pro tanto* the desirability of the action to which they refer, but we cannot measure the amount of the increase.

The 'risk' may be defined in some such way as follows. If A is the amount of good which may result, p its probability ($p+q=1$), and E the value of the 'mathematical expectation,' so that $E=pA$, then the 'risk' is R, where $R=p(A-E)=p(1-p)A=pqA=qE$. This may be put in another way: E measures the net immediate sacrifice which should be made in the hope of obtaining A; q is the probability that this sacrifice will be made in vain; so that qE is the 'risk.'[1] The ordinary theory supposes that the ethical value of an expectation is a function of E only and is entirely independent of R.

We could, if we liked, define a conventional coefficient c of weight and risk, such as $c=\frac{2pw}{(1+q)(1+w)}$, where w measures the 'weight,' which is equal to unity when $p=1$ and $w=1$, and to zero when $p=0$ or $w=0$, and has an intermediate value in other cases.[2] But if doubts as to the sufficiency of the conception of 'mathematical expectation' be sustained, it is not likely that the solution will lie, as D'Alembert suggests, and as has been exemplified above, in the discovery of some more

[1] The theory of *Risiko* is briefly dealt with by Czuber, *Wahrscheinlichkeitsrechnung*, vol. ii. pp. 219 *et seq.* If R measures the first insurance, this leads to a *Risiko* of the second order, $R_1=qR=q^2E$. This again may be insured against, and by a sufficient number of such reinsurances the risk can be completely shifted: $E+R_1+R_2+\ldots=E(1+q+q^2+\ldots)=\frac{E}{1-q}=\frac{E}{p}=A$.

[2] If $pA=p'A'$, $w>w'$, and $q=q'$, then $cA>c'A'$; if $pA=p'A'$, $w=w'$, and $q<q'$, then $cA>c'A'$; if $pA=p'A'$, $w>w'$, and $q<q'$, then $cA>c'A'$; but if $pA=p'A'$, $w=w'$, and $q>q'$, we cannot in general compare cA and $c'A'$.

complicated function of the probability wherewith to compound the proposed good. The judgment of goodness and the judgment of probability both involve somewhere an element of direct apprehension, and both are quantitative. We have raised a doubt as to whether the magnitude of the 'oughtness' of an action can be in all cases directly determined by simply multiplying together the magnitudes obtained in the two direct judgments; and a new direct judgment may be required, respecting the magnitude of the 'oughtness' of an action under given circumstances, which need not bear any simple and necessary relation to the two former.

The hope, which sustained many investigators in the course of the nineteenth century, of gradually bringing the moral sciences under the sway of mathematical reasoning, steadily recedes—if we mean, as they meant, by mathematics the introduction of precise numerical methods. The old assumptions, that all quantity is numerical and that all quantitative characteristics are additive, can be no longer sustained. Mathematical reasoning now appears as an aid in its symbolic rather than in its numerical character. I, at any rate, have not the same lively hope as Condorcet, or even as Edgeworth, "éclairer les Sciences morales et politiques par le flambeau de l'Algèbre." In the present case, even if we are able to range goods in order of magnitude, and also their probabilities in order of magnitude, yet it does not follow that we can range the products composed of each good and its corresponding probability in this order.

9. Discussions of the doctrine of Mathematical Expectation, apart from its directly ethical bearing, have chiefly centred round the classic Petersburg Paradox,[1] which has been treated by almost all the more notable writers, and has been explained by them in a great variety of ways. The Petersburg Paradox arises out of a game in which Peter engages to pay Paul one shilling if a head appears at the first toss of a coin, two shillings if it does not appear until the second, and, in general, 2^{r-1} shillings if no head appears until the rth toss. What is the value of Paul's expectation, and what sum must he hand over to Peter before the game commences, if the conditions are to be fair?

[1] For the history of this paradox see Todhunter. The name is due, he says, to its having first appeared in a memoir by Daniel Bernoulli in the *Commentarii* of the Petersburg Academy.

The mathematical answer is $\sum_1^n(\frac{1}{2})^r 2^{r-1}$, if the number of tosses is not in any case to exceed n in all, and $\sum_1^\infty(\frac{1}{2})^r 2^{r-1}$ if this restriction is removed. That is to say, Paul should pay $\frac{n}{2}$ shillings in the first case, and an infinite sum in the second. Nothing, it is said, could be more paradoxical, and no sane Paul would engage on these terms even with an honest Peter.

Many of the solutions which have been offered will occur at once to the reader. The conditions of the game *imply* contradiction, say Poisson and Condorcet; Peter has undertaken engagements which he cannot fulfil; if the appearance of heads is deferred even to the 100th toss, he will owe a mass of silver greater in bulk than the sun. But this is no answer. Peter has promised much and a belief in his solvency will strain our imagination; but it is imaginable. And in any case, as Bertrand points out, we may suppose the stakes to be, not shillings, but grains of sand or molecules of hydrogen.

D'Alembert's principal explanations are, first, that true expectation is not necessarily the product of probability and profit (a view which has been discussed above), and second, that very long runs are not only very improbable, but do not occur at all.

The next type of solution is due, in the first instance, to Daniel Bernoulli, and turns on the fact that no one but a miser regards the desirability of different sums of money as directly proportional to their amount; as Buffon says, "L'avare est comme le mathématicien: tous deux estiment l'argent par sa quantité numérique." Daniel Bernoulli deduced a formula from the assumption that the importance of an increment is inversely proportional to the size of the fortune to which it is added. Thus, if x is the 'physical' fortune and y the 'moral' fortune,

$$dy = k\frac{dx}{x},$$

or $y = k \log \frac{x}{a}$, where k and a are constants.

On the basis of this formula of Bernoulli's a considerable

theory has been built up both by Bernoulli[1] himself and by Laplace.[2] It leads easily to the further formula—

$$x=(a+x_1)p_1(a+x_2)p_2\ldots,$$

where a is the initial 'physical' fortune, p_1, etc., the probabilities of obtaining increments x_1, etc., to a, and x the 'physical' fortune whose present possession would yield the same 'moral' fortune as does the expectation of the various increments x_1, etc. By means of this formula Bernoulli shows that a man whose fortune is £1000 may reasonably pay a £6 stake in order to play the Petersburg game with £1 units. Bernoulli also mentions two solutions proposed by Cramer. In the first all sums greater than 2^{24} (16,777,116) are regarded as 'morally' equal; this leads to £13 as the fair stake. According to the other formula the pleasure derivable from a sum of money varies as the square root of the sum; this leads to £2 : 9s. as the fair stake. But little object is served by following out these arbitrary hypotheses.

As a solution of the Petersburg problem this line of thought is only partially successful: if increases of 'physical' fortune beyond a certain finite limit can be regarded as 'morally' negligible, Peter's claim for an infinite initial stake from Paul is, it is true, no longer equitable, but with any reasonable law of diminution for successive increments Paul's stake will still remain paradoxically large. Daniel Bernoulli's suggestion is, however, of considerable historical interest as being the first explicit attempt to take account of the important conception known to modern economists as the diminishing marginal utility of money,—a conception on which many important arguments are founded relating to taxation and the ideal distribution of wealth.

Each of the above solutions probably contains a part of the psychological explanation. We are unwilling to be Paul, partly because we do not believe Peter will pay us if we have good fortune in the tossing, partly because we do not know what we should do with so much money or sand or hydrogen if we won it, partly because we do not believe we ever should win it, and partly because we do not think it would be a rational act to risk

[1] "Specimen Theoriae Novae de Mensura Sortis," *Comm. Acad. Petrop.* vol. v. for 1730 and 1731, pp. 175-192 (published 1738). See Todhunter, pp. 213 *et seq.*

[2] *Théorie analytique*, chap. x. "De l'espérance morale," pp. 432-445.

an infinite sum or even a very large finite sum for an infinitely larger one, whose attainment is infinitely unlikely.

When we have made the proper hypotheses and have eliminated these elements of psychological doubt, the theoretic dispersal of what element of paradox remains must be brought about, I think, by a development of the theory of risk. It is primarily the great *risk* of the wager which deters us. Even in the case where the number of tosses is in no case to exceed a finite number, the risk R, as already defined, may be very great, and the relative risk $\frac{R}{E}$ will be almost unity. Where there is no limit to the number of tosses, the risk is infinite. A relative risk, which approaches unity, may, it has been already suggested, be a factor which must be taken into account in ethical calculation.

10. In establishing the doctrine, that all private gambling must be with certainty a losing game, precisely contrary arguments are employed to those which do service in the Petersburg problem. The argument that "you must lose if only you go on long enough" is well known. It is succinctly put by Laurent:[1] Two players A and B have a and b francs respectively. $f(a)$ is the chance that A will be ruined. Thus $f(a)=\frac{b}{a+b}$,[2] so that the poorer a gambler is, relatively to his opponent, the more likely he is to be ruined. But further, if $b=\infty$, $f(a)=1$, *i.e.* ruin is certain. The infinitely rich gambler is the public. It is against the public that the professional gambler plays, and his ruin is therefore certain.

Might not Poisson and Condorcet reply, The conditions of the game *imply* contradiction, for no gambler plays, as this argument supposes, for ever?[3] At the end of any *finite* quantity of play, the player, even if he is not the public, *may* finish with winnings of any finite size. The gambler is in a worse position if his capital is smaller than his opponents'—at poker, for instance, or on the Stock Exchange. This is clear. But our desire for moral improvement outstrips our logic if we tell him that he *must* lose. Besides it is paradoxical to say that everybody

[1] *Calcul des probabilités*, p. 129.

[2] This would possibly follow from the theorem of Daniel Bernoulli. The reasoning by which Laurent obtains it seems to be the result of a mistake.

[3] Cf. also Mr. Bradley, *Logic*, p. 217.

individually must lose and that everybody collectively must win. For every individual gambler who loses there is an individual gambler or syndicate of gamblers who win. The true moral is this, that poor men should not gamble and that millionaires should do nothing else. But millionaires gain nothing by gambling with one another, and until the poor man departs from the path of prudence the millionaire does not find his opportunity. If it be replied that in fact most millionaires are men originally poor who departed from the path of prudence, it must be admitted that the poor man is not doomed with certainty. Thus the philosopher must draw what comfort he can from the conclusion with which his theory furnishes him, that millionaires are often fortunate fools who have thriven on unfortunate ones.[1]

11. In conclusion we may discuss a little further the conception of 'moral' risk, raised in § **8** and at the end of § **9**. Bernoulli's formula crystallises the undoubted truth that the value of a sum of money to a man varies according to the amount he already possesses. But does the value of an amount of goodness also vary in this way? May it not be true that the addition of a given good to a man who already enjoys much good is less good than its bestowal on a man who has little? If this is the case, it follows that a smaller but relatively certain good is better than a greater but proportionately more uncertain good.

In order to assert this, we have only to accept a particular theory of organic goodness, applications of which are common enough in the mouths of political philosophers. It is at the root of all principles of equality, which do not arise out of an assumed diminishing marginal utility of money. It is behind the numerous arguments that an equal distribution of benefits is better than a very unequal distribution. If this is the case, it follows that, the sum of the goods of all parts of a community taken together being fixed, the organic good of the whole is greater the more equally the benefits are divided amongst the individuals. If the doctrine is to be accepted, moral risks, like financial risks, must not be undertaken unless they promise a profit actuarially.

[1] From the social point of view, however, this moral against gambling may be drawn—that those who start with the largest initial fortunes are most likely to win, and that a given increment to the wealth of these benefits them, on the assumption of a diminishing marginal utility of money, less than it injures those from whom it is taken.

There is a great deal which could be said concerning such a doctrine, but it would lead too far from what is relevant to the study of Probability. One or two instances of its use, however, may be taken from the literature of Probability. In his essay, "Sur l'application du calcul des probabilités à l'inoculation de la petite vérole,"[1] D'Alembert points out that the community would gain on the average if, by sacrificing the lives of one in five of its citizens, it could ensure the health of the rest, but he argues that no legislator could have the right to order such a sacrifice. Galton, in his *Probability, the Foundation of Eugenics*, employed an argument which depends essentially on the same point. Suppose that the members of a certain class cause an average detriment M to society, and that the mischiefs done by the several individuals differ more or less from M by amounts whose average is D, so that D is the average amount of the individual deviations, all regarded as positive, from M; then, Galton argued, the smaller D is, the stronger is the justification for taking such drastic measures against the propagation of the class as would be consonant to the feelings, if it were known that each individual member caused a detriment M. The use of such arguments seems to involve a qualification of the simple ethical doctrine that right action should make the sum of the benefits of the several individual consequences, each multiplied by its probability, a maximum.

On the other hand, the opposite view is taken in the *Port Royal Logic* and by Butler, when they argue that everything ought to be sacrificed for the hope of heaven, even if its attainment be thought infinitely improbable, since "the smallest degree of facility for the attainment of salvation is of higher value than all the blessings of the world put together."[2] The argument is, that we ought to follow a course of conduct which may with the slightest probability lead to an infinite good, until it is logically disproved that such a result of our action is impossible. The Emperor who embraced the Roman Catholic religion, not because

[1] *Opuscules mathématiques*, vol. ii.

[2] *Port Royal Logic* (Eng. trans.), p. 369: "It belongs to infinite things alone, as eternity and salvation, that they cannot be equalled by any temporal advantage; and thus we ought never to place them in the balance with any of the things of the world. This is why the smallest degree of facility for the attainment of salvation is of higher value than all the blessings of the world put together. . . ."

he believed it, but because it offered insurance against a disaster whose future occurrence, however improbable, he could not certainly disprove, may not have considered, however, whether the product of an infinitesimal probability and an infinite good might not lead to a finite or infinitesimal result. In any case the argument does not enable us to choose between different courses of conduct, unless we have reason to suppose that one path is *more* likely than another to lead to infinite good.

12. In estimating the risk, 'moral' or 'physical,' it must be remembered that we cannot necessarily apply to individual cases results drawn from the observation of a long series resembling them in some particular. I am thinking of such arguments as Buffon's when he names $\frac{1}{10,000}$ as the limit, beyond which probability is negligible, on the ground that, being the chance that a man of fifty-six *taken at random* will die within a day, it is practically disregarded by a man of fifty-six *who knows his health to be good*. "If a public lottery," Gibbon truly pointed out, "were drawn for the choice of an immediate victim, and if our name were inscribed on one of the ten thousand tickets, should we be perfectly easy?"

Bernoulli's second axiom,[1] that in reckoning a probability we must take everything into account, is easily forgotten in these cases of statistical probabilities. The statistical result is so attractive in its definiteness that it leads us to forget the more vague though more important considerations which may be, in a given particular case, within our knowledge. To a stranger the probability that I shall send a letter to the post unstamped may be derived from the statistics of the Post Office; for me those figures would have but the slightest bearing upon the question.

13. It has been pointed out already that no knowledge of probabilities, less in degree than certainty, helps us to know what conclusions are true, and that there is no direct relation between the truth of a proposition and its probability. Probability begins and ends with probability. That a scientific investigation pursued on account of its probability will generally lead to truth, rather than falsehood, is at the best only probable. The proposition that a course of action guided by the most probable considerations will generally lead to success, is not certainly true and has nothing to recommend it but its probability.

[1] See p. 76.

The importance of probability can only be derived from the judgment that it is *rational* to be guided by it in action; and a practical dependence on it can only be justified by a judgment that in action we *ought* to act to take some account of it. It is for this reason that probability is to us the "guide of life," since to us, as Locke says, "in the greatest part of our concernment, God has afforded only the Twilight, as I may so say, of Probability, suitable, I presume, to that state of Mediocrity and Probationership He has been pleased to place us in here."

PART V

THE FOUNDATIONS OF STATISTICAL INFERENCE

CHAPTER XXVII

THE NATURE OF STATISTICAL INFERENCE

1. THE Theory of Statistics, as it is now understood,[1] can be divided into two parts which are for many purposes better kept distinct. The first function of the theory is purely *descriptive.* It devises numerical and diagrammatic methods by which certain salient characteristics of large groups of phenomena can be briefly described; and it provides formulae by the aid of which we can measure or summarise the variations in some particular character which we have observed over a long series of events or instances. The second function of the theory is *inductive.* It seeks to extend its description of certain characteristics of observed events to the corresponding characteristics of other events which have not been observed. This part of the subject may be called the Theory of Statistical Inference; and it is this which is closely bound up with the theory of probability.

2. The union of these two distinct theories in a single science is natural. If, as is generally the case, the development of some inductive conclusion which shall go beyond the actually observed instances is our ultimate object, we naturally choose those modes of description, while we are engaged in our preliminary investigation, which are most capable of extension beyond the particular instances which they primarily describe. But this union is also the occasion of a great deal of confusion. The statistician, who is mainly interested in the technical methods of his science, is less concerned to discover the precise conditions in which a description can be legitimately extended by induction. He slips somewhat easily from one to the other, and having found a complete and satisfactory mode of description he

[1] See Yule, *Introduction to Statistics*, pp. 1-5, for a very interesting account of the evolution of the meaning of the term *statistics.*

may take less pains over the transitional argument, which is to permit him to use this description for the purposes of generalisation.

One or two examples will show how easy it is to slip from description into generalisation. Suppose that we have a series of similar objects one of the characteristics of which is under observation;—a number of persons, for example, whose age at death has been recorded. We note the proportion who die at each age, and plot a diagram which displays these facts graphically. We then determine by some method of curve fitting a mathematical frequency curve which passes with close approximation through the points of our diagram. If we are given the equation to this curve, the number of persons who are comprised in the statistical series, and the degree of approximation (whether to the nearest year or month) with which the actual age has been recorded, we have a very complete and succinct account of one particular characteristic of what may constitute a very large mass of individual records. In providing this comprehensive description the statistician has fulfilled his first function. But in determining the accuracy with which this frequency curve can be employed to determine the probability of death at a given age in the population at large, he must pay attention to a new class of considerations and must display a different kind of capacity. He must take account of whatever extraneous knowledge may be available regarding the sample of the population which came under observation, and of the mode and conditions of the observations themselves. Much of this may be of a vague kind, and most of it will be necessarily incapable of exact, numerical, or statistical treatment. He is faced, in fact, with the normal problems of inductive science, *one* of the data, which must be taken into account, being given in a convenient and manageable form by the methods of descriptive statistics.

Or suppose, again, that we are given, over a series of years, the marriage rate and the output of the harvest in a certain area of population. We wish to determine whether there is any apparent degree of correspondence between the variations of the two within this field of observation. It is technically difficult to measure such degree of correspondence as may appear to exist between the variations in two series, the terms of which are in some manner associated in couples,—by coincidence, in this case, of time and place. By the method of correlation tables and correlation coefficients the descriptive statistician is able to effect this object, and to present the inductive scientist with a highly significant part of his data in a compact and instructive form. But the statistician has not, in calculating these coefficients of observed correlation, covered the whole ground of which the inductive scientist must take cognisance. He has recorded the results of the observations in circumstances where they cannot be recorded so clearly without the aid of technical methods; but the precise nature of the conditions in which the observations took place and the numerous other considerations of one sort or another, of which we must take account when we wish to generalise, are not usually susceptible of numerical or statistical expression.

The truth of this is obvious; yet, not unnaturally, the more complicated and technical the preliminary statistical investigations become, the more prone inquirers are to mistake the statistical description for an inductive generalisation.[1] This tendency, which has existed in some degree, as, I think, the whole history of the subject shows, from the eighteenth century down to the present time, has been further encouraged by the terminology in ordinary use. For several statistical coefficients are given the same name when they are used for purely descriptive purposes, as when corresponding coefficients are used to measure the force or the precision of an induction. The term 'probable error,' for example, is used *both* for the purpose of supplementing and improving a statistical description, *and* for the purpose of indicating the precision of some generalisation. The term 'correlation' itself is used *both* to describe an observed characteristic of particular phenomena *and* in the enunciation of an inductive law which relates to phenomena in general.

3. I have been at pains to enforce this contrast between statistical description and statistical induction, because the chapters which follow are to be entirely about the latter, whereas nearly all statistical treatises are mainly concerned with the former. My object will be to analyse, so far as I can, the logical

[1] Cf. Whitehead, *Introduction to Mathematics*, p. 27: "There is no more common error than to assume that, because prolonged and accurate mathematical calculations have been made, the application of the result to some fact of nature is absolutely certain."

basis of statistical modes of argument. This involves a double task. To mark down those which are invalid amongst arguments having the support of authority is relatively easy. The other branch of our investigation, namely, to analyse the ground of validity in the case of those arguments the force of which all of us do in fact admit, presents the same kind of fundamental difficulties as we met with in the case of Induction.

4. The arguments with which we have to deal fall into three main classes:

(i.) Given the probability relative to certain evidence of each of a series of events, what are the probabilities, relative to the same evidence, of various proportionate frequencies of occurrence for the events over the whole series? Or more briefly, how often may we expect an event to happen over a series of occasions, given its probability on each occasion?

(ii.) Given the frequency with which an event has occurred on a series of occasions, with what probability may we expect it on a further occasion?

(iii.) Given the frequency with which an event has occurred on a series of occasions, with what frequency may we probably expect it on a further series of occasions?

In the first type of argument we seek to infer an unknown statistical frequency from an *à priori* probability. In the second type we are engaged on the inverse operation, and seek to base the calculation of a probability on an observed statistical frequency. In the third type we seek to pass from an observed statistical frequency, not merely to the probability of an individual occurrence, but to the probable values of other unknown statistical frequencies.

Each of these types of argument can be further complicated by being applied not simply to the occurrence of a simple event but to the concurrence under given conditions of two or more events. When this two or more dimensional classification replaces the one dimensional, the theory becomes what is sometimes termed Correlation, as distinguished from simple Statistical Frequency.

5. In Chapter XXVIII. I touch briefly on the observed phenomena which have given rise to the so-called Law of Great Numbers, and the discovery of which first set statistical

investigation going. In Chapter XXIX. the first type of argument, as classified above, is analysed, and the conditions which are required for its validity are stated. The crucial problem of attacking the second and third types of argument is the subject of my concluding chapters.

CHAPTER XXVIII

THE LAW OF GREAT NUMBERS

Natura quidem suas habet consuetudines, natas ex reditu causarum, sed non nisi ὡς ἐπὶ τὸ πολύ. Novi morbi inundant subinde humanum genus, quodsi ergo de mortibus quotcunque experimenta feceris, non ideo naturae rerum limites posuisti, ut pro futuro variare non possit.—LEIBNIZ *in a letter to Bernoulli, December* 3, 1703.

1. IT has always been known that, while some sets of events *invariably* happen together, other sets *generally* happen together. That experience shows one thing, while not always a sign of another, to be a usual or probable sign of it, must have been one of the earliest and most primitive forms of knowledge. If a dog is *generally* given scraps at table, that is sufficient for him to judge it reasonable to be there. But this kind of knowledge was slow to be made precise. Numerous experiments must be carefully recorded before we can know at all accurately *how* usual the association is. It would take a dog a long time to find out that he was given scraps except on fast days, and that there was the same number of these in every year.

The necessary kind of knowledge began to be accumulated during the seventeenth and eighteenth centuries by the early statisticians. Halley and others began to construct mortality tables; the proportion of the births of each sex were tabulated; and so forth. These investigations brought to light a new fact which had not been suspected previously—namely, that in certain cases of partial association the degree of association, *i.e.* the proportion of instances in which it existed, shows a very surprising regularity, and that this regularity becomes more marked the greater the number of the instances under consideration. It was found, for example, not merely that boys and girls are born on the whole in about equal proportions, but that the proportion, which is not one of complete equality, tends everywhere, when the number of recorded instances becomes large, to approximate towards a certain definite figure.

During the eighteenth century matters were not pushed much further than this, that in certain cases, of which comparatively few were known, there was this surprising regularity, increasing in degree as the instances became more numerous. Bernoulli, however, took the first step towards giving it a theoretical basis by showing that, if the *à priori* probability is known throughout, then (subject to certain conditions which he himself did not make clear) *in the long run* a certain determinate frequency of occurrence is to be expected. Süssmilch (*Die göttliche Ordnung in den Veränderungen des menschlichen Geschlechts*, 1741) discovered a theological interest in these regularities. Such ideas had become sufficiently familiar for Gibbon to characterise the results of probability as "so true in general, so fallacious in particular." Kant found in them (as many later writers have done) some bearing on the problem of Free Will.[1]

But with the nineteenth century came bolder theoretical methods and a wider knowledge of facts. After proving his extension of Bernoulli's Theorem,[2] Poisson applied it to the observed facts, and gave to the principle underlying these regularities the title of the *Law of Great Numbers.* "Les choses de toutes natures," he wrote,[3] "sont soumises à une loi universelle qu'on peut appeler la loi des grands nombres. . . . De ces exemples de toutes natures, il résulte que la loi universelle des grands nombres est déjà pour nous un fait général et incontestable, résultant d'expériences qui ne se démentent jamais." This is the language of exaggeration; it is also extremely vague. But

[1] In *Idee zu einer allgemeinen Geschichte in weltbürgerlicher Absicht*, 1784. For a discussion of this passage and for the connection between Kant and Süssmilch, see Lottin's *Quetelet*, pp. 367, 368.

[2] See p. 345.

[3] *Recherches*, pp. 7-12. Von Bortkiewicz (*Kritische Betrachtungen*, 1st part, pp. 655-660) has maintained that Poisson intended to state his principle in a less general way than that in which it has been generally taken, and that he was misunderstood by Quetelet and others. If we attend only to Poisson's contributions to *Comptes Rendus* in 1835 and 1836 and to the examples he gives there, it is possible to make out a good case for thinking that he intended his law to extend only to cases where certain strict conditions were fulfilled. But this is not the spirit of his more popular writings or of the passage quoted above. At any rate, it is the fashion, in which Poisson influenced his contemporaries, that is historically interesting; and this is certainly not represented by Von Bortkiewicz's interpretation.

it is exciting; it seems to open up a whole new field to scientific investigation; and it has had a great influence on subsequent thought. Poisson seems to claim that, in the whole field of chance and variable occurrence, there really exists, amidst the apparent disorder, a discoverable system. Constant causes are always at work and assert themselves in the long run, so that each class of event does eventually occur in a definite proportion of cases. It is not clear how far Poisson's result is due to *à priori* reasoning, and how far it is a natural law based on experience; but it is represented as displaying a certain harmony between natural law and the *à priori* reasoning of probabilities.

Poisson's conception was mainly popularised through the writings of Quetelet. In 1823 Quetelet visited Paris on an astronomical errand, where he was introduced to Laplace and came into touch with "la grande école française." "Ma jeunesse et mon zèle," he wrote in later years, "ne tardèrent pas à me mettre en rapport avec les hommes les plus distingués de cette époque; qu'on me permette de citer Fourier, Poisson, Lacroix, spécialement connus, comme Laplace, par leurs excellents écrits sur la théorie mathématique des probabilités. . . . C'est donc au milieu des savants, statisticiens, et économistes de ce temps que j'ai commencé mes travaux." [1] Shortly afterwards began his long series of papers, extending down to 1873, on the application of Probability to social statistics. He wrote a text-book on Probability in the form of letters for the instruction of the Prince Consort.

Before accepting in 1815 at the age of nineteen (with a view to a livelihood) a professorship of mathematics, Quetelet had studied as an art student and written poetry; a year later an opera, of which he was part-author, was produced at Ghent. The character of his scientific work is in keeping with these beginnings. There is scarcely any permanent, accurate contribution to knowledge which can be associated with his name. But suggestions, projects, far-reaching ideas he could both conceive and express, and he has a very fair claim, I think, to be regarded as the parent of modern statistical method.

Quetelet very much increased the number of instances of the

[1] For the details of the life of Quetelet and for a very full discussion of his writings with special reference to Probability, see Lottin's *Quetelet, statisticien et sociologue*.

Law of Great Numbers, and also brought into prominence a slightly variant type of it, of which a characteristic example is the law of height, according to which the heights of any considerable sample taken from any population tend to group themselves according to a certain well-known curve. His instances were chiefly drawn from social statistics, and many of them were of a kind well calculated to strike the imagination—the regularity of the number of suicides, "l'effrayante exactitude avec laquelle les crimes se reproduisent," and so forth. Quetelet writes with an almost religious awe of these mysterious laws, and certainly makes the mistake of treating them as being as adequate and complete in themselves as the laws of physics, and as little needing any further analysis or explanation.[1] Quetelet's sensational language may have given a considerable impetus to the collection of social statistics, but it also involved statistics in a slight element of suspicion in the minds of some who, like Comte, regarded the application of the mathematical calculus of probability to social science as "purement chimérique et, par conséquent, tout à fait vicieuse." The suspicion of quackery has not yet disappeared. Quetelet belongs, it must be admitted, to the long line of brilliant writers, not yet extinct, who have prevented Probability from becoming, in the scientific salon, perfectly respectable. There is still about it for scientists a smack of astrology, of alchemy.

The progress of the conception since the time of Quetelet has been steady and uneventful; and long strides towards this perfect respectability have been taken. Instances have been multiplied and the conditions necessary for the existence of statistical stability have been to some extent analysed. While the most fruitful applications of these methods have still been perhaps, as at first, in social statistics and in errors of observation, a number of uses for them have been discovered in quite recent times in the other sciences; and the principles of Mendelism have opened out for them a great field of application throughout biology.

[1] Compare, for instance, the following passage from *Recherches sur le penchant au crime*: "Il me semble que ce qui se rattache à l'espèce humaine, considérée en masse, est de l'ordre des faits physiques; plus le nombre des individus est grand, plus la volonté individuelle s'efface et laisse prédominer la série des faits généraux qui dépendent des causes générales. . . . Ce sont ces causes qu'il s'agit de saisir, et dès qu'on les connaîtra, on en déterminera les effets pour la société comme on détermine les effets par les causes dans les sciences physiques."

2. The existence of numerous instances of the Law of Great Numbers, or of something of the kind, is absolutely essential for the importance of Statistical Induction. Apart from this the more precise parts of statistics, the collection of facts for the prediction of future frequencies and associations, would be nearly useless. But the 'Law of Great Numbers' is not at all a good name for the principle which underlies Statistical Induction. The 'Stability of Statistical Frequencies' would be a much better name for it. The former suggests, as perhaps Poisson intended to suggest, but what is certainly false, that every class of event shows statistical regularity of occurrence if only one takes a sufficient number of instances of it. It also encourages the method of procedure, by which it is thought legitimate to take any observed degree of frequency or association, which is shown in a fairly numerous set of statistics, and to assume with insufficient investigation that, because the statistics are *numerous*, the observed degree of frequency is therefore *stable*. Observation shows that some statistical frequencies are, within narrower or wider limits, stable. But stable frequencies are not very common, and cannot be assumed lightly.

The gradual discovery, that there are certain classes of phenomena, in which, though it is impossible to predict what will happen in each individual case, there is nevertheless a regularity of occurrence if the phenomena be considered together in successive sets, gives the clue to the abstract inquiry upon which we are about to embark.

CHAPTER XXIX

THE USE OF *À PRIORI* PROBABILITIES FOR THE PREDICTION OF STATISTICAL FREQUENCY—THE THEOREMS OF BERNOULLI, POISSON, AND TCHEBYCHEFF

Hoc igitur est illud Problema, quod evulgandum hoc loco proposui, postquam jam per vicennium pressi, et cujus tum novitas, tum summa utilitas cum pari conjuncta difficultate omnibus reliquis hujus doctrinae capitibus pondus et pretium superaddere potest.—BERNOULLI.[1]

1. BERNOULLI'S Theorem is generally regarded as the central theorem of statistical probability. It embodies the first attempt to deduce the measures of statistical frequencies from the measures of individual probabilities, and it is a sufficient fruit of the twenty years which Bernoulli alleges that he spent in reaching his result, if out of it the conception first arose of general laws amongst masses of phenomena, in spite of the uncertainty of each particular case. But, as we shall see, the theorem is only valid subject to stricter qualifications, than have always been remembered, and in conditions which are the exception, not the rule.

The problem, to be discussed in this chapter, is as follows: Given a series of occasions, the probability[2] of the occurrence of a certain event at each of which is known relative to certain initial *data* h, on what proportion of these occasions may we reasonably anticipate the occurrence of the event? Given, that is to say, the individual probability of each of a series of events *à priori*, what statistical frequency of occurrence of these events is to be anticipated over the whole series? Beginning with Bernoulli's Theorem, we will consider the various solutions of this problem which have been propounded, and endeavour to

[1] *Ars Conjectandi*, p. 227.

[2] In the simplest cases, dealt with by Bernoulli, these probabilities are all supposed equal.

determine the proper limits within which each method has validity.

2. Bernoulli's Theorem in its simplest form is as follows: If the probability of an event's occurrence under certain conditions is p, then, if these conditions are present on m occasions, the most probable number of the event's occurrences is mp (or the nearest integer to this), *i.e.* the most probable *proportion* of its occurrences to the total number of occasions is p; further, the probability that the proportion of the event's occurrences will diverge from the most probable proportion p by less than a given amount b, increases as m increases, the value of this probability being calculable by a process of approximation.

The probability of the event's occurring n times and failing $m-n$ times out of the m occasions is (subject to certain conditions to be elucidated later) $p^n q^{m-n}$ multiplied by the coefficient of this expression in the expansion of $(p+q)^m$, where $p+q=1$. If we write $n=mp-h$, this term is $\frac{m!}{(mp-h)!\,(mq+h)!}p^n q^{m-n}$. It is easily shown that this is a maximum when $h=0$, *i.e.* when $n=mp$ (or the nearest integer to this, where mp is not integral). This result constitutes the first part of Bernoulli's Theorem.

For the second part of the theorem some method of approximation is required. Provided that m is large, we can simplify the expression $\frac{m!}{(mp-h)!\,(mq+h)!}p^n q^{m-n}$ by means of Stirling's Theorem, and obtain as its approximate value

$$\frac{1}{\sqrt{2\pi mpq}}e^{-\frac{h^2}{2mpq}}.$$

As before, this is a maximum when $h=0$, *i.e.* when $n=mp$.

It is possible, of course, by more complicated formulae to obtain closer approximations than this.[1] But there is an objection, which can be raised to this approximation, quite distinct from the fact that it does not furnish a result correct to as many places of decimals as it might. This is, that the approximation is independent of the sign of h, whereas the original expression is not thus independent. That is to say, the approximation implies a symmetrical distribution for different values of h about

[1] See, *e.g.*, Bowley, *Elements of Statistics*, p. 298. The objection about to be raised does not apply to these closer approximations.

the value for $h=0$; while the expression under approximation is unsymmetrical. It is easily seen that this want of symmetry is appreciable unless mpq is large. We ought, therefore, to have laid it down as a condition of our approximation, not only that m must be large, but also that mpq must be large. Unlike most of my criticisms, this is a mathematical, rather than a logical point. I recur to it in § 15.

"Par une fiction qui rendra les calculs plus faciles" (to quote Bertrand), we now replace the integer h by a continuous variable z and argue that the probability that the amount of the divergence from the most probable value mp will lie between z and $z+dz$, is

$$\frac{1}{\sqrt{2\pi mpq}}e^{-\frac{z^2}{2mpq}}\,dz.$$

This 'fiction' will do no harm so long as it is remembered that we are now dealing with a particular kind of approximation. The probability that the divergence h from the most probable value mp will be less than some given quantity a is, therefore,

$$\frac{1}{\sqrt{2\pi mpq}}\int_{-a}^{+a}e^{-\frac{z^2}{2mpq}}dz.$$

If we put $\frac{z}{\sqrt{2mpq}}=t$, this is equal to

$$\frac{2}{\sqrt{\pi}}\int_0^{\frac{a}{\sqrt{2mpq}}}e^{-t^2}dt.$$

Thus, if we write $a=\sqrt{2mpq}\,\gamma$, the probability[1] that the number of occurrences will lie between

$$mp+\sqrt{2mpq}\,\gamma \text{ and } mp-\sqrt{2mpq}\,\gamma$$

is measured by[2] $\frac{2}{\sqrt{\pi}}\int_0^{\gamma}e^{-t^2}dt.$ This same expression measures

[1] The replacement of the integer h by the continuous variable z may render the formula rather deceptive. It is certain, for example, that the error does not lie *between* h and $h+1$.

[2] The above proof follows the general lines of Bertrand's (*Calcul des probabilités*, chap. iv.). Some writers, using rather more precision, give the result as

$$\frac{2}{\sqrt{\pi}}\int_0^{\gamma}e^{-t^2}dt+\frac{e^{-\gamma^2}}{\sqrt{2\pi mpq}}$$

(*e.g.* Laplace, by the use of Euler's Theorem, and more recently Czuber, the probability that the *proportion* of occurrences will lie between

$$p+\sqrt{\frac{2pq}{m}}\gamma \text{ and } p-\sqrt{\frac{2pq}{m}}\gamma.$$

The different values of the integral $\frac{2}{\sqrt{\pi}}\int_0^t e^{-t^2}dt=\Theta(t)$ are given in tables.[1]

The probability that the proportion of occurrences will lie between given limits varies with the magnitude of $\sqrt{\frac{2pq}{m}}$, and this expression is sometimes used, therefore, to measure the 'precision' of the series. Given the *à priori* probabilities, the precision varies inversely with the *square root* of the number of instances. Thus, while the probability that the *absolute* divergence will be less than a given amount a decreases, the probability that the corresponding *proportionate* divergence (*i.e.* the absolute divergence divided by the number of instances) will be less than a given amount b, increases, as the number of instances increases. This completes the second part of Bernoulli's Theorem.

3. Bernoulli himself was not acquainted with Stirling's theorem, and his proof differs a good deal from the proof outlined in § 2. His final enunciation of the theorem is as follows: If in each of a given series of experiments there are r contingencies favourable to a given event out of a total number of contingencies t, so that $\frac{r}{t}$ is the probability of the event at each experiment, then, given any degree of probability c, it is possible to make such a number of experiments that the probability, that the proportionate number of the event's occurrences will lie between $\frac{r+1}{t}$ and $\frac{r-1}{t}$, is greater than c.[2]

Wahrscheinlichkeitsrechnung, vol. i. p. 121). As the whole formula is approximate, the simpler expression given in the text is probably not less satisfactory in practice. See also Czuber, *Entwicklung*, pp. 76, 77, and Eggenberger, *Beiträge zur Darstellung des Bernoullischen Theorems*.

[1] A list of the principal tables is given by Czuber, *loc. cit.* vol. i. p. 122.

[2] *Ars Conjectandi*, p. 236 (I have translated freely). There is a brief account of Bernoulli's proof in Todhunter's *History*, pp. 71, 72. The problem is dealt with by Laplace, *Théorie analytique*, livre ii. chap. iii. For an account of Laplace's proof see Todhunter's *History*, pp. 548-553.

4. We seem, therefore, to have proved that, if the *à priori* probability of an event under certain conditions is p, the proportion of times most probable *à priori* for the event's occurrence on a series of occasions where the conditions are satisfied is also p, and that if the series is a long one the proportion is very unlikely to differ widely from p. This amounts to the principle which Ellis [1] and Venn have employed as the defining axiom of probability, save that if the series is 'long enough' the proportion, according to them, will *certainly* be p. Laplace [2] believed that the theorem afforded a demonstration of a general law of nature, and in his second edition published in 1814 he replaces [3] the eloquent dedication, *A Napoléon-le-Grand*, which prefaces the edition of 1812, by an explanation that Bernoulli's Theorem must always bring about the eventual downfall of a great power which, drunk with the love of conquest, aspires to a universal domination,—"c'est encore un résultat du calcul des probabilités, confirmé par de nombreuses et funestes expériences."

5. Such is the famous Theorem of Bernoulli which some have believed [4] to have a universal validity and to be applicable to *all* 'properly calculated' probabilities. Yet the theorem exhibits algebraical rather than logical insight. And, for reasons about to be given, it will have to be conceded that it is only true of a special class of cases and requires conditions, before it can be legitimately applied, of which the fulfilment is rather the exception than the rule. For consider the case of a coin of which it is given that the two faces are either both heads or both tails: at every toss, provided that the results of the other tosses are unknown, the probability of heads is $\frac{1}{2}$ and the probability of tails is $\frac{1}{2}$; yet the probability of m heads and m tails in $2m$ tosses

1 *On the Foundation of the Theory of Probabilities*: "If the probability of a given event be correctly determined, the event will on a long run of trials tend to recur with frequency proportional to this probability. This is generally proved mathematically. It seems to me to be true *à priori*. . . . I have been unable to sever the judgment that one event is more likely to happen than another from the belief that in the long run it will occur more frequently."

2 *Essai philosophique*, p. 53: "On peut tirer du théorème précédent cette conséquence qui doit être regardée comme une loi générale, savoir, que les rapports des effets de la nature, sont à fort peu près constans, quand ces effets sont considérés en grand nombre."

3 Introduction, pp. liii, liv.

4 Even by Mr. Bradley, *Principles of Logic*, p. 214. After criticising Venn's view he adds: "It is false that the chances must be realised in a series. It is, however, true that they most probably will be, and true again that this probability is increased, the greater the length we give to our series."

is zero, and it is certain *à priori* that there will be either $2m$ heads or none. Clearly Bernoulli's Theorem is inapplicable to such a case. And this is but an extreme case of a normal condition.

For the first stage in the proof of the theorem assumes that, if p is the probability of one occurrence, p^r is the probability of r occurrences running. Our discussion of the theorems of multiplication will have shown how considerable an assumption this involves. It assumes that a *knowledge* of the fact that the event has occurred on every one of the first $r-1$ occasions does not in any degree affect the probability of its occurrence on the rth. Thus Bernoulli's Theorem is only valid if our initial *data* are of such a character that additional knowledge, as to the proportion of failures and successes in one part of a series of cases is altogether irrelevant to our expectation as to the proportion in another part. If, for example, the initial probability of the occurrence of an event under certain circumstances is one in a million, we may only apply Bernoulli's Theorem to evaluate our expectation over a million trials, if our original *data* are of such a character that, even after the occurrence of the event in every one of the first million trials, the probability in the light of this additional knowledge that the event will occur on the next occasion is still no more than one in a million.

Such a condition is very seldom fulfilled. If our initial probability is partly founded upon experience, it is clear that it is liable to modification in the light of further experience. It is, in fact, difficult to give a concrete instance of a case in which the conditions for the application of Bernoulli's Theorem are completely fulfilled. At the best we are dealing in practice with a good approximation, and can assert that no realised series of moderate length can much affect our initial probability. If we wish to employ the expression $\frac{2}{\sqrt{\pi}}\int_0^\gamma e^{-t^2}dt$ we are in a worse position. For this is an approximate formula which requires for its validity that the series should be *long*; whilst it is precisely in this event, as we have seen above, that the use of Bernoulli's Theorem is more than usually likely to be illegitimate.

6. The conditions, which have been described above, can be expressed precisely as follows:

Let ${}_m x_n$ represent the statement that the event has occurred on m out of n occasions and has not occurred on the others; and let ${}_1x_1/h=p$, where h represents our *à priori data*, so that p is the *à priori* probability of the event in question. Bernoulli's Theorem then requires a series of conditions, of which the following is typical: ${}_{m+1}x_{n+1}/{}_m x_n \cdot h={}_1x_1/h$, *i.e.* the probability of the event on the $n+1$th occasion must be unaffected by our knowledge of its proportionate frequency on the first n occasions, and must be exactly equal to its *à priori* probability before the first occasion.

Let us select one of these conditions for closer consideration. If y_r represents the statement that the event has occurred on each of r successive occasions, $y_r/h=y_r/y_{r-1}h \cdot y_{r-1}/h$ and so on, so that $y_r/h=\prod_{s=1}^{s=r} y_s/y_{s-1}h$. Hence if we are to have $y_r/h=p^r$, we must have $y_s/y_{s-1}h=p$ for all values of s from 1 to r. But in many particular examples $y_s/y_{s-1}h$ increases with s, so that $y_r/h>p^r$. Bernoulli's Theorem, that is to say, tends, if it is carelessly applied, to exaggerate the rate at which the probability of a given divergence from the most probable decreases as the divergence increases. If we are given a penny of which we have no reason to doubt the regularity, the probability of heads at the first toss is $\frac{1}{2}$; but if heads fall at every one of the first 999 tosses, it becomes reasonable to estimate the probability of heads at the thousandth toss at much more than $\frac{1}{2}$. For the *à priori* probability of its being a conjurer's penny, or otherwise biassed so as to fall heads almost invariably, is not usually so infinitesimally small as $(\frac{1}{2})^{1000}$. We can only apply Bernoulli's Theorem with rigour for a prediction as to the penny's behaviour over a series of a thousand tosses, if we have *à priori* such exhaustive knowledge of the penny's constitution and of the other conditions of the problem that 999 heads running would not cause us to modify in any respect our prediction *à priori*.

7. It seldom happens, therefore, that we can apply Bernoulli's Theorem with reference to a long series of natural events. For in such cases we seldom possess the exhaustive knowledge which is necessary. Even where the series is short, the perfectly rigorous application of the Theorem is not likely to be legitimate, and some degree of approximation will be involved in utilising its results.

Not so infrequently, however, artificial series can be devised in which the assumptions of Bernoulli's Theorem are relatively legitimate.[1] Given, that is to say, a proposition a_1, *some* series $a_1 a_2 \ldots$ can be found, which satisfies the conditions:

$$\text{(i.) } a_1/h = a_2/h \ldots = a_r/h.$$
$$\text{(ii.) } a_r/a_s \ldots \bar{a}_t \ldots h = a_r/h.$$

Adherents of the Frequency Theory of Probability, who use the principal conclusion of Bernoulli's Theorem as the defining property of *all* probabilities, sometimes seem to mean no more than that, relative to given evidence, every proposition belongs to *some* series, to the members of which Bernoulli's Theorem is rigorously applicable. But the *natural* series, the series, for example, in which we are most often interested, where the a's are *alike* in being accompanied by certain specified conditions c, is not, as a rule, rigorously subject to the Theorem. Thus 'the probability of a in certain conditions c is $\frac{1}{2}$' is *not* in general equivalent, as has sometimes been supposed, to 'It is 500 to 1 that in 90,000 occurrences of c, a will not occur more than 20,200 times, and 500 to 1 that it will not occur less than 19,800 times.'

8. Bernoulli's Theorem supplies the simplest formula by which we can attempt to pass from the *à priori* probabilities of each of a series of events to a prediction of the statistical frequency of their occurrence over the whole series. We have seen that Bernoulli's Theorem involves two assumptions, one (in the form in which it is usually enunciated) tacit and the other explicit. It is assumed, first, that a knowledge of what has occurred at some of the trials would not affect the probability of what may occur at any of the others; and it is assumed, secondly, that these probabilities are all *equal à priori*. It is assumed, that is to say, that the probability of the event's occurrence at the rth trial is equal *à priori* to its probability at the nth trial, and, further, that it is unaffected by a knowledge of what may actually have occurred at the nth trial.

A formula, which dispenses with the explicit assumption of equal *à priori* probabilities at every trial, was proposed by Poisson,[2] and is usually known by his name. It does *not* dispense,

[1] In the discussion in Chapter XVI., p. 170, of the probability of a divergence from an equality of heads and tails in coin-tossing, an example has been given of the construction of an artificial series in which the application of Bernoulli's Theorem is more legitimate than in the natural series.

[2] *Recherches*, pp. 246 *et seq.*

however, with the other inexplicit assumption. The difference between Poisson's Theorem and Bernoulli's is best shown by reference to the ideal case of balls drawn from an urn. The typical example for the valid application of Bernoulli's Theorem is that of balls drawn from a single urn, containing black and white balls in a known proportion, and replaced after each drawing, or of balls drawn from a series of urns, each containing black and white balls in the *same* known proportion. The typical example for Poisson's Theorem is that of balls drawn from a series of urns, each containing black and white balls in *different* known proportions.

Poisson's Theorem may be enunciated as follows:[1] Let s trials be made, and at the λth trial ($\lambda = 1, 2 \ldots s$) let the probabilities for the occurrence and non-occurrence of the event be p_λ, q_λ respectively. Then, if $\frac{\Sigma p_\lambda}{s} = p$, the probability that the number of occurrences m of the event in the s trials will lie between the limits $sp \pm l$ is given by

$$P = \frac{2}{k\sqrt{\pi s}} \int_0^l e^{-\frac{x^2}{k^2 s}} dx + \frac{e^{-\frac{l^2}{k^2 s}}}{k\sqrt{\pi s}},$$

where $k = \sqrt{\frac{2\Sigma p_\lambda q_\lambda}{s}}$.

By substituting $\frac{x}{k\sqrt{s}} = t$ and $\frac{l}{k\sqrt{s}} = \gamma$, this may be written in a form corresponding to that of Bernoulli's Theorem,[2] namely:

The probability that the number of occurrences of the event will lie between $sp \pm \gamma k\sqrt{s}$ is given by

$$P = \frac{2}{\sqrt{\pi}} \int_0^\gamma e^{-t^2} dt + \frac{e^{-\gamma^2}}{k\sqrt{\pi s}}.$$

9. This is a highly ingenious theorem and extends the application of Bernoulli's results to some important types of cases. It embraces, for example, the case in which the successive terms of a series are drawn from distinct populations known to be characterised by differing statistical frequencies; no further com-

[1] For the proof see Poisson, *Recherches*, *loc. cit.*, or Czuber, *Wahrscheinlichkeitsrechnung*, vol. i. pp. 153-159.

[2] For the analogous form of Bernoulli's Theorem see p. 339 (footnote).

plication being necessary beyond the calculation of two simple functions of these frequencies and of the number of terms in the series. But it is important not to exaggerate the degree to which Poisson's method has extended the application of Bernoulli's results. Poisson's Theorem leaves untouched all those cases in which the probabilities of some of the terms in the series of events can be influenced by a knowledge of how some of the other terms in the series have turned out.

Amongst these cases two types can be distinguished. In the first type such knowledge would lead us to discriminate between the conditions to which the different instances are subject. If, for example, balls are drawn from a bag, containing black and white balls in known proportions, and not replaced, the knowledge whether or not the first ball drawn was black affects the probability of the second ball's being black because it tells us how the conditions in which the second ball is drawn differ from those in which the first ball was drawn. In the second type such knowledge does not lead us to discriminate between the conditions to which the different instances are subject, but it leads us to modify our opinion as to the nature of the conditions which apply to all the terms alike. If, for instance, balls are drawn from a bag, which is one, but it is not certainly known which, out of a number of bags containing black and white balls in differing proportions, the knowledge of the colour of the first ball drawn affects the probabilities at the second drawing, because it throws some light upon the question as to which bag is being drawn from.

This last type is that to which most instances conform which are drawn from the real world. A knowledge of the characteristics of some members of a population may give us a clue to the general character of the population in question. Yet it is this type, where there is a change in knowledge but *no change in the material conditions* from one instance to the next, which is most frequently overlooked.[1] It will be worth while to say something further about each of these two types.[2]

[1] Numerous instances could be quoted. To take a recent English example, reference may be made to Yule, *Introduction to the Theory of Statistics*, p. 251. Mr. Yule thinks that the condition of independence is satisfied if "the *result* of any one throw or toss does not affect, and is unaffected by, the results of the preceding and following tosses," and does not allow for the cases in which *knowledge* of the result is relevant apart from any change in the physical conditions.

[2] The types which I distinguish under four heads (the Bernoullian, the

10. For problems of the first type, where there is physical or material dependence between the successive trials, it is not possible, I think, to propose any general solution; since the probabilities of the successive trials may be modified in all kinds of different ways. But for particular problems, if the conditions are precise enough, solutions can be devised. The problem, for instance, of an urn, containing black and white balls in known proportions, from which balls are drawn successively and *not replaced*,[1] is ingeniously solved by Czuber[2] with the aid of Stirling's Theorem. If σ is the number of balls and s the number of drawings, he reaches the interesting conclusion (assuming that σ, s and $\sigma - s$ are all large) that the probability of the number of black balls lying within given limits is the same as it would be if the balls were replaced after each drawing and the number of drawings were $\frac{\sigma - s}{\sigma} s$ instead of s.

In addition to the assumptions already stated, Professor Czuber's solution applies only to those cases where the limits, for which we wish to determine the probability, are narrow compared with the total number of black balls $p\sigma$. Professor Pearson[3] has worked out the same problem in a much more general manner, so as to deal with the *whole* range, *i.e.* the frequency or probability of all possible ratios of black balls, even where $s > p\sigma$. The various forms of curve, which result, according to the different relations existing between p, s, and σ, supply examples of each of the different types of frequency curve which arise out of a

---

Poissonian, and the two described above) Bachelier (*Calcul des probabilités*, p. 155) classifies as follows:

(i.) When the conditions are identical throughout, the problem has *uniformité*;

(ii.) When they vary from stage to stage, but according to a law given from the beginning and in a manner which does not depend upon what has happened at the earlier stages, it has *indépendance*;

(iii.) When they vary in a manner which depends upon what has happened at the earlier stages, it has *connexité*.

Bachelier gives solutions for each type on the assumption that the number of trials is very great, and that the number of successes or failures can be regarded as a continuous variable. This is the same kind of assumption as that made in the proof of Bernoulli's Theorem given in § 2, and is open to the same objections,—or rather the value of the results is limited in the same way.

[1] It is of no consequence whether the balls are drawn successively and not replaced, or are drawn simultaneously.

[2] *Loc. cit.* vol. i. pp. 163, 164.

[3] "Skew Variation in Homogeneous Material," *Phil. Trans.* (1895), p. 360.

classification according to (i.) skewness or symmetry, (ii.) limitation of range in one, both or neither direction; and he designates, therefore, the curves which are thus obtained as *generalised probability curves*. His discussion of the properties of these curves is interesting, however, to the student of descriptive statistics rather than to the student of probability. The most generalised and, mathematically, by far the most elegant treatment of this problem, with which I am acquainted, is due to Professor Tschuprow.[1]

Poisson, in attempting a somewhat similar problem,[2] arrives at a result, which seems obviously contrary to good sense, by a curious, but characteristic, misapprehension of the meaning of 'independence' in probability. His problem is as follows: If l balls be taken out from an urn, containing c black and white balls in known proportions, and not replaced, and if a further number of balls μ be then taken out, the probability that a given proportion $\frac{m}{m+n}$ of these μ balls will be black *is independent of the number and the colour of the l balls originally drawn out.* For, he argues, if $l+\mu$ balls are drawn out, the probability of a combination, which is made up of l black and white balls in given proportions followed by μ balls, of which m are white and n black, must be the same as that of a similar combination in which the μ balls precede the l balls. Hence the probability of m white balls in μ drawings, given that the l balls have already been drawn out, must be equal to the probability of the same result, when no balls have been previously drawn out. The reader will perceive that Poisson, thinking only of physical dependence, has been led to his paradoxical conclusion by a failure to distinguish between the cases where the proportion of black and white balls amongst the l balls originally drawn is *known* and where it is not. The *fact* of their having been drawn in certain proportions, provided that only the total number drawn is known and the proportions are *unknown*, does not influence the probability. Poisson states in his conclusion that the probability is independent of the number and colour of the l balls originally drawn. If he had added—as he ought—'provided the number of each colour is

[1] "Zur Theorie der Stabilität statistischer Reihen," p. 216, published in the *Skandinavisk Aktuarietidskrift* for 1919.

[2] *Loc. cit.* pp. 231, 232.

unknown,' the air of paradox disappears. This is an exceedingly good example of the failure to perceive that a probability cannot be influenced by the *occurrence* of a material event but only by such *knowledge*, as we may have, respecting the occurrence of the event.[1]

11. For problems of the second type, where knowledge of the result of one trial is capable of influencing the probability at the next apart from any change in the material conditions, there is, likewise, no general solution. The following artificial example, however, will illustrate the sort of considerations which are involved.

In the cases where Bernoulli's Theorem is applied to practical questions, the *à priori* probability is generally obtained empirically by reference to the statistical frequency of each alternative in past experience under apparently similar conditions. Thus the *à priori* probability of a male birth is estimated by reference to the recorded proportion of male births in the past.[2] The validity of estimating probabilities in this manner will be discussed later. But for the purposes of this example let us assume that the *à priori* probability has been calculated on this basis. Thus the *à priori* probability $p\left(=\frac{r}{s}\right)$ of an event is based on the observation of its occurrence r times out of s occasions on which the given conditions were present. Now, according to Bernoulli's Theorem directly applied, the probability of the event's occurring n times running is p^n or $\left(\frac{r}{s}\right)^n$. But, if the event occurs at the first trial, the probability at the second

1 For an attempt to solve other problems of this type see Bachelier, *Calcul des probabilités*, chap. ix. (*Probabilités connexes*). I think, however, that the solutions of this chapter are vitiated by his assuming in the course of them both that certain quantities are very large, and also, at a later stage, that the same quantities are infinitesimal. On this account, for example, his solution of the following difficult problem breaks down: Given an urn A with m white and n black balls and an urn B with m' white and n' black balls, if at each move a ball is taken from A and put into B, and at the same time a ball is taken from B and put into A, what is the probability after x moves that the urns A and B shall have a given composition?

2 Cf. Yule, *Theory of Statistics*, p. 258: "We are not able to assign an *à priori* value to the chance p (*i.e.* of a male birth) as in the case of dice-throwing, but it is quite sufficiently accurate for practical purposes to use the proportion of male births actually observed if that proportion be based on a moderately large number of observations."

becomes $\frac{r+1}{s+1}$, and so on. Hence the probability P, properly calculated, of n successive occurrences is

$$\frac{r}{s}\cdot\frac{r+1}{s+1}\cdot\frac{r+2}{s+2}\cdots\frac{r+n-1}{s+n-1}.$$

Hence

$$P=\frac{(r+n-1)!\,(s-1)!}{(s+n-1)!\,(r-1)!}$$

$$=\frac{(r+n-1)^{r+n-\frac{1}{2}}e^{-(r+n-1)}s^{s-\frac{1}{2}}e^{-(s-1)}}{(s+n-1)^{s+n-\frac{1}{2}}e^{-(s+n-1)}r^{r-\frac{1}{2}}e^{-(r-1)}}$$ by Stirling's Theorem, provided that r and s are large;

$$=\left(\frac{r}{s}\right)^n\frac{\left(1+\frac{n-1}{r}\right)^{r+n-\frac{1}{2}}}{\left(1+\frac{n-1}{s}\right)^{s+n-\frac{1}{2}}}.$$

$$=p^nQ^n,\text{ where } Q=\frac{\left(1+\frac{n-1}{r}\right)^{\frac{r-\frac{1}{2}}{n}+1}}{\left(1+\frac{n-1}{s}\right)^{\frac{s-\frac{1}{2}}{n}+1}}.$$

Thus, in this case, the assumption of Bernoulli's Theorem is approximately correct, only if Q is nearly unity. This condition is not satisfied unless n is small both compared with r and compared with s. It is very important to notice that *two* conditions are involved. Not only must the experience, upon which the *à priori* probability is based, be extensive in comparison with the number of instances to which we apply our prediction; but also the number of previous instances multiplied by the probability based upon them, *i.e.* sp $(=r)$, must be large in comparison with the number of new instances. Thus, even where the prior experience, upon which we found the initial probability P, is very extensive, we must not, if P is very small, say that the probability of n successive occurrences is approximately p^n, unless n is also small. Similarly if we wish to determine, by the methods of Bernoulli, the probability of n occurrences and m failures on $m+n$ occasions, it is necessary that we should have m and n small compared with s, n small compared with r, and m small compared with $s-r$.[1]

The case solved above is the simplest possible. The general problem is as follows: If an event has occurred x times in the first y trials, its probability at the $y+1$th is $\frac{r+x}{s+y}$; determine the *à priori* probability of the event's occurring p times in q trials. If the *à priori* probability in question is represented by $\phi(p, q)$, we have $\phi(p, q)=\frac{r+p-1}{s+q-1}\phi(p-1, q-1)+\frac{s+q-1-r-p}{s+q-1}\phi(p, q-1)$. I know of no solution of this, even approximate. But we may say that the conditions are those of supernormal dispersion as compared with Bernoulli's conditions. That is to say, the probability of a proportion differing widely from $\frac{r}{s}$ is greater than in Bernoullian conditions; for when the proportion begins to diverge it becomes more probable that it will continue to diverge in the same direction. If, on the other hand, the conditions of the problem had been such, that when the proportion begins to diverge it becomes more probable that it will recover itself and tend back towards $\frac{r}{s}$ (as when we draw balls without replacing them from a bag of known composition), we should have subnormal dispersion.[2]

12. The condition elucidated in the preceding paragraph is frequently overlooked by statisticians. The following example from Czuber[3] will be sufficient for the purpose of illustration. Czuber's argument is as follows:

In the period 1866–1877 there were registered in Austria

$$\begin{aligned} m &= 4{,}311{,}076 \text{ male births} \\ n &= 4{,}052{,}193 \text{ female births} \\ \hline s &= 8{,}363{,}269\,; \end{aligned}$$

[1] This paragraph is concerned with a different point from that dealt with in Professor Pearson's article "On the Influence of Past Experience on Future Expectation," to which it bears a superficial resemblance. Professor Pearson's article which deals, not with Bernoulli's Theorem, but with Laplace's "Rule of Succession," will be referred to in § 16 of this chapter and in § 12 of the next.

[2] Bachelier (*Calcul des probabilités*, p. 201) classifies these two kinds of conditions as *conditions accélératrices* and *conditions retardatrices*.

[3] *Loc. cit.* vol. ii. p. 15. I choose my example from Professor Czuber because he is usually so careful an exponent of theoretical statistics.

for the succeeding period, 1877–1899, we are given only

$$m' = 6{,}533{,}961 \text{ male births};$$

what conclusion can we draw as to the number n' of female births? We can conclude, according to Czuber, that the most probable value

$$n_0' = \frac{nm'}{m} = 6{,}141{,}587,$$

and that there is a probability $P = \cdot 9999779$ that n' will lie between the limits 6,118,361 and 6,164,813.

It seems in plain opposition to good sense that on such evidence we should be able with practical certainty $\left(P = \cdot 9999779 = 1 - \frac{1}{45250}\right)$ to estimate the number of female births within such narrow limits. And we see that the conditions laid down in § 11 have been flagrantly neglected. The number of cases, over which the prediction based on Bernoulli's Theorem is to extend, actually *exceeds* the number of cases upon which the *à priori* probability has been based. It may be added that for the period, 1877–1894, the actual value of n' did lie between the estimated limits, but that for the period, 1895–1905, it lay *outside* limits to which the same method had attributed practical certainty.

That Professor Czuber should have thought his own argument plausible, is to be explained, I think, by his tacitly taking account in his own mind of evidence not stated in the problem. He was relying upon the fact that there is a great mass of evidence for believing that the ratio of male to female births is peculiarly stable. But he has not brought this into the argument, and he has not used as his *à priori* probability and as his coefficient of dispersion the values which the whole mass of this evidence would have led him to adopt. Would not the argument have seemed very preposterous if m had been the number of males called George, and n the number of females called Mary? Would it not have seemed rather preposterous if m had been the number of legitimate births and n the number of illegitimate births? Clearly we must take account of other considerations than the mere numerical values of m and n in estimating our *à priori* probability. But this question belongs to the subject-matter of later chapters, and, quite apart from the manner of calculation of the *à priori* probability, the argument is invalidated by the fact than an *à priori* probability founded on 8,363,269 instances, without corroborative evidence of a non-statistical character, cannot be assumed stable through a calculation which extends over 12,700,000 instances.

13. Before we leave the theorems of Bernoulli and Poisson, it is necessary to call attention to a very remarkable theorem by Tchebycheff, from which both of the above theorems can be derived as special cases. This result is reached rigorously and without approximation, by means of simple algebra and without the aid of the differential calculus. Apart from the beauty and simplicity of the proof, the theorem is so valuable and so little known that it will be worth while to quote it in full:[1]

Let $x, y, z \ldots$ represent certain magnitudes, of which x can take the values $x_1 x_2 \ldots x_k$ with probabilities $p_1 p_2 \ldots p_k$ respectively, y the values $y_1 y_2 \ldots y_l$ with probabilities $q_1 q_2 \ldots q_l$, z the values $z_1 z_2 \ldots z_m$ with probabilities $r_1 r_2 \ldots r_m$ and so on, so that

$$\sum_1^k p = 1, \ \sum_1^l q = 1, \ \sum_1^m r = 1, \text{ etc.}$$

Write
$$\sum_1^k p_\kappa x_\kappa = a, \ \sum_1^l q_\lambda y_\lambda = b, \ \sum_1^m r_\mu z_\mu = c, \text{ etc.,}$$

and
$$\sum_1^k p_\kappa x_\kappa^2 = a_1, \ \sum_1^l q_\lambda y_\lambda^2 = b_1, \ \sum_1^m r_\mu z_\mu^2 = c, \text{ etc.,}$$

so that we can describe a as the mathematical expectation or average value of x, and a_1 as the mathematical expectation or average value of x^2, etc.

The probability that the sum $x + y + z + \ldots$ will have for its value $x_\kappa + y_\lambda + z_\mu + \ldots$ is $p_\kappa q_\lambda r_\mu \ldots$ (provided that the values of $x, y, z \ldots$ are independent). Hence

[1] From *Journ. Liouville* (2), xii., 1867, "Des valeurs moyennes," an article translated from the Russian of Tchebycheff. This proof is also quoted by Czuber, *loc. cit.* p. 212, through whom I first became acquainted with it. Most of Tchebycheff's work was published previous to 1870 and appeared originally in Russian. It was not easily accessible, therefore, until the publication at Petrograd in 1907 of the collected edition of his works in French. His theorems are, consequently, not nearly so well known as they deserve to be, although his most important theorems were reproduced from time to time in the journals of Euler and Liouville. For full references see the Bibliography.

$$\Sigma(x_\kappa + y_\lambda + z_\mu + \ldots - a - b - c - \ldots)^2 p_\kappa q_\lambda r_\mu \ldots$$

summed for all values of κ, λ, μ is the average expectation for

$$(x_\kappa + y_\lambda + z_\mu + \ldots - a - b - c - \ldots)^2.$$

Now $$\sum_1^k (x_\kappa^2 - 2ax_\kappa + a^2)p_\kappa = \Sigma p_\kappa x_\kappa^2 - 2a\Sigma p_\kappa x_\kappa + a^2 \Sigma p_\kappa$$
$$= a_1 - 2a^2 + a^2 = a_1 - a^2.$$

Also $\Sigma q_\lambda r_\mu \ldots = 1$, summed for all values of $\lambda, \mu \ldots$, and

$$\sum_1^k 2(x_\kappa - a)(y_\lambda - b)p_\kappa = \sum_1^k 2(x_\kappa y_\lambda - bx_\kappa - ay_\lambda + ab)p_\kappa$$
$$= 2(y_\lambda \Sigma p_\kappa x_\kappa - b\Sigma p_\kappa x_\kappa - ay_\lambda \Sigma p_\kappa + ab\Sigma p_\kappa)$$
$$= 2(ay_\lambda - ab - ay_\lambda + ab) = 0.$$

Therefore $\Sigma(x_\kappa + y_\lambda + z_\mu t + \ldots - a - b - c - \ldots)^2 p_\kappa q_\lambda r_\mu \ldots$
$$= a_1 + b_1 + c_1 + \ldots - a^2 - b^2 - c^2 - \ldots,$$

whence $$\frac{\Sigma(x_\kappa + y_\lambda + z_\mu + \ldots - a - b - c \ldots)^2 p_\kappa q_\lambda r_\mu \ldots}{a^2(a_1 + b_1 + c_1 + \ldots - a^2 - b^2 - c^2 - \ldots)} = \frac{1}{a^2},$$

where the summation extends over all values of $\kappa, \lambda, \mu \ldots$ and a is some arbitrary number greater than unity.

If we omit those terms of the sum on the left-hand side of the above equation for which

$$\frac{(x_\kappa + y_\lambda + z_\mu + \ldots - a - b - c - \ldots)^2}{a^2(a_1 + b_1 + c_1 + \ldots - a^2 - b^2 - c^2 - \ldots)} < 1,$$

and write unity for this expression in the remaining terms, both these processes diminish the magnitude of the left-hand side. Hence $\Sigma p_\kappa q_\lambda r_\mu \ldots < \frac{1}{a^2}$, where the summation covers those sets of values only for which

$$\frac{(x_\kappa + y_\lambda + z_\mu + \ldots - a - b - c \ldots)^2}{a^2(a_1 + b_1 + c_1 + \ldots - a^2 - b^2 - c^2 \ldots)} \geqq 1.$$

If P is the probability that

$$\frac{(x_\kappa + y_\lambda + z_\mu + \ldots - a - b - c \ldots)^2}{a^2(a_1 + b_1 + c_1 + \ldots - a^2 - b^2 - c^2 - \ldots)}$$

is equal to or less than unity, it follows that

$$1 - P < \frac{1}{a^2},$$

i.e.

$$P > 1 - \frac{1}{a^2}.$$

Hence the probability that the sum

$x_\kappa + y_\lambda + z_\mu + \ldots$ lies between the limits

$$a + b + c + \ldots - a\sqrt{a_1 + b_1 + c_1 + \ldots - a^2 - b^2 - c^2 - \ldots}$$

and $$a + b + c + \ldots + a\sqrt{a_1 + b_1 + c_1 + \ldots - a^2 - b^2 - c^2 - \ldots}$$

is greater than $1 - \frac{1}{a^2}$, where a is some number greater than unity.

This result constitutes Tchebycheff's Theorem. It may also be written in the following form:

Let n be the number of the magnitudes $x, y, z \ldots$, and write $a = \frac{\sqrt{n}}{t}$; then the probability that the arithmetic mean $\frac{x_\kappa + y_\lambda + z_\mu + \ldots}{n}$ lies between the limits

$$\frac{a + b + c + \ldots}{n} \pm \frac{1}{t}\sqrt{\frac{a_1 + b_1 + c_1 + \ldots}{n} - \frac{a^2 + b^2 + c^2 + \ldots}{n}}$$

is greater than $1 - \frac{t^2}{n}$.

It is also easy to show [1] as a deduction from Tchebycheff's Theorem that, if an amount A is won when an event of probability $p[p = 1 - q]$ occurs and an amount B lost when it fails, then in s trials the probability that the total winnings (or losses) will lie between the limits

$$s(pA - qB) \pm a(A + B)\sqrt{spq}$$

is greater than $1 - \frac{1}{a^2}$.

14. From this very general result for the probable limits of a sum composed of a number of independently varying magnitudes, Bernoulli's Theorem is easily derived. For let there be

[1] For a proof see Czuber, *loc. cit.* vol. i. p. 216.

s observations or trials, and s magnitudes $x_1 x_2 \ldots x_s$ corresponding, such that $x=1$ when the event under consideration occurs, and $x=0$ when it fails. If the probability of the event's occurrence is p, we have $a=p$, $b=p$, etc., and $a_1=p$, $b_1=p$, etc. Hence the probability P that the number of the event's occurrences will lie between the limits $sp \pm a\sqrt{sp-sp^2}$, *i.e.* between the limits $sp \pm a\sqrt{spq}$ where $q=1-p$, is $>1-\frac{1}{a^2}$. If we compare this formula with the formula for Bernoulli's Theorem already given, we find that, where this formula gives $P>1-\frac{1}{a^2}$, Bernoulli's Theorem with greater precision gives $P=\Theta\left(\frac{a}{\sqrt{2}}\right)$. The degree of superiority in the matter of precision supplied by the latter can be illustrated by the following table:

a^2.	$\Theta\frac{a}{\sqrt{2}}$.	$1-\frac{1}{a^2}$.
1·5	·7788	·333
2	·8427	·5
4·5	·9661	·7778
8	·9953	·875
12·5	·9996	·92
18	·99998	·9445

Thus when the limits are narrow and a is small, Bernoulli's formula gives a value of P very much in excess of $1-\frac{1}{a^2}$. But Bernoulli's formula involves a process of approximation which is only valid when s is large. Tchebycheff's formula involves no such process and is equally valid for all values of s. We have seen in § 11 that there are numerous cases in which for a different reason Bernoulli's formula exaggerates the results, and, therefore, Tchebycheff's more cautious limits may sometimes prove useful.

The deduction of a corresponding form of Poisson's Theorem from Tchebycheff's general formula obviously follows on similar lines. For we put[1] $a=p_1$, $b=p_2$, etc., and $a_1=p_1$, $b_1=p_2$, etc.,

[1] I am using the same notation as that used for Poisson's Theorem in § 8.

and find that the probability that the number of the event's occurrences will lie between the limits

$$\sum_1^\lambda p_\lambda \pm a\sqrt{\sum_1^\lambda p_\lambda - \sum_1^\lambda p_\lambda{}^2},$$

i.e. between the limits $sp \pm a\sqrt{\sum^\lambda p_\lambda q_\lambda}$,

i.e. between the limits $sp \pm \sqrt{2ak}\sqrt{s}$,

is greater than $t - \dfrac{1}{a^2}$.

In *Crelle's Journal*[1] Tchebycheff proves Poisson's Theorem directly by a method similar to his general method, and also obtains several supplementary results such as the following:

I. If the chances of an event E in μ consecutive trials are $p_1 p_2 \ldots p_\mu$ respectively, and their sum is s, the probability that E will occur at least m times is less than

$$\frac{1}{2(m-s)}\sqrt{\frac{m(\mu-m)}{\mu}}\left(\frac{s}{\mu}\right)^m\left(\frac{\mu-s}{\mu-m}\right)^{\mu-m+1}$$

provided that $m > s+1$;

II. and the probability that E will not occur more than n times is less than

$$\frac{1}{2(s-n)}\sqrt{\frac{\mu(\mu-n)}{\mu}}\left(\frac{\mu-s}{\mu-n}\right)^{\mu-n}\left(\frac{s}{n}\right)^{n+1}$$

provided that $n < s-1$.

III. Hence the probability that E will occur less than m times and more than n is greater than

$$1 - \frac{1}{2(m-s)}\sqrt{\frac{m(\mu-m)}{\mu}}\left(\frac{s}{m}\right)^m\left(\frac{\mu-s}{\mu-m}\right)^{\mu-m+1}$$
$$-\frac{1}{2(s-n)}\sqrt{\frac{n(\mu-n)}{\mu}}\left(\frac{s}{n}\right)^{n+1}\left(\frac{\mu-s}{\mu-n}\right)^{n-\mu}$$

provided $m > s+1$, $n < s-1$.

15. Tchebycheff's methods have been set out and his results admirably extended by A. A. Markoff.[2] And some develop-

[1] Vol. 33 (1846), *Démonstration élémentaire d'une proposition générale de la théorie des probabilités.*

[2] The reader is referred to Markoff's *Wahrscheinlichkeitsrechnung*, and particularly to p. 67, for a striking development, along mathematical lines, of

ments along the same lines by Tschuprow ("Zur Theorie der Stabilität statistischer Reihen," *Skandinavisk Aktuarietidskrift*, 1919) have convinced me that Tchebycheff's discovery is far more than a technical device for solving a special problem, and points the way to the fundamental method for attacking these questions on the mathematical side. The Laplacian mathematics, although it still holds the field in most text-books, is really obsolete, and ought to be replaced by the very beautiful work which we owe to these three Russians.

16. There is one other investigation relating to Bernoulli's Theorem which deserves remark. I have already pointed out, in § 2, that the dispersion about the most probable value, even when the conditions for the applicability of Bernoulli's Theorem in its non-approximate form are strictly fulfilled, is unsymmetrical. The fact, that the usual approximation for the probability of a divergence h from the most probable number of occurrences (the notation is that of § 2 above) takes the form $\frac{1}{\sqrt{2\pi mpq}} e^{-\frac{h^2}{2mpq}}$, which is the same for $+h$ as for $-h$, has led to this want of symmetry being very generally overlooked; and it is not uncommon to assume that the probability of a given divergence less than pm is equal to that of the same divergence in excess of pm, and, in general, that the probability of the frequency's exceeding pm in a set of m trials is *equal* to that of its falling short of pm.

That this is not strictly the case is obvious. If a die is cast 60 times, the most probable number of appearances of the ace is 10; but the ace is more likely to appear 9 times than 11 times; and much more likely (about 5 times as likely) not to appear at all than to appear exactly 20 times. That this must be so will be clear to the reader (without his requiring to trouble himself with the algebra), when he reflects that the ace cannot appear less often than not at all, whereas it may well appear more than 20 times, so that the smallness of the possible divergence in defect from the most probable value 10, as compared with the possible divergence in excess, must be made up for by the greater

---

Tchebycheff's leading idea. Further references to later memoirs, which, being in the Russian language, are inaccessible to me, will be found in the Bibliography.

frequency of any given defection as compared with the corresponding excess. Thus the actual frequency in a series of trials of an event, of which the probability at each trial is less than $\frac{1}{2}$, is likely to fall short of its most probable value more often than it exceeds it. What is in fact true is that the mathematical expectation of deficiency is equal to the mathematical expectation of excess, *i.e.* that the sum of the possible deficiencies each multiplied by its probability is equal to the sum of the possible excesses each multiplied by its probability.

The actual measurement of this want of symmetry and the determination of the conditions, in which it can be safely neglected, involves laborious mathematics, of which I am only acquainted with one direct investigation, that published in the *Proceedings of the London Mathematical Society* by Mr. T. C. Simmons.[1]

For the details of the proof I must refer the reader to Mr. Simmons's article. His principal theorem[2] is as follows: If $\frac{1}{a+1}$ is the probability of the event at each trial and $n(a+1)$ the number of trials, n and a being integers,[3] the probability that the frequency of occurrence will fall short of n is always greater than the probability that it will exceed n; the difference between the two probabilities being a maximum when $n=1$, constantly diminishing as n increases, lying always between $\frac{1}{3}\frac{a-1}{a+1}$ times the greatest term in $\left(\frac{a}{a+1}+\frac{1}{a+1}\right)^{n(a+1)}$ and $\frac{1}{3}\frac{a-1}{a+1}$ times the

1 "A New Theorem in Probability." Mr. Simmons claimed novelty for his investigation, and so far as I know this claim is justified; but recent investigations obtaining closer approximations to Bernoulli's Theorem by means of the Method of Moments are essentially directed towards the same problem.

A somewhat analogous point has, however, been raised by Professor Pearson in his article (*Phil. Mag.*, 1907) on "The Influence of Past Experience on Future Expectation." He brings out an exactly similar want of symmetry in the probabilities of the various possible frequencies about the most probable frequency, when the calculation is based, not on Bernoulli's Theorem as in Mr. Simmons's investigation, but on Laplace's rule of succession (see next chapter). The want of symmetry has also been pointed out by Professor Lexis (*Abhandlungen*, p. 120).

2 I am not giving his own enunciation of it.

3 Mr. Simmons does not seem to have been able to remove this restriction on the generality of his theorem, but there does not seem much reason to doubt that it can be removed.

greatest term in $\left(\frac{a}{a+1}+\frac{1}{a+1}\right)^{(n+1)(a+1)}$, and being approximately equal, when n is very large, to $\frac{1}{3}\frac{a-1}{\sqrt{2\pi na(a+1)}}$.

The following table gives the value of the excess Δ of the probability of a frequency less than pm over the probability of a frequency greater than pm for various values of p the probability and m the number of trials $\left[p=\frac{1}{a+1},\ m=n(a+1)\right]$, as calculated by Mr. Simmons:

p.	m.	Δ.
$\frac{1}{3}$	3	·037037
$\frac{1}{3}$	15	·02243662
$\frac{1}{3}$	24	·0182706
$\frac{1}{4}$	4	·054687
$\frac{1}{4}$	20	·03201413
$\frac{1}{10}$	10	·084777
$\frac{1}{10}$	20	·068673713
$\frac{1}{100}$	100	·101813
$\frac{1}{100}$	200	·081324387
$\frac{1}{1000}$	1000	·103454

Thus unless not only m but mp also is large the want of symmetry is likely to be appreciable. Thus it is easily found that in 100 sets of 4 trials each, where $p=\frac{1}{4}$, the actual frequency is likely to exceed the most probable 26 times and to fall short of it 31 times; and in 100 sets of 10 trials each, where $p=\frac{1}{10}$, to exceed 26 times and to fall short 34 times.

Mr. Simmons was first directed to this investigation through noticing in the examination of sets of random digits that "each digit presented itself, with unexpected frequency, *less* than $\frac{1}{10}$ of the number of times. For instance, in 100 sets of 150 digits each, I found that a digit presented itself in a set more frequently under 15 times than over 15 times; similarly in the case of 80 sets each of 250 digits, and also in other aggregations." Its possible bearing on such experiments with dice and roulette, as are described at the end of this chapter, is clear. But apart from these artificial experiments, it is sometimes worth the statistician's while to bear in mind this appreciable want of symmetry in the distribution about the mode or most probable value in many even of those cases in which Bernoullian conditions are strictly fulfilled.

17. I will conclude this chapter by an account of some of the attempts which have been made to verify *à posteriori* the conclusions of Bernoulli's Theorem. These attempts are nearly useless, first, because we can seldom be certain *à priori* that the conditions assumed in Bernoulli's Theorem are fulfilled, and, secondly, because the theorem predicts not what will happen but only what is, on certain evidence, likely to happen. Thus even where our results do not verify Bernoulli's Theorem, the theorem is not thereby discredited. The results have bearing on the conditions in which the experiments took place, rather than upon the truth of the theorem. In spite, therefore, of the not unimportant place which these attempts have in the history of probability, their scientific value is very small. I record them, because they have a good deal of historical and psychological interest, and because they satisfy a certain idle curiosity from which few students of probability are altogether free.[1]

18. The *data* for these investigations have been principally drawn from four sources—coin-tossing, the throw of dice, lotteries, and roulette; for in such cases as these the conditions for Bernoulli's Theorem seem to be fulfilled most nearly. The earliest recorded experiment was carried out by Buffon,[2] who, assisted

[1] Mr. Yule (*Introduction to Statistics*, p. 254) recommends its indulgence: "The student is strongly recommended to carry out a few series of such experiments personally, in order to acquire confidence in the use of the theory." Mr. Yule himself has indulged moderately.

[2] *Essai d'arithmétique morale* (see Bibliography), published 1777, said to have been composed about 1760.

by a child tossing a coin into the air, played 2048 *partis* of the Petersburg game, in which a coin is thrown successively until the *parti* is brought to an end by the appearance of heads. The same experiment was repeated by a young pupil of De Morgan's 'for his own satisfaction.' [1] In Buffon's trials there were 1992 tails to 2048 heads; in Mr. H.'s (De Morgan's pupil) 2044 tails to 2048 heads. A further experiment, due to Buffon's example, was carried out by Quetelet [2] in 1837. He drew 4096 balls from an urn, replacing them each time, and recorded the result at different stages, in order to show that the precision of the result tended to increase with the number of the experiments. He drew altogether 2066 white balls and 2030 black balls. Following in this same tradition is the experiment of Jevons,[3] who made 2048 throws of ten coins at a time, recording the proportion of heads at each throw and the proportion of heads altogether. In the whole number of 20,480 single throws, he obtained heads 10,353 times. More recently Weldon [4] threw twelve dice 4096 times, recording the proportion of dice at each throw which showed a number greater than three.

All these experiments, however, are thrown completely into the shade by the enormously extensive investigations of the Swiss astronomer Wolf, the earliest of which were published in 1850 and the latest in 1893.[5] In his first set of experiments Wolf completed 1000 sets of tosses with two dice, each set continuing until every one of the 21 possible combinations had occurred at least once. This involved altogether 97,899 tosses, and he then completed a total of 100,000. These *data* enabled him to work out a great number of calculations, of which Czuber quotes the following, namely a proportion of ·83533 of unlike pairs, as against the theoretical value ·83333, *i.e.* $\frac{5}{6}$. In his second set of experi-

[1] *Formal Logic*, p. 185, published 1847. De Morgan gives Buffon's results, as well as his pupil's, in full. Buffon's results are also investigated by Poisson, *Recherches*, pp. 132-135.

[2] *Letters on the Theory of Probabilities* (Eng. trans.), p. 37.

[3] *Principles of Science* (2nd ed.), p. 208.

[4] Quoted by Edgeworth, "Law of Error" (*Ency. Brit.* 10th ed.), and by Yule, *Introduction to Statistics*, p. 254.

[5] See Bibliography. Of the earliest of these investigations I have no first-hand knowledge and have relied upon the account given by Czuber, *loc. cit.* vol. i. p. 149. For a general account of empirical verifications of Bernoulli's Theorem reference may be made to Czuber, *Wahrscheinlichkeitsrechnung*, vol. i. pp. 139-152, and Czuber, *Entwicklung der Wahrscheinlichkeitstheorie*, pp. 88-91.

ments Wolf used two dice, one white and one red (in the first set the dice were indistinguishable), and completed 20,000 tosses, the details of each result being recorded in the *Vierteljahrsschrift der Naturforschenden Gesellschaft in Zürich.* He studied particularly the number of sequences with each die, and the relative frequency of each of the 36 possible combinations of the two dice. The sequences were somewhat fewer than they ought to have been, and the relative frequency of the different combinations very different indeed from what theory would predict.[1] The explanation of this is easily found; for the records of the relative frequency of each face show that the dice must have been very irregular, the six face of the white die, for example, falling 38 per cent more often than the four face of the same die. This, then, is the sole conclusion of these immensely laborious experiments,—that Wolf's dice were very ill made. Indeed the experiments could have had no bearing except upon the accuracy of his dice. But ten years later Wolf embarked upon one more series of experiments, using *four* distinguishable dice,—white, yellow, red, and blue,—and tossing this set of four 10,000 times. Wolf recorded altogether, therefore, in the course of his life 280,000 results of tossing individual dice. It is not clear that Wolf had any well-defined object in view in making these records, which are published in curious conjunction with various astronomical results, and they afford a wonderful example of the pure love of experiment and observation.[2]

19. Another series of calculations have been based upon the ready-made *data* provided by the published results of lotteries and roulette.[3]

[1] Czuber quotes the principal results (*loc. cit.* vol. i. pp. 149-151). The frequencies of only 4, instead of 18, out of the 36 combinations lay within the probable limits, and the standard deviation was 76·8 instead of 23·2.

[2] The latest experiment of the kind, of which I am aware, is that of Otto Meissner ("Würfelversuche," *Zeitschrift für Math. und Phys.* vol. 62 (1913), pp. 149-156), who recorded 24 series of 180 throws each with four distinguishable dice.

[3] For the publication of such returns there has always been a sufficient demand on the part of gamblers. An *Almanach romain sur la loterie royale de France* was published at Paris in 1830, which contained all the drawings of the French lottery (two or three a month) from 1758 to 1830. Players at Monte Carlo are provided with cards and pins with which to record the results of successive coups, and the results at the tables are regularly published in *Le Monaco.* Gamblers study these returns on account of the belief, which they usually hold, that as the number of cases is increased the *absolute* deviation from the most probable proportion becomes less, whereas at the best Bernoulli's

Czuber[1] has made calculations based on the lotteries of Prague (2854 drawings) and Brünn (2703 drawings) between the years 1754 and 1886, in which the actual results agree very well with theoretical predictions. Fechner[2] employed the lists of the ten State lotteries of Saxony between the years 1843 and 1852. Of a rather more interesting character are Professor Karl Pearson's investigations[3] into the results of Monte Carlo Roulette as recorded in *Le Monaco* in the course of eight weeks. Applying Bernoulli's Theorem, on the hypothesis of the equiprobability of all the compartments throughout the investigation, he found that the actually recorded proportions of red and black were not unexpected, but that alternations and long runs were so much in excess that, on the assumption of the exact accuracy of the tables, the *à priori* odds were at least a thousand millions to one against some of the recorded deviations. Professor Pearson concluded, therefore, that Monte Carlo Roulette is not objectively a game of chance in the sense that the tables on which it is played are absolutely devoid of bias. Here also, as in the case of Wolf's dice, the conclusion is solely relevant, not to the theory or philosophy of Chance, but to the material shapes of the tools of the experiment.

Professor Pearson's investigations into Roulette, which dealt with 33,000 Monte Carlo coups, have been overshadowed, just

---

Theorem shows that the *proportionate* deviation decreases while the absolute deviation *increases*. Cf. Houdin's *Les Tricheries des Grecs dévoilées*: "In a game of chance, the oftener the same combination has occurred in succession, the nearer we are to the certainty that it will not recur at the next cast or turn-up. This is the most elementary of the theories on probabilities; it is termed the *maturity of the chances*." Laplace (*Essai philosophique*, p. 142) quotes an amusing instance of the same belief not drawn from the annals of gambling: "J'ai vu des hommes désirant ardemment d'avoir un fils, n'apprendre qu'avec peine les naissances des garçons dans le mois où ils allaient devenir pères. S'imaginant que le rapport de ces naissances à celles des filles devait être le même à la fin de chaque mois, ils jugaient que les garçons déjà nés rendaient plus probables les naissances prochaines des filles."

The literature of gambling is very extensive, but, so far as I am acquainted with it, excessively lacking in variety, the maturity of the chances and the martingale continually recurring in one form or another. The curious reader will find tolerable accounts of such topics in Proctor's *Chance and Luck*, and Sir Hiram Maxim's *Monte Carlo Facts and Fallacies*.

[1] *Zum Gesetz der grossen Zahlen*. The results are summarised in his *Wahrscheinlichkeitsrechnung*, vol. i. p. 139.

[2] *Kollektivmasslehre*, p. 229. These results also are summarised by Czuber, *loc. cit.*

[3] *The Chances of Death*, vol. i.

as all other tosses of coins and dice have been outdone by Wolf, by Dr. Karl Marbe,[1] who has examined 80,000 coups from Monte Carlo and elsewhere. Dr. Marbe arrived at exactly opposite conclusions; for he claims to have shown that long runs, so far from being in excess, were greatly in defect. Dr. Marbe introduces this experimental result in support of his thesis that the world is so constituted that long runs do not as a matter of fact occur in it.[2] Not merely are long runs very improbable. They do not, according to him, occur at all. But we may doubt whether roulette can tell us very much either of the laws of logic or of the constitution of the universe.

Dr. Marbe's main thesis is identical, as he himself recognises, with one of the heterodox contentions of D'Alembert.[3] But this principle of variety, precisely opposite to the usual principle of Induction, can have no claim to be accepted *à priori* and, as a *general* principle, there is no adequate evidence to support it from experience. Its origin is to be found, perhaps, in the fact that

[1] *Naturphilosophische Untersuchungen zur Wahrscheinlichkeitstheorie.*

[2] Dr. Marbe's monograph has given rise in Germany to a good deal of discussion, not directed towards showing what a preposterous method this is for demonstrating a natural law, but because the experimental result itself does not really follow from the *data* and is due to a somewhat subtle error in Marbe's reasoning, by which he has been led into an incorrect calculation of the probable proportions *à priori* of the various sequences. The problem is discussed by Von Bortkiewicz, Brömse, Bruns, Grimsehl, and Grünbaum (for exact references to these see the Bibliography), and by Lexis (*Abhandlungen*, pp. 222-226) and Czuber (*Wahrscheinlichkeitsrechnung*, vol. i. pp. 144-149). Largely as a result of this controversy, Von Bortkiewicz has lately devoted a complete treatise (*Die Iterationen*) to the mathematics of 'runs.' Dr. Marbe has been given far more attention by his colleagues in Germany than he conceivably deserves.

[3] D'Alembert's principal contributions to Probability are most accessible in the volumes of his *Opuscules mathématiques* (1761). Works on Probability usually contain some reference to D'Alembert, but his sceptical opinions, rejected rather than answered by the orthodox school of Laplace, have not always received full justice. D'Alembert has three main contentions to which in his various papers he constantly recurs:

(1) That a probability very small mathematically is really zero;

(2) That the probabilities of two successive throws with a die are not independent;

(3) That 'mathematical expectation' is not properly measured by the product of the probability and the prize.

The first and third of these were partly advanced in explanation of the Petersburg paradox (see p. 316). The second is connected with the first, and was also used to support his incorrect evaluation of the probability of heads twice running; but D'Alembert, in spite of many of his results being wrong, does not altogether deserve the ridicule which he has suffered at the hands of writers, who accepted without sceptical doubts the hardly less incorrect conclusions of the orthodox theory of that time.

in a certain class of cases, especially where conscious human agency comes in, it may contain some element of truth. The fact of an act's having been done in a particular way once may be a special reason for thinking that it will not be performed on the next occasion in precisely the same manner. Thus in many so-called random events some slight degree of causal and material dependence between successive occurrences may, nevertheless, exist. In these cases 'runs' may be fewer and shorter than those which we should predict, if a complete absence of such dependence is assumed. If, for example, a pack of cards be dealt, collected, and shuffled, to the extent that card-players do as a rule shuffle, there may be a greater presumption against the second hand's being identical with the first than against any other particular distribution. In the case of croupiers long experience might possibly suggest some psychological generalisation,—that they are very mechanical, giving an excess of numbers belonging to a particular section of the wheel, or, on the other hand, that when a croupier sees a run beginning, he tends to vary his spin more than usual, thus bringing runs to an end sooner than he ought.[1] At any rate, it is worth emphasising once more that from such experiments as these this is the only *kind* of knowledge which we can hope to obtain,—knowledge of the material construction of a die or of the psychology of a croupier.

[1] A good roulette table is, however, so delicate an instrument that no probable degree of regularity of habit on the part of the spinner could be sufficient to produce regularity in the result.

CHAPTER XXX

THE MATHEMATICAL USE OF STATISTICAL FREQUENCIES FOR THE DETERMINATION OF PROBABILITY *À POSTERIORI*—THE METHODS OF LAPLACE

Utilissima est aestimatio probabilitatum, quanquam in exemplis juridicis politicisque plerumque non tam subtili calculo opus est, quam accurata omnium circumstantiarum enumeratione.—LEIBNIZ.

1. IN the preceding chapter we have assumed that the probability of an event at each of a series of trials is given, and have considered how to infer from this the probabilities of the various possible frequencies of the event over the whole series, without discussing in detail by what method the initial probability had been determined. In statistical inquiries it is generally the case that this initial probability is based, not upon the Principle of Indifference, but upon the statistical frequencies of similar events which have been observed previously. In this chapter, therefore, we must commence the complementary part of our inquiry,—namely, into the method of deriving a measure of probability from an observed statistical frequency.

I do not myself believe that there is any direct and simple method by which we can make the transition from an observed numerical frequency to a numerical measure of probability. The problem, as I view it, is part of the general problem of founding judgments of probability upon experience, and can only be dealt with by the general methods of induction expounded in Part III. The nature of the problem precludes any other method, and direct mathematical devices can all be shown to depend upon insupportable assumptions. In the next chapters we will consider the applicability of general inductive methods to this problem, and in this we will endeavour to discredit the mathematical charlatanry by which, for a hundred years past, the basis of theoretical statistics has been greatly undermined.

2. Two direct methods have been commonly employed, theoretically inconsistent with one another, though not in every case noticeably discrepant in practice. The first and simplest of these may be termed the Inversion of Bernoulli's Theorem, and the other Laplace's Rule of Succession.

The earliest discussion of this problem is to be found in the Correspondence of Leibniz and Jac. Bernoulli,[1] and its true nature cannot be better indicated than by some account of the manner in which it presented itself to these very illustrious philosophers. The problem is tentatively proposed by Bernoulli in a letter addressed to Leibniz in the year 1703. We can determine from *à priori* considerations, he points out, by how much it is more probable that we shall throw 7 rather than 8 with two dice, but we cannot determine by such means the probability that a young man of twenty will outlive an old man of sixty. Yet is it not possible that we might obtain this knowledge *à posteriori* from the observation of a great number of similar couples, each consisting of an old man and a young man ? Suppose that the young man was the survivor in 1000 cases and the old man in 500 cases, might we not conclude that the young man is twice as likely as the old man to be the survivor ? For the most ignorant persons seem to reason in this way by a sort of natural instinct, and feel that the risk of error is diminished as the number of observations is increased. Might not the solution tend asymptotically to some determinate degree of probability with the increase of observations ? *Nescio, Vir Amplissime, an speculationibus istis soliditatis aliquid inesse Tibi videatur.*

Leibniz's reply goes to the root of the difficulty. The calculation of probabilities is of the utmost value, he says, but in statistical inquiries there is need not so much of mathematical subtlety as of a precise statement of all the circumstances. The possible contingencies are too numerous to be covered by a finite number of experiments, and exact calculation is, therefore, out of the question. Although nature has her habits, due to the recurrence of causes, they are general, not invariable. Yet empirical calculation, although it is inexact, may be adequate in affairs of practice.[2]

[1] For the exact references see Bibliography.

[2] Leibniz's actual expressions (in a letter to Bernoulli, December 3, 1703) are as follows: Utilissima est aestimatio probabilitatum, quanquam in exemplis juridicis politicisque plerumque non tam subtili calculo opus est, quam accurata omnium circumstantiarum enumeratione. Cum empirice aestimamus proba-

Bernoulli in his answer fell back upon the analogy of balls drawn from an urn, and maintained that without estimating each separate contingency we might determine within narrow limits the *proportion* favouring each alternative. If the true proportion were 2 : 1, we might estimate it with moral certainty *à posteriori* as lying between 201 : 100 and 199 : 100. "Certus sum," he concluded the controversy, "Tibi placituram demonstrationem, cum publicavero." But whether he was impressed by the just caution of Leibniz, or whether death intercepted him, he advances matters no further in the *Ars Conjectandi*. After dealing with some of Leibniz's objections[1] and seeming to promise some mode of estimating probabilities *à posteriori* by an inversion of his theorem, he proves the direct theorem only and the book is suddenly at an end.

3. In dealing with the correspondence of Leibniz and Bernoulli, I have not been mainly influenced by the historical interest of it. The view of Leibniz, dwelling mainly on considerations of analogy, and demanding "not so much mathematical subtlety as a precise statement of all the circumstances," is, substantially, the view which will be supported in the following chapters. The desire of Bernoulli for an exact formula, which would derive from the numerical frequency of the experimental results a numerical measure of their probability, preludes the exact formulas of later and less cautious mathematicians, which will be examined immediately.

4. During the greater part of the eighteenth century there is no trace, I think, of the explicit use of the Inversion of Bernoulli's Theorem. The investigations carried out by D'Alembert, Daniel Bernoulli, and others relied upon the type of argument examined in Chapter XXV. They showed, that is to say, that certain observed series of events would have been very improbable, if we had supposed independence between some two factors or if

---

bilitates per experimenta successuum, quaeris an ea via tandem aestimatio perfecte obtineri possit. Idque a Te repertum scribis. Difficultas in eo mihi inesse videtur, quod contingentia seu quae infinitis pendent circumstantiis, per finita experimenta determinari non possunt; natura quidem suas habet consuetudines, natas ex reditu causarum, sed non nisi ὡς ἐπὶ τὸ πολύ. Novi morbi inundant subinde humanum genus, quodsi ergo de mortibus quotcunque experimenta feceris, non ideo naturae rerum limites posuisti, ut pro futuro variare non possit. Etsi autem empirice non posset haberi perfecta aestimatio, non ideo minus empirica aestimatio in praxi utilis et sufficiens foret.

[1] The relevant passages are on pp. 224-227 of the *Ars Conjectandi*.

some occurrence had been assumed to be as likely as not, and they inferred from this that there was in fact a measure of dependence or that the occurrence had probability in its favour. But they did not endeavour to pass from the observed frequency of occurrence to an exact measure of the probability. With the advent of Laplace more ambitious methods took the field.

Laplace began by assuming without proof a direct inversion of Bernoulli's Theorem. Bernoulli's Theorem, in the form in which Laplace proved it, states that, if p is the probability *à priori*, there is a probability P that the proportion of times $\frac{m}{m+n}$ of the event's occurrence in $\mu\,(=m+n)$ trials will lie between $p \pm \gamma\sqrt{\frac{2pq}{\mu}}$, where $P = \frac{2}{\sqrt{\pi}}\int_0^\gamma e^{-t^2}dt + \frac{1}{\sqrt{2\pi\mu pq}}e^{-\gamma^2}$. The inversion of the theorem, which he assumes without proof, states that, if the event is observed to happen m times in μ trials, there is a probability P that the probability of the event p will lie between $\frac{m}{\mu} \pm \gamma\sqrt{\frac{2mn}{\mu^3}}$, where $P = \frac{2}{\sqrt{\pi}}\int_0^\gamma e^{-t^2}dt + \frac{1}{\sqrt{2\pi\mu\frac{m}{\mu^2}}}e^{-\gamma^2}$. The same result is also given by Poisson.[1] Thus, given the frequency of occurrence in μ trials, these writers infer the probability of occurrence at subsequent trials within certain limits, just as, given the *à priori* probability, Bernoulli's Theorem would enable them to predict the frequency of occurrence in μ trials within corresponding limits.

[1] For an account of the treatments of this topic both by Laplace and by Poisson, see Todhunter's *History*, pp. 554-557. Both of them also obtain a formula slightly different from that given above by a method analogous to the first part of the proof of Laplace's Rule of Succession; *i.e.* by an application of the inverse principle of probability to the assumption that the probability of the probability's lying within any interval is proportional to the length of the interval. This discrepancy has given rise to some discussion. See Todhunter, *loc. cit.*; De Morgan, *On a Question in the Theory of Probabilities*; Monro, *On the Inversion of Bernoulli's Theorem in Probabilities*; and Czuber, *Entwicklung*, pp. 83, 84. But this is not the important distinction between the two mathematical methods by which this question has been approached, and this minor point, which is of historical interest mainly, I forbear to enter into.

If the number of trials is at all numerous, these limits are narrow and the purport of the inversion of Bernoulli's Theorem may therefore be put briefly as follows. By the direct theorem, if p measures the probability, p also measures the most probable value of the frequency; by the inversion of the theorem, if $\frac{m}{m+n}$ measures the frequency, $\frac{m}{m+n}$ also measures the most probable value of the probability. The simplicity of the process has recommended it, since the time of Laplace, to a great number of writers. Czuber's argument, criticised on p. 351, with reference to the proportions of male and female births in Austria, is based upon an unqualified use of it. But examples abound throughout the literature of the subject, in which the theorem is employed in circumstances of greater or less validity.

The theorem was originally given without proof, and is indeed incapable of it, unless some illegitimate assumption has been introduced. But, apart from this, there are some obvious objections. We have seen in the preceding chapter that Bernoulli's Theorem itself cannot be applied to all kinds of *data* indiscriminately, but only when certain rather stringent conditions are fulfilled. Corresponding conditions are required equally for the inversion of the theorem, and it cannot possibly be inferred from a statement of the number of trials and the frequency of occurrence merely, that these have been satisfied. We must know, for instance, that the examined instances are similar in the main relevant particulars, both to one another and to the unexamined instances to which we intend our conclusion to be applicable. An unanalysed statement of frequency cannot tell us this.

This method of passing from statistical frequencies to probabilities is not, however, like the method to be discussed in a moment, radically false. With due qualifications it has its place in the solution of this problem. The conditions in which an inversion of Bernoulli's Theorem is legitimate will be elucidated in Chapter XXXI. In the meantime we will pass on to Laplace's second method, which is more powerful than the first and has obtained a wider currency. The more extreme applications of it are no longer ventured upon, but the theory which underlies it is still widely adopted, especially by French writers upon probability, and seldom repudiated.

5. The formula in question, which Venn [1] has called the *Rule of Succession*, declares that, if we know no more than that an event has occurred m times and failed n times under given conditions, then the probability of its occurrence when those conditions are next fulfilled is $\frac{m+1}{m+n+2}$. It is necessary, however, before we examine the proof of this formula, to discuss in detail the reasoning which leads up to it.

This preliminary reasoning involves the Laplacian theory of 'unknown probabilities.' The postulate, upon which it depends, is introduced to supplement the Principle of Indifference, and is in fact the extension of this principle from the probabilities of arguments, when we know nothing about the arguments, to the probabilities that the probabilities of arguments have certain values, when we know nothing about the probabilities. Laplace's enunciation is as follows: "Quand la probabilité d'un événement simple est inconnue, on peut lui supposer également toutes les valeurs depuis zéro jusqu'à l'unité. La probabilité de chacune de ces hypothèses tirée de l'événement observé est . . . une fraction dont le numérateur est la probabilité de l'événement dans cette hypothèse, et dont le dénominateur est la somme des probabilités semblables relatives à toutes les hypothèses. . . ." [2]

Thus when the probability of an event is unknown, we may suppose all possible values of the probability between 0 and 1 to be equally likely *à priori*. The probability, *after* the event has occurred, that the probability *à priori* was $\frac{1}{r}$ (say), is measured by a fraction of which $\frac{1}{r}$ is the numerator and the sums of all the possible *à priori* values the denominator. The origin of this rule is evident. If we consider the problem in which a ball is drawn from a bag containing an infinite number of black and white balls in unknown proportions, we have hypotheses, corresponding to each of the possible constitutions of the bag, the assumption of which yields in turn every value between 0 and 1 as the *à priori* probability of drawing a white ball. If we could assume that these constitutions are equally probable *à priori*, we should obtain probabilities for each of them *à posteriori* according to Laplace's rule.

[1] *Logic of Chance*, p. 190.

[2] *Essai philosophique*, p. 16.

On the analogy of this Laplace assumes in general that, where everything is unknown, we may suppose an infinite number of possibilities, each of which is equally likely, and each of which leads to the event in question with a *different* degree of probability, so that for every value between 0 and 1 there is one and only one hypothetical constitution of things, the assumption of which invests the event with a probability of that value.

6. It might be an almost sufficient criticism of the above to point out that these assumptions are entirely baseless. But the theory has taken so important a place in the development of probability that it deserves a detailed treatment.

What, in the first place, does Laplace mean by an *unknown* probability? He does not mean a probability, whose value is in fact unknown *to us*, because we are unable to draw conclusions which *could be drawn* from the *data*; and he seems to apply the term to any probability whose value, according to the argument of Chapter III., is numerically indeterminate. Thus he assumes that *every* probability has a numerical value and that, in those cases where there seems to be no numerical value, this value is not non-existent but unknown; and he proceeds to argue that where the numerical value is unknown, or as I should say where there is *no* such value, every value between 0 and 1 is equally probable. With the possible interpretations of the term 'unknown probability,' and with the theory that every probability can be measured by one of the real numbers between 0 and 1, I have dealt, as carefully as I can, in Chapter III. If the view taken there is correct, Laplace's theory breaks down immediately. But even if we were to answer these questions, not as they have been answered in Chapter III., but in a manner favourable to Laplace's theory, it remains doubtful whether we could legitimately attribute a value to the probability of an unknown probability's having such and such a value. If a probability is unknown, surely the probability, relative to the same evidence, that this probability has a given value, is also unknown; and we are involved in an infinite regress.

7. This point leads on to the second objection; Laplace's theory requires the employment of both of two inconsistent methods. Let us consider a number of alternatives a_1, a_2, etc., having probabilities p_1, p_2, etc.; if we do not know anything about a_1, we do not know the value of its probability p_1, and we

must consider the various possible values of p_1, namely b_1, b_2, etc., the probabilities of these possible values being q_1, q_2, etc. respectively. There is no reason why this process should ever stop. For as we do not know anything about b_1, we do not know the value of its probability q_1, and we must consider the various possible values of q_1, namely c_1, c_2, etc., the probabilities of these possible values being r_1, r_2, etc. respectively; and so on. This method consists in supposing that, when we do not know anything about an alternative, we must consider all the possible values of the probability of the alternative; these possible values can form in their turn a set of alternatives, and so on. But this method *by itself* can lead to no final conclusion. Laplace superimposes on it, therefore, his other method of determining the probabilities of alternatives about which we know nothing,—namely, the Principle of Indifference. According to this method, when we know nothing about a set of alternatives, we suppose the probabilities of each of them to be *equal*. In some parts of his writings—and this is true also of most of his followers—he applies this method from the beginning. If, that is to say, we know nothing about a_1, since a_1 and its contradictory form a pair of exhaustive alternatives two in number, the probability of these alternatives is *equal* and each is $\frac{1}{2}$. But in the reasoning which leads up to the Law of Succession he chooses to apply this method at the second stage, having used the other method at the first stage. If, that is to say, we know nothing about a_1, its probability p_1 may have any of the values b_1, b_2, etc. where b_1 is any fraction between 0 and 1; and, as we know nothing about the probabilities q_1, q_2, etc. of these alternatives b_1, b_2, etc., we may by the Principle of Indifference suppose them to be *equal*. This account may seem rather confused; but it is not easy to give a lucid account of so confused a doctrine.

8. Turning aside from these considerations, let us examine the theory, for a moment, from another side. When we reach the Rule of Succession, it will be seen that the hypothetical *à priori* probabilities are treated as if they were possible *causes* of the event. It is assumed, that is to say, that the number of possible sets of antecedent conditions is proportional to the number of real numbers between 0 and 1; and that these fall into equal groups, each group corresponding to one of the real numbers

between 0 and 1, this number measuring the degree of probability with which we could predict the event, if we knew that an antecedent condition belonging to that group was fulfilled. It is then assumed that all of these possible antecedent conditions are *à priori* equally likely. The argument has arisen by false analogy from the problem in which a ball is drawn from an urn containing an infinite number of black and white balls. But for the assumption that we have *in general* the kind of knowledge which is necessary about the possible antecedents, no reasonable foundation has been suggested.

De Morgan endeavoured to deal with the difficulty in much the same way in the following passage:[1] "In determining the chance which exists (under known circumstances) for the happening of an event a number of times which lies between certain limits, we are involved in a consideration of some difficulty, namely, the *probability of a probability*, or, as we have called it, the presumption of a probability. To make this idea more clear, remember that any state of probability may be immediately made the expression of the result of a set of circumstances, which being introduced into the question, the difficulty disappears. The word presumption refers distinctly to an act of the mind, or a state of the mind, while in the word probability we feel disposed rather to think of the external arrangements on the knowledge of which the strength of our presumption ought to depend, than of the presumption itself." The point of this explanation lies in the assumption that "any state of probability may be immediately made the expression of the result of a set of circumstances." It cannot be allowed that this is generally true;[2] and even in those cases in which it is true we are thrown back on the *à priori* probabilities of the various sets of circumstances which need not be, as De Morgan assumes, either equal or exhaustive alternatives.

9. The proof of the Rule of Succession, which is based upon this theory of unknown probabilities, is, briefly, as follows:

If x stands for the *à priori* probability of an event in given conditions, then the probability that the event will occur m times and fail n times in these conditions is $x^m(1-x)^n$. If, however, x is unknown, all values of it between 0 and

[1] *Cabinet Encyclopaedia*, p. 87.

[2] For instance, it is not true even in the standard instance of balls drawn from an urn containing black and white in unknown proportions, unless the number of balls is infinite.

1 are *à priori* equally probable. It follows from these two sets of considerations that, if the event has been observed to occur m times out of $m+n$, the probability *à posteriori* that x lies between x and $x+dx$ is proportional to $x^m(1-x)^n dx$, and is equal, therefore, to $Ax^m(1-x)^n dx$ where A is a constant. Since the event has in fact occurred, and since x must have one of its possible values, A is determined by the equation

$$\int_0 Ax^m(1-x)^n dx = 1 \quad \therefore A = \frac{\Gamma(m+n+2)}{\Gamma(m+1)\Gamma(n+1)}.$$

Hence the probability that the event will occur at the $(m+n+1)$th trial, when we know that it has occurred m times in $m+n$ trials, is

$$A\int_0^1 x^{m+1}(1-x)^n dx.$$

If we substitute the value of A found above, this is equal to $\dfrac{m+1}{m+n+2}$.[1]

The class of problem to which the theorem is supposed to apply is the following: There are certain conditions such that we are ignorant *à priori* as to whether they do or do not lead to the occurrence of a particular event; on m out of $m+n$ occasions, however, on which these conditions have been observed, the event has occurred; what is the probability in the light of this experience that the event will occur on the next occasion? The answer to all such problems is $\dfrac{m+1}{m+n+2}$. In the cases where $n=0$, *i.e.* when the event has invariably occurred, the formula

[1] The theorem is sometimes enunciated by contemporary writers in a much more guarded form, *e.g.* by Czuber, *Wahrscheinlichkeitsrechnung*, vol. i. p. 197, and by Bachelier, *Calcul des probabilités*, p. 487. Bachelier, instead of assuming that the *à priori* probabilities of all possible values of the probability of the event are equal, writes $\hat{\omega}(y)dy$ as the *à priori* probability that the probability is y, so that after m occurrences is $m+n$ trials the probability that the probability lies between y and $y+dy$ is $\dfrac{y^m(1-y)^n\hat{\omega}(y)dy}{\int y^m(1-y)^n\hat{\omega}(y)dy}$. If one has no idea of $\hat{\omega}$ *à priori*, he suggests that the simplest hypothesis is to put $\hat{\omega}=1$, which leads, as above, to Laplace's Law of Succession. He also proposes the hypothesis $\hat{\omega}(y)=a+a_1y+a_2y^2+\ldots$, in which case the denominator is a series of Eulerian integrals. There is a discussion of the Law of Succession, and of the contradictions and paradoxes to which it leads, by E. T. Whittaker and others in Part VI. vol. viii. (1920) of the *Transactions of the Faculty of Actuaries in Scotland*.

yields the result $\frac{m+1}{m+2}$. In the case where the conditions have been observed once only and the event has occurred on that occasion, the result is $\frac{2}{3}$. If the conditions have *never* been met with at all, the probability of the event is $\frac{1}{2}$. And even in the case where on the only occasion on which the conditions were observed, the event did *not* occur, the probability is $\frac{1}{3}$.

Some of the flaws in this proof have been already explained. One minor objection may be pointed out in addition. It is assumed that, if x is the *à priori* probability of the event's happening once, then x^n is the *à priori* probability of its happening n times in succession, whereas by the theorem's own showing the knowledge that the event has happened once modifies the probability of its happening a second time; its successive occurrences are not, therefore, independent. If the *à priori* probability of the event is $\frac{1}{2}$, and if, after it has been observed once, the probability that it will occur a second time is $\frac{2}{3}$, then it follows that the *à priori* probability of its occurring twice is not $\frac{1}{2} \times \frac{1}{2}$, but $\frac{1}{2} \times \frac{2}{3}$, *i.e.* $\frac{1}{3}$; and in general the *à priori* probability of its happening n times in succession is not $\left(\frac{1}{2}\right)^n$ but $\frac{1}{n+1}$.

10. But refinements of disproof are hardly needed. The principle's conclusion is inconsistent with its premisses. We begin with the assumption that the *à priori* probability of an event, about which we have no information and no experience, is unknown, and that all values between 0 and 1 are equally probable. We end with the conclusion that the *à priori* probability of such an event is $\frac{1}{2}$. It has been pointed out in § 7 that this contradiction was latent, as soon as the Principle of Indifference was superimposed on the principle of unknown probabilities.

The theorem's conclusions, moreover, are a *reductio ad absurdum* of the reasoning upon which it is based. Who could suppose that the probability of a purely hypothetical event, of whatever complexity, in favour of which no positive argument exists, the like of which has *never* been observed, and which has failed to occur on the one occasion on which the hypothetical conditions were fulfilled, is no less than $\frac{1}{3}$? Or if we do suppose it, we are involved in contradictions,—for it is easy to imagine more than three *incompatible* events which satisfy these conditions.

11. The theorem was first suggested by the problem of the urn which contains black and white balls in unknown proportions: *m* white and *n* black balls have been successively drawn and replaced; what is the probability that the next draw will yield a white ball? It is supposed that all compositions of the urn are equally probable, and the proof then proceeds precisely as in the case of the more general rule of succession. The rule of succession has been, sometimes, directly deduced from the case of the urn, by assimilating the occurrence of the event to the drawing of a white ball and its non-occurrence to the drawing of a black ball.

On the hypothesis that all compositions of the urn are equally probable, an hypothesis to which in general there is nothing corresponding, and on the further hypothesis that the number of balls is infinite, this solution is correct.[1] But the rule of succession does not apply, as it is easy to demonstrate, even to the case of balls drawn from an urn, if the number of balls is finite.[2]

12. If the Rule of Succession is to be adopted by adherents of the Frequency Theory of Probability,[3] it is necessary that they should make some modification in the preliminary reasoning on which it is based. By Dr. Venn, however, the rule has been

[1] This second condition is often omitted (*e.g.* Bertrand, *Calcul des probabilités*, p. 172).

[2] The correct solution for the case of a finite number of balls, on the hypothesis that each possible ratio is equally likely, is as follows: The probability of a black ball at a further trial, after black balls have been successively withdrawn and replaced p times, is $\frac{1}{n}\frac{s_{p+1}}{s_p}$ where there are n balls and s_r represents the sum of the rth powers of the first n natural numbers. This reduces to $\frac{p+1}{p+2}$,—the solution usually given,—when n is infinite. More generally, if p black balls and q white balls have been drawn and replaced, the chance that the next ball will be black is $\frac{1}{n}\frac{\sum_{r=o}^{r=n} r^{p+1}(n-r)^b}{\sum_{r=o}^{r=n} r^p(n-r)^b}$.

[3] See Chapter VIII.

explicitly rejected on the ground that it does not accord with experience.[1] But Professor Karl Pearson, who accepts it, has made the necessary restatement,[2] and it will be worth while to examine the reasoning when it is put in this form. Professor Pearson's proof of the Rule of Succession is as follows:

"I start, as most mathematical writers have done, with 'the equal distribution of ignorance,' or I assume the truth of Bayes' Theorem. I hold this theorem not as rigidly demonstrated, but I think with Edgeworth [3] that the hypothesis of the equal distribution of ignorance is, within the limits of practical life, justified by our experience of statistical ratios, which *à priori* are unknown, *i.e.* such ratios do not tend to cluster markedly round any particular value. 'Chances' lie between 0 and 1, but our experience does not indicate any tendency of actual chances to cluster round any particular value in this range. The ultimate basis of the theory of statistics is thus not mathematical but observational. Those who do not accept the hypothesis of the equal distribution of ignorance and its justification in observation are compelled to produce definite evidence of the clustering of chances, or to drop all application of past experience to the judgment of probable future statistical ratios. . . .

"Let the chance of a given event occurring be supposed to lie between x and $x+dx$, then if on $n=p+q$ trials an event has been observed to occur p times and fail q times, the probability that the true chance lies between x and $x+dx$ is, on the equal distribution of our ignorance,

$$P_x = \frac{x^p(1-x)^q dx}{\int_0^1 x^p(1-x)^q dx}.$$

"This is Bayes' Theorem. . . .[4]

[1] *Logic of Chance*, p. 197.

[2] "On the Influence of Past Experience on Future Experience on Future Expectation," *Phil. Mag.*, 1907, pp. 365-378. The quotations given below are taken from this article.

[3] This reference is, no doubt, to Edgeworth's "Philosophy of Chance" (*Mind*, 1884, p. 230), when he wrote: "The assumption that any probability-constant about which we know nothing in particular is as likely to have one value as another is grounded upon the rough but solid experience that such constants do, as a matter of fact, as often have one value as another." See also Chapter VII. § 6, above.

[4] Professor Pearson's use of this title for the above formula is not, I think, historically correct. Bayes' Theorem is the Inverse Principle of Probability itself, and not this extension of it.

"Now suppose that a second trial of $m=r+s$ instances be made, then the probability that the given event will occur r times and fail s, is on the *à priori* chance being between x and $x+dx$

$$=P_x\frac{\Gamma m}{\Gamma r\Gamma s}n^r(1-x)^s,$$

and accordingly the total chance C_r, whatever x may be of the event occurring r times in the second series, is

$$C_r=\frac{\Gamma m}{\Gamma r\Gamma s}\frac{\int_0^1 x^{p+r}(1-x)^{q+s}dx}{\int_0^1 x^p(1-x)^q dx}.$$

This is, with a slight correction, Laplace's extension of Bayes' Theorem." [1]

13. This argument can be restated as follows. Of all the objects which satisfy $\phi(x)$, let us suppose that a proportion p also satisfy $f(x)$. In this case p measures the probability that any object, of which we know only that it is ϕ, is in fact also f. Now if we do not know the value of p and have no relevant information which bears upon it, we can assume *à priori* that all values of p between 0 and 1 are equally likely. This assumption, which is termed the 'equal distribution of ignorance,' is justified by our experience of statistical ratios. Our experience, that is to say, leads us to suppose that of all the theories, which could be propounded, there are just as many which are always true as there are which are always false, just as many which are true once in fifty times as there are which are true once in three times, and so on. Professor Pearson challenges those who do not accept this assumption to produce definite evidence to the contrary.

The challenge is easily met. It would not be difficult to produce 10,000 positive theories which are always false corresponding to every one which is always true, and 10,000 correlations of posi-

[1] The rest of the article is concerned with the determination of the probable error when Laplace's Rule of Succession is used not simply to yield the probability of a single additional occurrence, but to predict the probable limits within which the frequency will lie in a considerable series of additional trials. Professor Pearson's method applies more rigorous methods of approximation to the fundamental formulae given above than have been sometimes used. As my main purpose in this chapter is to dispute the general validity of the fundamental formulae, it is not worth while to consider these further developments here. If the validity of the fundamental formula were to be granted, Professor Pearson's methods of approximation would, I think, be satisfactory.

tive qualities which hold less often than once in three times for every one we can name which holds more often than once in three times. And the converse is the case for negative theories and correlations between negative qualities; for corresponding to every positive theory which is true there is a negative theory which is false, and so on. Thus experience, if it shows anything, shows that there is a very marked clustering of statistical ratios in the neighbourhoods of zero and unity,—of those for positive theories and for correlations between positive qualities in the neighbourhood of zero, and of those for negative theories and for correlations between negative qualities in the neighbourhood of unity. Moreover, we are seldom in so complete a state of ignorance regarding the nature of the theory or correlation under investigation as not to know whether or not it is a positive theory or a correlation between positive qualities. In general, therefore, whenever our investigation is a practical one, experience, if it tells us anything, tells us not only that the statistical ratios cluster in the neighbourhood of zero and unity, but in which of these two neighbourhoods the ratio in this particular case is most likely *à priori* to be found. If we seek to discover what proportion of the population suffer from a certain disease, or have red hair, or are called Jones, it is preposterous to suppose that the proportion is as likely *à priori* to exceed as to fall short of (say) fifty per cent. As Professor Pearson applies this method to investigations where it is plain that the qualities involved are positive, he seems to maintain that experience shows that there are as many positive attributes which are shared by more than half of any population as there are which are shared by less than half.

It is also worth while to point out that it is formally impossible that it should be true of all characters, simple and complex, that they are as likely to have any one frequency as any other. For let us take a character c which is compound of two characters a and b, between which there is no association, and let us suppose that a has a frequency x in the population in question and that b has a frequency y, so that, in the absence of association, the frequency z of c is equal to xy. Then it is easy to show that, if all values of x and y between 0 and 1 are equally probable, all values of z between 0 and 1 are *not* equally probable. For the value $\frac{1}{2}$ is more probable than any other, and the possible values of

z become increasingly improbable as they differ more widely from $\frac{1}{2}$.

It may be added that the conclusions, which Professor Pearson himself derives from this method, provide a *reductio ad absurdum* of the arguments upon which they rest. He considers, for example, the following problem: A sample of 100 of a population shows 10 per cent affected with a certain disease. What percentage may be reasonably expected in a second sample of 100? By approximation he reaches the conclusion that the percentage of the character in the second sample is as likely to fall inside as outside the limits, 7·85 and 13·71. Apart from the preceding criticisms of the reasoning upon which this depends, it does not seem reasonable upon general grounds that we should be able on so little evidence to reach so certain a conclusion. The argument does not require, for example, that we have any knowledge of the manner in which the samples are chosen, of the positive and negative analogies between the individuals, or indeed anything at all beyond what is given in the above statement. The method is, in fact, much too powerful. It invests any positive conclusion, which it is employed to *support*, with far too high a degree of probability. Indeed this is so foolish a theorem that to entertain it is discreditable.

14. The Rule of Succession has played a very important part in the development of the theory of probability. It is true that it has been rejected by Boole [1] on the ground that the hypotheses on which it is based are arbitrary, by Venn [2] on the ground that it does not accord with experience, by Bertrand [3] because it is ridiculous, and doubtless by others also. But it has been very widely accepted,—by De Morgan,[4] by Jevons,[5] by Lotze,[6] by Czuber,[7] and by Professor Pearson,[8]—to name some representative writers of successive schools and periods. And, in any case, it

[1] *Laws of Thought*, p. 369.

[2] *Logic of Chance*, p. 197.

[3] *Calcul des probabilités*, p. 174.

[4] Article in *Cabinet Encyclopaedia*, p. 64.

[5] *Principles of Science*, p. 297.

[6] *Logic*, pp. 373, 374; Lotze propounds a "simple deduction" "as convincing" to him "as the more obscure analysis, by which it is usually obtained." The proof is among the worst ever conceived, and may be commended to those who seek instances of the profound credulity of even considerable thinkers.

[7] *Wahrscheinlichkeitsrechnung*, vol. i. p. 199,—though much more guardedly and with more qualifications than in the form discussed above.

[8] *Loc. cit.*

is of interest as being one of the most characteristic results of a way of thinking in probability introduced by Laplace, and never thoroughly discarded to this day. Even amongst those writers who have rejected or avoided it, this rejection has been due more to a distrust of the particular applications of which the law is susceptible than to fundamental objections against almost every step and every presumption upon which its proof depends.

Some of these particular applications have certainly been surprising. The law, as is evident, provides a numerical measure of the probability of any simple induction, provided only that our ignorance of its conditions is sufficiently complete, and, although, when the number of cases dealt with is small, its results are incredible, there is, when the number dealt with is large, a certain plausibility in the results it gives. But even in these cases paradoxical conclusions are not far out of sight. When Laplace proves that, account being taken of the experience of the human race, the probability of the sun's rising to-morrow is 1,826,214 to 1, this large number may seem in a kind of way to represent our state of mind about the matter. But an ingenious German, Professor Bobek,[1] has pushed the argument a degree further, and proves by means of these same principles that the probability of the sun's rising every day for the next 4000 years, is not more, approximately, than two-thirds,—a result less dear to our natural prejudices.

[1] *Lehrbuch der Wahrscheinlichkeitsrechnung*, p. 208.

CHAPTER XXXI

THE INVERSION OF BERNOULLI'S THEOREM

1. I CONCLUDE, then, that the application of the mathematical methods, discussed in the preceding chapter, to the general problem of statistical inference is invalid. Our state of knowledge about our material must be positive, not negative, before we can proceed to such definite conclusions as they purport to justify. To apply these methods to material, unanalysed in respect of the circumstances of its origin, and without reference to our general body of knowledge, merely on the basis of arithmetic and of those of the characteristics of our material with which the methods of descriptive statistics are competent to deal, can only lead to error and to delusion.

But I go further than this in my opposition to them. Not only are they the children of loose thinking, and the parents of charlatanry. Even when they are employed by wise and competent hands, I doubt whether they represent the most fruitful form in which to apply technical and mathematical methods to statistical problems, except in a limited class of special cases. The methods associated with the names of Lexis, Von Bortkiewicz, and Tschuprow (of whom the last named forms a link, to some extent, between the two schools), which will be briefly described in the next chapter, seem to me to be much more clearly consonant with the principles of sound induction.

2. Nevertheless it is natural to suppose that the fundamental ideas, from which these methods have sprung, are not wholly *égarés*. It is reasonable to presume that, subject to suitable conditions and qualifications, an inversion of Bernoulli's Theorem must have validity. If we *knew* that our material could be likened to a game of chance, we might expect to infer chances from frequencies, with the same sort of confidence as that with

which we infer frequencies from chances. This part of our inquiry will not be complete, therefore, until we have endeavoured to elucidate the conditions for the validity of an Inversion of Bernoulli's Theorem.

3. The problem is usually discussed in terms of the happening of an event under certain conditions, that is to say, of the co-existence of the conditions, as affecting a particular event, with that event. The same problem can be dealt with more generally and more conveniently as an investigation of the correlation between two characters A(x) and B(x), which, as in Part III., are propositional functions which may be said to concur or co-exist when they are both true of the same argument x. Given that, within the field of our knowledge, B(x) is true for a certain proportion of the values of x for which A(x) is true, what is the probability for a further value a of x that, if A(a) holds, B(a) will hold also?

Let us suppose that the occurrence of an instance of A(x) is a sign of one of the events $e_1(x)$, $e_2(x)$ . . . or $e_m(x)$, and that these are exhaustive, exclusive, and ultimate alternatives. By *exhaustive* it is meant that, whenever there is an instance of A(x), one of the e's is present; by *exclusive*, that the presence of one of the e's is not a sign of the presence of any other, but not that the concurrence of two or more of the e's is in fact impossible; by *ultimate*, that no one of the e's is a disjunction of two or more alternatives which might themselves be members of the e's. Let us assume that these alternatives are initially and *throughout the argument* equally probable, which, subject to the above conditions, is justified by the Principle of Indifference. We have no reason, that is to say, and no part of our evidence ever gives us one, for thinking that A(a) is more likely to be a sign of one of the e's than of any other, or even for thinking that some e's, although we do not know which, are more likely to occur than others. Let us also assume that, out of $e_1(x)$, $e_2(x)$ . . . $e_m(x)$, the set $e_1(x)$, $e_2(x)$ . . . $e_l(x)$, and these only, are signs or occasions of B(x); and further that we have no evidence bearing on the actual magnitude of the integers l and m, so that the *ratio* l/m is the only factor of which the probability varies as the evidence accumulates. Let us assume, lastly, that our knowledge of the several instances of B(x) is adequate to establish a perfect analogy between them; the instances a, etc., of B(x), that is to say, must not have anything in common except B, unless we have reason to know that the additional resemblances are immaterial. Even by these considerable simplifications not every difficulty has been avoided. But a development along the usual lines with the assistance of Bernoulli's Theorem is now possible.

Let $l/m = q$. If the value of q were known, the problem would be solved. For this numerical ratio would represent the probability that A is, in any random instance, a sign of B; and no further evidence, which satisfies the conditions of the preceding hypothesis, can possibly modify it. But in the inverse problem q is not known; and our problem is to determine whether evidence can be forthcoming of such a kind, that, as this evidence is increased in quantity, the probability that A will be in any instance a sign of B, tends to a limit which lies between two determinate ratios, just as the probability of an inductive generalisation may tend towards certainty, when the evidence is increased in a manner satisfying given conditions.

Let $f(q)$ represent the proposition that q is the true value of l/m. Let q' represent the ratio of the number of instances actually before us in which A has been accompanied by B to that of the instances in which A has not been accompanied by B; and let $f'(q')$ be the proposition which asserts this. Now if the ratio q is known, then, subject to the assumptions already stated, the number q must also represent the *à priori* probability in any instance, both before and after the results of other instances are known, that A, if it occurs, will be accompanied by B. We have, in fact, the conditions as set forth in Chapter XXIX., in which Bernoulli's Theorem can be validly applied, so that this theorem enables us to give a numerical value, for all numerical values of q and q', to the probability $f'(q')/h\,.\,f(q)$,—which expression represents the likelihood *à priori* of the frequency q', given q.

An application of the inverse formula allows us to infer from the above the *à posteriori* probability of q, given q', namely:

$$\frac{f(q)/h\,.\,f(q')/h\,.\,f(q)}{\Sigma f(q)/h\,.\,f(q')/h\,.\,f(q)}$$

where the summation in the denominator covers all possible values of q. In rough applications of this inverse of Bernoulli's Theorem it has been usual to suppose that $f(q)/h$ is constant for all values of q,—that, in other words, all possible values of the ratio q are *à priori* equally likely. If this supposition were legitimate, the formula could be reduced to the algebraical expression

$$\frac{f(q')/h \,.\, f(q)}{\Sigma f(q')/h \,.\, f(q)},$$

all the terms of which can be determined numerically by Bernoulli's Theorem. It is easy to show that it is a maximum when $q = q'$, *i.e.* that q' is the most probable value of l/m, and that, when the instances are very numerous, it is very improbable that l/m differs from q' widely. If, therefore, the number of instances is increased in such a manner that the ratio continues in the neighbourhood of q', the probability that the true value of l/m is nearly q' tends to certainty; and, consequently, the probability, that A is in any instance a sign of B, also tends to a magnitude which is measured by q'.

I see, however, no justification for the assumption that all possible values of the ratio q are *à priori* equally likely. It is not even equivalent to the assumptions that all integral values of l and m respectively are equally probable. I am not satisfied *either* that different values of q, *or* that different values of m, satisfy the conditions which have been laid down in Part I. for alternatives which are equal before the Principle of Indifference. There seem, for instance, to be relevant differences between the statement that A can arise in exactly two ways and the statement that it can arise in exactly a thousand ways. We must, therefore, be content with some lesser assumption and with a less precise form for our final conclusion.

4. Since, in accordance with our hypothesis, m cannot exceed some finite number, and since l must necessarily be less than m, the possible values of m, and therefore of q, are finite in number. Perhaps we can assume, therefore, as one of our fundamental assumptions, that there is *à priori* a finite probability in favour of each of these possible values. Let μ be the finite number which m cannot exceed. Then there is a *finite* probability for each of the intervals [1]

$$\frac{1}{\mu} \text{ to } \frac{2}{\mu}, \quad \frac{2}{\mu} \text{ to } \frac{3}{\mu}, \quad \ldots \quad \frac{\mu-1}{\mu} \text{ to } 1$$

[1] The intervals are supposed to include their lower but not their upper limit.

that q lies in this interval; but we cannot assume that there is an *equal* probability for each interval.

We must now return to the formula

$$\frac{f(q)/h \,.\, f(q')/hf(q)}{\Sigma f(q)/h \,.\, f(q')/hf(q)},$$

which represents the *à posteriori* probability of q, given q'. Since by sufficiently increasing the number of instances, the sum of terms $f(q')/hf(q)$ for possible values of q within a certain finite interval in the neighbourhood of q' can be made to exceed the other terms by any required amount, and since the sum of the values of $f(q)/h$ for possible values of q within this interval is finite, it clearly follows that a finite number of instances can make the probability, that q lies in an interval of magnitude $1/\mu$ in the neighbourhood of q', to differ from certainty by less than any finite amount however small.

5. We have, therefore, reached the main part of the conclusion after which we set out—namely, that as the number of instances is increased the probability, that q is in the neighbourhood of q', tends towards certainty; and hence that, subject to certain specified conditions, if the frequency with which B accompanies A is found to be q' in a great number of instances, then the probability that A will be accompanied by B in any further instance is also approximately q'. But we are left with the same vagueness, as in the case of generalisation, respecting the value of μ and the number of instances that we require. We know that we can get as near certainty as we choose by a finite number of instances, but what this number is we do not know. This is not very satisfactory, but it accords very well, I think, with what common sense tells us. It would be very surprising, in fact, if logic could tell us exactly how many instances we want, to yield us a given degree of certainty in empirical arguments.

Nobody supposes that we can measure exactly the probability of an induction. Yet many persons seem to believe that in the weaker and much more difficult type of argument, where the association under examination has been in our experience, not invariable, but merely in a certain proportion, we can attribute a definite measure to our future expectations and can claim practical certainty for the results of predictions which lie within relatively narrow limits. Coolly considered, this is a preposterous claim, which would have been universally rejected long ago, if those who made it had not so successfully concealed themselves from the eyes of common sense in a maze of mathematics.

6. Meantime we are in danger of forgetting that, in order to reach even our modified conclusion, material assumptions have been introduced. In the first place, we are faced with exactly the same difficulties as in the case of universal induction dealt with in Part III., and our original starting-point must be the same. We have the same difficulty as to how our *initial* probability is to be obtained; and I have no better suggestion to offer in this than in the former case—namely, the supposed principle of a limitation of independent variety in experience. We have to suppose that if A and B occur together (*i.e.* are true of the same object), this is some just appreciable reason for supposing that in *this* instance they have a common cause; and that, if A occurs again, this is a just appreciable reason for supposing that it is due to the *same* cause as on the former occasion. But in addition to the usual inductive hypothesis, the argument has rested on two particularly important assumptions, first, that we have no reason for supposing that some of the events of which A may be a sign are more likely to be exemplified in some of the particular instances than in others, and secondly, that the analogy amongst the examined B's is perfect. The first assumption amounts, in the language of statisticians, to an assumption of *random sampling* from amongst the A's. The second assumption corresponds precisely to the similar condition which we discussed fully in connection with inductive generalisation. The instances of $A(x)$ may be the result of random sampling, and yet it may still be the case that there are material circumstances, common to all the examined instances of $B(x)$, yet not covered by the statement $A(x)B(x)$. In so far as these two assumptions are not justified, an element of doubt and vagueness, which is not easily measured, assails the argument. It is an element of doubt precisely similar to that which exists in the case of generalisation. But we are most likely to forget it. For having overcome the difficulties peculiar to correlation,[1] it is, possibly, not un-

[1] I am here using this term in distinction to *generalisation*; that is to say, I call the statement that $A(x)$ is always accompanied by $B(x)$ a *generalisation*, and the statement that $A(x)$ is accompanied by $B(x)$ in a certain proportion of cases a *correlation*. This is not quite identical with its use by modern statisticians.

natural for a statistician to feel as if he had overcome *all* the difficulties.

In practice, however, our knowledge, in cases of correlation just as in cases of generalisation, will seldom justify the assumption of perfect analogy between the B's; and we shall be faced by precisely the same problems of analysing and improving our knowledge of the instances, as in the general case of induction already examined. If B has invariably accompanied A in 100 cases, we have all kinds of difficulties about the exact character of our evidence before we can found on this experience a valid generalisation. If B has accompanied A, not invariably, but only 50 times in the 100 cases, clearly we have just the same kind of difficulties to face, and more too, before we can announce a valid correlation. Out of the mere analysed statement that B has accompanied A as often as not in 100 cases, without precise particulars of the cases, or even if there were 1,000,000 cases instead of 100, we can conclude very little indeed.

CHAPTER XXXII

THE INDUCTIVE USE OF STATISTICAL FREQUENCIES FOR THE DETERMINATION OF PROBABILITY *À POSTERIORI* — THE METHODS OF LEXIS

1. No one supposes that a good induction can be arrived at merely by counting cases. The business of strengthening the argument chiefly consists in determining whether the alleged association is *stable*, when the accompanying conditions are varied. This process of improving the Analogy, as I have called it in Part III., is, both logically and practically, of the essence of the argument.

Now in statistical reasoning (or inductive correlation) that part of the argument, which corresponds to counting the cases in inductive generalisation, may present considerable technical difficulty. This is especially so in the particularly complex cases of what in the next chapter (§ 9) I shall term *Quantitative Correlation*, which have greatly occupied the attention of English statisticians in recent years. But clearly it would be an error to suppose that, when we have successfully overcome the mathematical or other technical difficulties, we have made any greater progress towards establishing our conclusion than when, in the case of inductive generalisation, we have counted the cases but have not yet analysed or compared the descriptive and non-numerical differences and resemblances. In order to get a good scientific argument we still have to pursue precisely the same scientific methods of experiment, analysis, comparison, and differentiation as are recognised to be necessary to establish any scientific generalisation. These methods are not reducible to a precise mathematical form for the reasons examined in Part III. of this treatise. But that is no reason for ignoring them, or for pretending that the calculation of a probability, which takes into account nothing whatever except the numbers of the instances, is a rational proceeding. The passage already quoted from Leibniz (*In exemplis juridicis politicisque plerumque non tamen subtili calculo opus est, quam accurata omnium circumstantiarum enumeratione*) is as applicable to scientific as to political inquiries.

Generally speaking, therefore, I think that the business of statistical technique ought to be regarded as strictly limited to preparing the numerical aspects of our material in an intelligible form, so as to be ready for the application of the usual inductive methods. Statistical technique tells us how to 'count the cases' when we are presented with complex material. It must not proceed also, except in the exceptional case where our evidence furnishes us from the outset with data of a particular kind, to turn its results into probabilities; not, at any rate, if we mean by probability a measure of rational belief.

2. There is, however, one type of technical, statistical investigation not yet discussed, which seems to me to be a valuable aid to inductive correlation. This method consists in breaking up a statistical series, according to appropriate principles, into a number of sub-series, with a view to analysing and measuring, not merely the frequency of a given character over the aggregate series, but the *stability* of this frequency amongst the sub-series; that is to say, the series as a whole is divided up by some principle of classification into a set of sub-series, and the *fluctuation* of the statistical frequency under examination between the various sub-series is then examined. It is, in fact, a technical method of increasing the Analogy between the instances, in the sense given to this process in Part III.

3. The method of analysing statistical series, as opposed to the Laplacian or *mathematical* method, one might designate the *inductive* method. Independently of the investigations of Bernoulli or Laplace, practical statisticians began at least as early as the end of the seventeenth century[1] to pay attention to the *stability* of statistical series when analysed in this manner. Throughout the eighteenth century, students of mortality statistics, and of the ratio of male to female births (including Laplace himself), paid attention to the degree of constancy of the

[1] Graunt in his *Natural and Political Observations upon the Bills of Mortality* has been quoted as one of the earliest statisticians to pay attention to these considerations.

ratios over different parts of their series of instances as well as to their average value over the whole series. And in the early part of the nineteenth century, Quetelet, as we have already noticed, widely popularised the notion of the stability of various social statistics from year to year. Quetelet, however, sometimes asserted the existence of stability on insufficient evidence, and involved himself in theoretical errors through imitating the methods of Laplace too closely; and it was not until the last quarter of the nineteenth century that a school of statistical theory was founded, which gave to this way of approaching the problem the system and technique which it had hitherto lacked, and at the same time made explicit the contrast between this analytical or inductive method and the prevailing mathematical theory. The sole founder of this school was the German economist, Wilhelm Lexis, whose theories were expounded in a series of articles and monographs published between the years 1875 and 1879. For some years Lexis's fundamental ideas did not attract much notice, and he himself seems to have turned his attention in other directions. But more recently a considerable literature has grown up round them in Germany, and their full purport has been expressed with more clearness than by Lexis himself—although no one, with the exception of Ladislaus von Bortkiewicz, has been able to make additions to them of any great significance.[1] Lexis devised his theory with an immediate view to its practical application to the problems of sex ratio and mortality. The fact that his general theory is so closely intermingled with these particular applications of it is, probably, a part explanation of the long interval which elapsed before the general theoretical importance of his ideas was widely realised. I cannot help doubting how fully Lexis himself realised it in the first instance. It would certainly be easy to read his earlier contributions to the question without appreciating their generalised significance. After 1879 Lexis added nothing substantial to his earlier work, and later developments are mainly due to Von

[1] A list of Lexis's principal writings on these topics will be found in the Bibliography. There is little of first-rate importance which is not contained either in the volume, *Zur Theorie der Massenerscheinungen in der menschlichen Gesellschaft*, or in the *Abhandlungen zur Theorie der Bevölkerungs- und Moral-Statistik*. In this latter volume the two important articles on "Die Theorie der Stabilität statistischer Reihen" and on "Das Geschlechtsverhältnis der Geborenen und die Wahrscheinlichkeitsrechnung," originally published in Conrad's *Jahrbüche*, are reprinted.

Bortkiewicz. Those of the latter's writings, which have an important bearing on the relation between probability and statistics, are given in the Bibliography.[1]

On the logic and philosophy of Probability writers of the school of Lexis are in general agreement with Von Kries; but this seems to be due rather to the reaction which is common both to him and to them against the Laplacian tradition, than to any very intimate theoretical connection between Von Kries's main contributions to Probability and those of Lexis, though it is true that both show a tendency to find the ultimate basis of Probability in physical rather than in logical considerations. I am not acquainted with much work, which has been appreciably influenced by Lexis, written in other languages than German (including with Germans, that is to say, those Russians, Austrians, and Dutch who usually write in German, and are in habitual connection with the German scientific world). In France Dormoy[2] published independently and at about the same time as Lexis some not dissimilar theories, but subsequent French writers have paid little attention to the work of either. Such typical French treatises as that of Bertrand, or, more recently, that of Borel, contain no reference to them.[3] In Italy there has been some discussion recently on the work of Von Bortkiewicz. Among Englishmen Professor Edgeworth has shown a close acquaintance with the work of the German school,[4] he providing for nearly forty years past, on this as on other matters where the realms of

[1] The reader may be specially referred to the *Kritische Betrachtungen zur theoretischen Statistik* (*first* instalment—the later instalments being of less interest to the student of Probability), the *Anwendungen der Wahrscheinlichkeitsrechnung auf Statistik*, and *Homogeneität und Stabilität in der Statistik*. Of other German and Russian writers it will be sufficient to mention here Tschuprow, who in "Die Aufgaben der Theorie der Statistik" (Schmoller's *Jahrbuch*, 1905) and "Zur Theorie der Stabilität statistischer Reihen" (*Skandinavisk Aktuarietidskrift*) gives by far the best and most lucid general accounts that are available of the doctrines of the school, he alone amongst these authors writing in a style from which the foreign reader can derive pleasure, and Czuber, who in his *Wahrscheinlichkeitsrechnung* (vol. ii. part iv. section 1) supplies a useful mathematical commentary.

[2] *Journal des actuaires français*, 1874, and *Théorie mathématique des assurances sur la vie*, 1878; on the question of priority see Lexis, *Abhandlungen*, p. 130.

[3] Though both these writers touch on closely cognate matters, where Lexis's investigations would be highly relevant—Bertrand, *Calcul*, pp. 312-314; Borel, *Éléments*, p. 160.

[4] See especially his "Methods of Statistics" in the *Jubilee Volume of the Stat. Journ.*, 1885, and "Application of the Calculus of Probabilities to Statistics," *International Statistical Institute Bulletin*, 1910.

Statistics and Probability overlap, almost the only connecting link between English and continental thought.

Nevertheless, an account in English of the main doctrines of this school is still lacking. It would be outside the plan of the present treatise to attempt such an account here. But it may be useful to give a short summary of Lexis's fundamental ideas. After giving this account I shall find it convenient, in proceeding to my own incomplete observations on the matter, to approach it from a rather different standpoint from that of Lexis or of Von Bortkiewicz, though not for that reason the less influenced or illuminated by their eminent contributions to this problem.

4. It will be clearer to begin with some analysis due to Von Bortkiewicz,[1] and then to proceed to the method of Lexis himself, although the latter came first in point of time.

A group of observations may be made up of a number of sub-groups, to which different frequencies for the character under investigation are properly applicable. That is to say, a proportion $\frac{z_1}{z}$ of the observations may belong to a group, for which, given the frequency, the *à priori* probability of the character under observation in a particular instance would be p_1, a proportion $\frac{z_2}{z}$ may belong to a second group for which p_2 is the probability, and so on. In this case, given the frequencies for the sub-groups, the probability p for the group as a whole would be made up as follows:

$$p=\frac{z_1}{z}p_1+\frac{z_2}{z}p_2+\ldots.$$

We may call p a *general probability*, and p, etc., *special probabilities*. But the special probabilities may in their turn be general probabilities, so that there may be more than one way of resolving a general probability into special probabilities.

If $p_1=p_2=\ldots=p$, then p, for that particular way of resolving the total group into partial groups, is, in Bortkiewicz's terminology, *indifferent*. If p is indifferent for all conceivable resolutions into partial groups,[2] then, borrowing a phrase from Von Kries, Bortkiewicz says of it that it has a *definitive interpretation*. In

[1] What follows is a free rendering of some passages in his *Kritische Betrachtungen*.

[2] This is clearly a very loose statement of what Bortkiewicz really means.

dealing with *à priori* probabilities, we can resolve a total probability until we reach the special probabilities of each individual case ; and if we find that all these special probabilities are equal, then, clearly, the general probability satisfies the condition for definitive interpretation.

So far we have been dealing with *à priori* probabilities. But the object of the analysis has been to throw light on the inverse problem. We want to discover in what conditions we can regard an observed frequency as being an adequate approximation to a definitive general probability.

If p' is the empirical value of p (or, as I should prefer to call it, the frequency) given by a series of n observations, we may have

$$p' = \frac{n_1}{n}p_1' + \frac{n_2}{n}p_2' + \ldots .$$

Even if this particular way of resolving the series of observations is indifferent, the *actually observed* frequencies p_1', p_2', etc., may nevertheless be unequal, since they may fluctuate round the norm p' through the operation of 'chance' influences. If, however, n_1, n_2, etc., are large, we can apply the usual Bernoullian formula to discover whether, *if* there was a norm p', the divergences of p_1', p_2', etc., from it are within the limits reasonably attributable on Bernoullian hypotheses to 'chance' influences. We can, however, only base a sound argument in favour of the existence of a 'definitive' probability p' by resolving our aggregate of instances into sub-series in a great variety of ways, and applying the above calculations each time. Even so, some measure of doubt must remain, just as in the case of other inductive arguments.

Bortkiewicz goes on to say that probabilities having definitive interpretation (*definitive Bedeutung*) may be designated elementary probabilities (*Elementarwahrscheinlichkeiten*). But the probabilities which usually arise in statistical inquiries are not of this type, and may be termed *average probabilities* (*Durchschnittswahrscheinlichkeiten*). That is to say, a series of observed frequencies (or, as he calls them, empirical probabilities) does not, as a rule, group itself as it would if the series was in fact subject to an elementary probability.

5. This exposition is based on a philosophy of Probability different from mine ; but the underlying ideas are capable of translation. Suppose that one is endeavouring to establish an inductive correlation, *e.g.* that the chance of a male birth is m. The conclusion, which we are seeking to establish, takes no account of the place or date of birth or the race of the parents, and assumes that these influences are irrelevant. Now, if we had statistics of birth ratios for all parts of the world throughout the nineteenth century, and added them all up and found that the average frequency of male births was m, we should not be justified in arguing from this that the frequency of male births in England next year is very unlikely to diverge widely from m. For this would involve the unwarranted assumption, in Bortkiewicz's terminology, that the empirical probability m is elementary for any resolution dependent on time or place, and is not an average probability compounded out of a series of groups, relating to different times or places, to each of which a distinct special probability is applicable. And, in my terminology, it would assume that variations of time and place were irrelevant to the correlation, without any attempt having been made to employ the methods of positive and negative Analogy to establish this.

We must, therefore, break up our statistical material into groups by date, place, and any other characteristic which our generalisation proposes to treat as irrelevant. By this means we shall obtain a number of frequencies m_1', m_2', m_3', ... m_1'', m_2'', m_3'', ... etc., which are distributed round the average frequency m. For simplicity let us consider the series of frequencies m_1', m_2', m_3', ... obtained by breaking up our material according to the *date* of the birth. If the observed divergences of these frequencies from their mean are not significant, we have the beginnings of an inductive argument for regarding *date* as being in this connection irrelevant.

6. At this point Lexis's fundamental contribution to the problem must be introduced. He concentrated his attention on the nature of the dispersion of the frequencies m_1', m_2', m_3' . . . . round their mean value m; and he sought to devise a technical method for measuring the degree of stability displayed by the series of sub-frequencies, which are yielded by the various possible criteria for resolving the aggregate statistical material into a number of constituent groups.

For this purpose he classified the various types of dispersion which could occur. It may be the case that some of the sub-frequencies show such wide and discordant variations from the mean as to suggest that some significant Analogy has been overlooked. In this event the lack of symmetry, which characterises the oscillations, may be taken to indicate that some of the subgroups are subject to a relevant influence, of which we must take account in our generalisation, to which some of the other subgroups are not subject.

But amongst the various types of dispersion Lexis found one class clearly distinguishable from all the others, the peculiarity of which is that the individual values fluctuate in a 'purely chance' manner about a constant fundamental value. This type he called typical (*typische*) dispersion. He meant by this that the dispersion conformed approximately to the distribution which would be given by some normal law of error.

The next stage of Lexis's argument [1] was to point out that series of frequencies which are typical in character may have as their foundation either a constant probability,[2] or one which is itself subject to chance variations about a mean. The first case is typified by the example of a series of sets of drawings of balls, each set being drawn from a similar urn; the second case by the example of a series of sets of drawings, the urns from which each set is drawn being not similar, but with constitutions which vary in a chance manner about a mean.

As his *measure* of dispersion Lexis introduces a formula, which is evidently in part conventional (as is the case with so many other statistical formulae, the particular shape of which is often determined by mathematical convenience rather than by any more fundamental criterion). He expresses himself as follows. Where the underlying probability is constant, the probable error in a particular frequency *à priori* is $r=\rho\sqrt{\frac{2v(1-v)}{g}}$, where $\rho=\cdot 4769$, v is the underlying probability, and g is the number of instances to which the frequency refers. This follows from the usual Bernoullian assumptions. Now let R be the corresponding expression derived *à posteriori* by reference to the actual deviations of a series of observed frequencies from their mean, so that

[1] I am here following fairly closely his paper, "Über die Theorie der Stabilität statisticher Reihen," reprinted in his *Abhandlungen zur Theorie der Bevölkerungs- und Moral-Statistik*, pp. 170-212.

[2] This mode of expression, which is not in accurate conformity with my philosophy of Probability, is Lexis's, not mine. His meaning is intelligible.

$R = \rho\sqrt{\frac{2[\delta^2]}{n-1}}$, where $[\delta^2]$ is the sum of the squares of the deviations of the individual frequencies from their mean and n is their number. Now, if the observed facts are due to merely chance variations about a constant v, we must have approximately $R = r$, though, if g is small, comparatively wide deviations between R and r will not be significant. If, on the other hand, v itself is not constant but is subject to chance variations, the case stands differently. For the fluctuations of the observed frequencies are now due to two components. The one which would be present, even if the underlying probability were constant, Lexis terms the ordinary or unessential component; the other he terms the physical component. If p is the probable deviation of the various values of v from their mean, then, on the same assumptions and as a deduction from the same theory as before, R will tend to equal not r but $\sqrt{r^2+p^2}$. In this event R cannot be less than r. If, therefore, $R<r$, one must suppose that the individual instances of each several series on which each frequency is based are not independent of one another. Such a series Lexis terms an organic or dependent (*gebundene*) series, and explains that it cannot be handled by purely statistical methods.

Since, therefore, we have three types of series, differing fundamentally from one another according as $R = r$, $>r$, or $<r$, Lexis puts $\frac{R}{r} = Q$, and takes Q as his measure of dispersion.[1] If $Q = 1$, we have normal dispersion; if $Q>1$, we have supernormal dispersion; and if $Q<1$, we have subnormal dispersion, which is an indication that the series is 'organic.'

If the number of instances on which the frequencies are based is very great, r becomes negligible in comparison with p (the physical component), and, therefore, $R = \sqrt{r^2+p^2}$ becomes approximately $R = p$. On the other hand, if p is not very large and the base number of instances is small, p becomes negligible

[1] In Tschuprow's notation (*Die Aufgaben der Theorie der Statistik*, p. 45), $Q = P/C$, where P (the Physical modulus) $= \sqrt{\frac{2\sum_{k=1}^{k=n}(p_k-p)^2}{n}}$ and C (the Combinatorial modulus) $= \sqrt{\frac{2p(1-p)}{M}}$, M being the number of instances in each set, n the number of sets, p_k the frequency for set k, and p the mean of the n frequencies.

in comparison with r, and we have a delusive appearance of normal dispersion.[1] Lexis well illustrates the former point by the example that the statistics of the ratio of male to female births for the forty-five registration districts of England over the years 1859–1871 approximately satisfy the relation $R=r$. But if we take the figures for all England over those thirteen years, although the extreme limits of the fluctuation of the ratio about its mean 1·042 are 1·035 and 1·047, nevertheless $R=2·6$ and $r=1·6$, so that $Q=1·625$; the explanation being that the base number of instances, namely 730,000, is so large that r is very small, with the result that it is swamped by the physical component p. And he illustrates the latter point by the assertion that, if in 20 or 30 series each of 100 draws from an urn containing black and white balls equally, the number of black balls drawn each time were only to vary between 49 and 51, he would have confidence that the game was in some way falsified and that the draws were not independent. That is to say, undue regularity is as fatal to the assumption of Bernoullian conditions as is undue dispersion.

7. In a characteristic passage[2] Professor Edgeworth has applied these theories to the frequency of dactyls in successive extracts from the *Aeneid*. The mean for the line is 1·6, exclusive of the fifth foot, thus sharply distinguishing the Virgilian line from the Ovidian, for which the corresponding figure is 2·2. But there is also a marked stability. "That the Mean of any five lines should differ from the general Mean by a whole dactyl is proved to be an exceptional phenomenon, about as rare as an Englishman measuring 5 feet, or 6 feet 3 inches. An excess of two dactyls in the Mean of five lines would be as exceptional as an Englishman measuring 6 feet 10 inches." But not only so—the stability is *excessive*, and the fluctuation is less "than that which is obtained upon the hypothesis of pure sortition. If we could imagine dactyls and spondees to be mixed up in the poet's brain in the proportion of 16 to 24 and shaken out at random, the modulus in the number of dactyls would be 1·38, whereas we have constantly obtained a smaller number, on an average (the square root of the average fluctuation) 1·2." On Lexian principles these statistical results would support the hypothesis that the

[1] This is part of the explanation of Bortkiewicz's *Law of Small Numbers*. See also p. 401.

[2] "On Methods of Statistics," *Jubilee Volume of the Royal Statistical Society*, p. 211.

series under investigation is 'organic' and not subject to Bernoullian conditions, an hypothesis in accordance with our ideas of poetry. That Edgeworth should have put forward this example in criticism of Lexis's conclusions, and that Lexis [1] should have retorted that the explanation was to be found in Edgeworth's series' not consisting of an adequate number of separate observations, indicates, if I do not misapprehend them, that these authorities are at fault in the principles, if not of Probability, of Poetry.

The dactyls of the Virgilian hexameter are, in fact, a very good example of what has been termed *connexité*, leading to subnormal dispersion. The quantities of the successive feet are not independent, and the appearance of a dactyl in one foot *diminishes* the probability of another dactyl in that line. It is like the case of drawing black and white balls out of an urn, where the balls are not replaced. But Lexis is wrong if he supposes that a *supernormal* dispersion cannot also arise out of *connexité*, or organic connection between the successive terms. It might have been the case that the appearance of a dactyl in one foot *increased* the probability of another dactyl in that line. He should, I think, have contemplated the result $R > r$ as possibly indicating a non-typical, organic series, and should not have assumed that, where R is greater than r, it is of the form $\sqrt{r^2+p^2}$.

In short, Lexis has not pushed his analysis far enough, and he has not fully comprehended the character of the underlying conditions. But this does not affect the fact that it was he who made the vital advance of taking as the unit, not the single observation, but the frequency in given conditions, and of conceiving the nature of statistical induction as consisting in the examination, and if possible the measurement, of the stability of the frequency when the conditions are varied.

8. There is one special piece of work illustrative of the above methods, due to Von Bortkiewicz, which must not be overlooked, and which it is convenient to introduce in this place—the so-called *Law of Small Numbers*.[2]

Quetelet, as we have seen in Chapter XXVIII., called attention

[1] "Über die Wahrscheinlichkeitsrechnung," p. 444 (see Bibliography).

[2] There are numerous references to this phenomenon in periodical literature; but it is sufficient to refer the reader to Von Bortkiewicz's *Das Gesetz der kleinen Zahlen*.

to the remarkable regularity of comparatively *rare* events. Von Bortkiewicz has enlarged Quetelet's catalogue with modern instances out of the statistical records of bureaucratic Germany. The classic instance, perhaps, is the number of Prussian cavalrymen killed each year by the kick of a horse. The table is worth giving as a statistical curiosity. (The period is from 1875 to 1894; G stands for the Corps of Guards, and I.-XV. for the 15 Army Corps.)

	75	76	77	78	79	80	81	82	83	84	85	86	87	88	89	90	91	92	93	94
G.	..	2	2	1	..	..	1	1	..	3	..	2	1	..	..	1	..	1	..	1
I.	..	..	..	2	..	3	..	2	..	..	..	1	1	1	..	2	..	3	1	..
II.	..	..	..	2	..	2	..	..	1	1	..	..	2	1	1	..	..	2	..	..
III.	..	..	..	1	1	1	2	..	2	..	..	..	1	..	1	2	1	..	..	..
IV.	..	1	..	1	1	1	1	..	..	..	..	1	..	..	..	..	1	1	..	..
V.	..	..	..	..	2	1	..	..	1	..	..	1	..	1	1	1	1	1	1	..
VI.	..	..	1	..	2	..	..	1	2	..	1	1	3	1	1	1	..	3	..	..
VII.	1	..	1	..	..	..	1	..	1	1	..	..	2	..	..	2	1	..	2	..
VIII.	1	..	..	..	1	..	..	1	..	..	..	..	1	..	..	..	1	1	..	1
IX.	..	..	..	..	..	2	1	1	1	..	2	1	1	..	1	2	..	1	.	..
X.	..	..	1	1	..	1	..	2	..	2	..	..	..	..	2	1	3	..	1	1
XI.	..	..	..	..	2	4	..	1	3	..	1	1	1	1	2	1	3	1	3	1
XIV.	1	1	2	1	1	3	..	4	..	1	..	3	2	1	..	2	1	1	..	..
XV.	..	1	..	..	..	..	..	1	..	1	1	..	..	..	2	2	..	..	..	..

The agreement of this table with the theoretical results of a random distribution of the total number of casualties is remarkably close:[1]

Casualties in a Year.	Number of Occasions on which the Annual Casualties in a Corps reach the Figure in Column 1.	
	Actual.	Theoretical.
0	144	143·1
1	91	92·1
2	32	33·3
3	11	8·9
4	2	2·0
5 and more	..	0·6

Other instances are furnished by the numbers of child suicides in Prussia, and the like.

It is Von Bortkiewicz's thesis that these observed regularities

[1] Bortkiewicz, *op. cit.* p. 24.

have a good theoretical explanation behind them, which he dignifies with the name of the *Law of Small Numbers*.

The reader will recall that, according to the theory of Lexis, his measure of stability Q is, in the more general case, made up of two components r and p, combined in the expression $\sqrt{r^2+p^2}$, of which one is due to fluctuations from the average of the conditions governing all the members of a series, which furnishes us with one of our observed frequencies, and of which the other is due to fluctuations in the individual members of the series about the true norm of the series. Bortkiewicz carries the same analysis a little further, and shows that Lexis's Q is of the form $\sqrt{1+(n-1)c^2}$, where n is the number of times that the event occurs in each series.[1] That is to say, Q increases with n, and, when n is small, Q is likely to exceed unity to a less extent than when n is large. To postulate that n is small, is, when we are dealing with observations drawn from a wide field, the same thing as to say that the event we are looking for is a comparatively rare one. This, in brief, is the mathematical basis of the Law of Small Numbers.

In his latest published work on these topics,[2] Von Bortkiewicz builds his mathematical structure considerably higher, without, however, any further underpinning of the logical foundations of it. He has there worked out further statistical constants, arising out of the conceptions on which Lexis's Q is based (the precise bearing of which is not made any clearer by his calling them *coefficients of syndromy*), which are explicitly dependent on the value of n; and he elaborately compares the theoretical value of the coefficients with the observed value in certain actual statistical material. He concludes with the thesis, that Homogeneity and Stability (defined as he defines them) are opposed conceptions, and that it is not correct to premise, that the larger statistical mass is as a rule more stable than the smaller, unless

[1] I refer the reader to the original, *op. cit.* pp. 29-31, for the interpretation of c (which is a function of the mean square errors arising in the course of the investigation) and for the mathematical argument by which the above result is justified.

[2] "Homogeneität und Stabilität in der Statistik," published in the *Skandinavisk Aktuarietidskrift*, 1918. Those readers, who look up my references, will, I think, agree with me that Von Bortkiewicz does not get any less obscure as he goes on. The mathematical argument is right enough, and often brilliant. But what it is all really about, what it all really amounts to, and what the premisses are, it becomes increasingly perplexing to decide.

we also assume that the larger mass is less homogeneous. At this point, it would have helped, if Von Bortkiewicz, excluding from his vocabulary homogeneity, paradromy, γ'_M, and the like, had stopped to tell in plain language where his mathematics had led him, and also whence they had started. But like many other students of Probability he is eccentric, preferring algebra to earth.

9. Where, then, though an admirer, do I criticise all this? I think that the argument has proceeded so far from the premisses, that it has lost sight of them. If the limitations prescribed by the premisses are kept in mind, I do not contest the mathematical accuracy of the results. But many technical terms have been introduced, the precise signification and true limitations of which will be misunderstood if the conclusion of the argument is allowed to detach itself from the premisses and to stand by itself. I will illustrate what I mean by two examples from the work of Von Bortkiewicz described above.

Von Bortkiewicz enunciates the seeming paradox that the larger statistical mass is only, as a rule, more stable if it is less homogeneous. But an illustration which he himself gives shows how misleading his aphorism is. The opposition between stability and homogeneity is borne out, he says, by the judgment of practical men. For actuaries have always maintained that their results average out better, if their cases are drawn from a wide field subject to *variable* conditions of risk, whilst they are chary of accepting too much insurance drawn from a single homogeneous area which means a concentration of risk. But this is really an instance of Von Bortkiewicz's own distinction between a general probability p and special probabilities p_1 etc., where

$$p=\frac{z_1}{z}p_1+\frac{z_2}{z}p_2+\ \ldots$$

If we are basing our calculations on p and do not know p_1, p_2, etc., then these calculations are more likely to be borne out by the result if the instances are selected by a method which spreads them over all the groups 1, 2, etc., than if they are selected by a method which concentrates them on group 1. In other words, the actuary does not like an undue proportion of his cases to be drawn from a group which may be subject to a common relevant influence *for which he has not allowed*. If the *à priori* calculations are based on the average over a field which is not homogeneous in all its parts, greater stability of result will be obtained if the instances are drawn from all parts of the non-homogeneous total field, than if they are drawn now from one homogeneous sub-field and now from another. This is not at all paradoxical. Yet I believe, though with hesitation, that this is all that Von Bortkiewicz's elaborately supported mathematical conclusion really amounts to.

My second example is that of the Law of Small Numbers. Here also we are presented with an apparent paradox in the statement that the regularity of occurrence of rare events is more stable than that of commoner events. Here, I suspect, the paradoxical result is really latent in the particular measure of stability which has been selected. If we look back at the figures, which I have quoted above, of Prussian cavalrymen killed by the kick of a horse, it is evident that a measure of stability could be chosen according to which exceptional instability would be displayed by this particular material; for the frequency varies from 0 to 4 round a mean somewhat less than unity, which is a very great *percentage* fluctuation. In fact, the particular measure of stability which Von Bortkiewicz has adopted from Lexis has about it, however useful and convenient it may be, especially for mathematical manipulation, a great deal that is arbitrary and conventional. It is only one out of a great many possible formulae which might be employed for the numerical measurement of the conception of stability, which, quantitatively at least, is not a perfectly precise one. The so-called Law of Small Numbers is, therefore, little more than a demonstration that, where rare events are concerned, the Lexian measure of stability does not lead to satisfactory results. Like some other formulae which involve a use of Bernoullian methods in an approximative form, it does not lead to reliable results in all circumstances. I should add that there is one other element which may contribute to the total psychological reaction of the reader's mind to the Law of Small Numbers, namely, the surprising and *piquant* examples which are cited in support of it. It is startling and even amusing to be told that horses kick cavalrymen with the same sort of regularity as characterises the rainfall. But our surprise at this particular example's fulfilling the Law of Great Numbers has little or nothing to do with the exceptional stability about which the Law of Small Numbers purports to concern itself.

CHAPTER XXXIII

OUTLINE OF A CONSTRUCTIVE THEORY

1. There is a great difference between the proposition " It is probable that *every* instance of this generalisation is true " and the proposition " It is probable of *any* instance of this generalisation taken at random that it is true." The latter proposition may remain valid, even if it is certain that some instances of the generalisation are false. It is more likely than not, for example, that any number will be divisible either by two or by three, but it is not more likely than not that all numbers are divisible either by two or by three.

The first type of proposition has been discussed in Part III. under the name of *Universal Induction*. The latter belongs to *Inductive Correlation* or *Statistical Induction*, an attempt at the logical analysis of which must be my final task.

2. What advocates of the Frequency Theory of Probability wrongly believe to be characteristic of *all* probabilities, namely, that they are essentially concerned not with single instances but with series of instances, is, I think, a true characteristic of statistical induction. A statistical induction either asserts the probability of an instance *selected at random* from a series of propositions, or else it assigns the probability of the assertion, that the truth frequency of a series of propositions (*i.e.* the proportion of true propositions in the series) is in the neighbourhood of a given value. In either case it is asserting a characteristic of a *series* of propositions, rather than of a particular proposition.

Whilst, therefore, our unit in the case of Universal Induction is a single instance which satisfies both the condition and the conclusion of our generalisation, our unit in the case of Statistical

Induction is not a single instance, but a set or series of instances, all of which satisfy the condition of our generalisation but which satisfy the conclusion only in a certain proportion of cases. And whilst in Universal Induction we build up our argument by examining the known positive and negative Analogy shown in a series of single instances, the corresponding task in Statistical Induction consists in examining the Analogy shown in a *series of series* of instances.

3. We are presented, in problems of Statistical Induction, with a set of instances all of which satisfy the conditions of our generalisation, and a proportion f of which satisfy its conclusion; and we seek to generalise as to the probable proportion in which further instances will satisfy the conclusion.

Now it is useless merely to pay attention to the proportion (or frequency) f discovered in the aggregate of the instances. For any collection whatever, comprising a definite number of objects, must, if the objects be classified with reference to the presence or absence of any specified characteristic whatever, show some definite proportion or statistical frequency of occurrence; so that a mere knowledge of what this frequency is can have no appreciable bearing on what the corresponding frequency will be for some other collection of objects, or on the probability of finding the characteristic in an object which does not belong to the original collection. We should be arguing in the same sort of way as if we were to base a universal induction as to the concurrence of two characteristics on a single observation of this concurrence, and without any analysis of the accompanying circumstances.

Let the reader be clear about this. To argue from the *mere* fact that a given event has occurred invariably in a thousand instances under observation, without any analysis of the circumstances accompanying the individual instances, that it is likely to occur invariably in future instances, is a feeble inductive argument, because it takes no account of the Analogy. Nevertheless an argument of this kind is not entirely worthless, as we have seen in Part III. But to argue, without analysis of the instances, from the mere fact that a given event has a frequency of 10 per cent in the thousand instances under observation, or even in a million instances, that its probability is 1/10 for the next instance, or that it is likely to have a frequency near to 1/10 in a further set of observations, is a far feebler argument; indeed it is hardly an argument at all. Yet a good deal of statistical argument is not free from this reproach;—though persons of common sense often conclude better than they argue, that is to say, they select for credence, from amongst arguments similar in form, those in favour of which there is in fact other evidence tacitly known to them though not explicit in the premisses as stated.

4. The analysis of statistical induction is not fundamentally different from that of universal induction already attempted in Part III. But it is much more intricate; and I have experienced exceptional difficulty, as the reader may discover for himself in the following pages, both in clearing up my own mind about it and in expounding my conclusions precisely and intelligibly. I propose to begin with a few examples of what commonly impresses us as good arguments in this field, and also of the attendant circumstances which, if they were known to exist, might be held to justify such a mode of reasoning; and, having thus attempted to bring before the reader's mind the character of the subject-matter, to proceed to an abstract analysis.

Example One.—Let us investigate the generalisation that the proportion of male to female births is m. The fact that the aggregate statistics for England during the nineteenth century yield the proportion m would go no way at all towards justifying the statement that the proportion of male births in Cambridge next year is likely to approximate to m. Our argument would be no better if our statistics, instead of relating to England during the nineteenth century, covered all the descendants of Adam. But if we were able to break up our aggregate series of instances into a series of sub-series, classified according to a great variety of principles, as for example by date, by season, by locality, by the class of the parents, by the sex of previous children, and so forth, and if the proportion of male births throughout these sub-series showed a significant stability in the neighbourhood of m, then indeed we have an argument worth something. Otherwise we must either abandon our generalisation, amplify its conditions, or modify its conclusion.

Example Two.—Let us take a series of objects s all alike in some specified respect, this resemblance constituting membership of the class F; let us determine of how many members of the series a certain property ϕ is true, the frequency of which is to be the subject of our generalisation; and if a proportion f of the series s have the property ϕ, we may say that the series s has a frequency f for the property ϕ.

Now if the whole field F has a finite number of constituents, it must have some determinate frequency p, and if, therefore, we increase the comprehensiveness of s until eventually it includes the whole field, f must come in the end to be equal to p. This is obvious and without interest and not what we mean by the law of great numbers and the stability of statistical frequency.

Let us now divide up the field F, according to some determinate principle of division D, into subfields F_1, F_2, etc.; and let the series s_1 be taken from F_1, s_2 from F_2, and so on. Where F_1, F_2, etc., have a finite number of constituents, s_1, s_2, etc., may possibly coincide with them; if s_1, s_2, etc., do not coincide with F_1, F_2, etc., but are chosen from them, let us suppose that they are chosen according to some principle of random or unbiassed selection—s_1, that is to say, will be a random sample from F_1. Now it may happen that the frequencies f_1, f_2, etc., of the series s_1, s_2, etc., thus selected cluster round some mean frequency f. If the frequencies show this characteristic (the measurement and precise determination of which I am not now considering), then the series of series s_1, s_2, etc., has a stable frequency for the classification D. 'Great numbers' only come in because it is difficult to ascertain the existence of stable frequency unless the series s_1, s_2, etc., are themselves numerous and unless each of these comprises numerous individual instances.

Let us then apply a different principle of division D′, leading to series s_1', s_2', etc., and to frequencies f_1', f_2', etc.; and then again a third principle of division D″ leading to frequencies f_1'', f_2'', etc.; and so on, to the full extent that our knowledge of the differences between the individual instances permits us. If the frequencies f_1, f_2, etc., f_1', f_2', etc., f_1'', f_2'', etc., and so on are all stable about f, we have an inductive ground of some weight for asserting a statistical generalisation.

Let the field F, for example, comprise all Englishmen in their sixtieth year, and let the property ϕ, about the frequency of which we are generalising, be their death in that year of their age. Now the field F can be divided into subfields F_1, F_2, etc., on innumerable different principles. F_1 might represent Englishmen in their sixtieth year in 1901, F_2 in 1902, and so on; or we might classify them according to the districts in which they live; or according to the amount of income tax they pay; or according as they are in workhouses, in hospitals, in asylums, in prisons, or at large. Let us take the second of these classifications and let the subfields F_1, F_2, etc., be constituted by the districts in which they live. If we take large random selections s_1, s_2, etc., from F_1, F_2, etc., respectively, and find that the frequencies f_1, f_2, etc., fluctuate closely round a mean value f, this can be expressed by the statement that there is a stable frequency f for death in the sixtieth year in different English districts. We might also find a similar stability for all the other classifications. On the other hand, for the third and fourth classifications we might find no stability at all, and for the first a greater or less degree of stability than for the second. In the latter case the form of our statistical generalisation must be modified or the argument in its favour weakened.

Example Three.—Let us return to the example given in Chapter XXVII. of the dog which is fed sometimes by scraps at table and so judges it reasonable to be there. From one year to another, let us assume, the dog gets scraps on a proportion of days more or less stable. What sorts of explanation might there be of this? First, it might be the case that he was fed on the movable feasts of the Church; there would be the same number of these in each year, but it would not be easy for any one who had not the clue to discover any regularity in the occasions of their individual occurrence. Second, it might be the case that he was given scraps whenever he looked thin, and that the scraps were withheld whenever he looked fat, so that if he was given scraps on one day, this diminished the likelihood of his getting scraps on the next day, whilst if they were withheld this would increase the likelihood; the dog's constitution remaining constant, the number of days for scraps would tend to fluctuate from year to year about a stable value. Third, it might be the case that the company at table varied greatly from day to day, and that some days people were there of the kind who give dogs scraps and other days not; if the set of people from whom the company was drawn remained more or less the same from year to year, and it was a matter of chance (in the objective sense defined in § 8 of Chapter XXIV. above) which of them were there from day to day, the proportion of days for scraps might again show some degree of stability from year to year. Lastly, a combination between the first and third type of circumstance gives rise to a variant deserving separate mention. It might be the case that the dog was only given scraps by his master, that his master generally went away for Saturday and Sunday, and was at home the rest of the week unless something happened to the contrary, and that "chance" causes would sometimes intervene to keep him at home for the week-end and away in the week; in this case the frequency of days for scraps would probably fluctuate in the neighbourhood of five-sevenths. In circumstances of this third type, however, the degree of stability would probably be less than in circumstances of the first two types; and in order to get a really stable frequency it might be necessary to take a longer period than a year as the basis for each series of observations, or even to take the average for a number of dogs placed in like circumstances instead of one dog only.

It has been assumed so far that we have an opportunity of observing what happens on *every* day of the year. If this is not the case and we have knowledge only of a random sample from the days of each year, then the stability, though it will be less in degree, may be nevertheless observable, and will increase as the number of days included in each sample is increased. This applies equally to each of the three types.

5. What is the correct logical analysis of this sort of reasoning? If an inductive generalisation is a true one, the conclusion which it asserts about the instance under inquiry is, so far as it goes, definite and final, and cannot be modified by the acquisition of more detailed knowledge about the particular instance. But a statistical induction, when applied to a particular instance, is not like this; for the acquisition of further knowledge might render the statistical induction, though not in itself less probable than before, *inapplicable* to that particular instance.

This is due to the fact that a statistical induction is not really about the particular instance at all, but has its subject, about which it generalises, a *series*; and it is only applicable to the particular instance, in so far as the instance is relative to our knowledge, a *random member* of the series. If the acquisition of new knowledge affords us additional relevant information about the particular instance, so that it ceases to be a random member of the series, then the statistical induction ceases to be applicable; but the statistical induction does not for that reason become any less probable than it was—it is simply no longer indicated by our data as being the statistical generalisation appropriate to the instance under inquiry. The point is illustrated by the familiar example that the probability of an unknown individual posting a letter unaddressed can be based on the statistics of the Post Office, but *my* expectation that *I* shall act thus, cannot be so determined.

Thus a statistical generalisation is always of the form: 'The probability, that an instance taken at random from the series S will have the characteristic ϕ, is p;' or, more precisely, if a is a random member of $S(x)$, the probability of $\phi(a)$ is p.

It will be convenient to recapitulate from Chapter XXIV. § 11 the definition of 'an instance taken at random': Let $\phi(x)$ stand for 'x has the characteristic ϕ,' and $S(x)$ for 'x is a member of the class S'; then, on evidence h, a is a random member of the class S for characteristic ϕ, if 'x is a' is irrelevant to $\phi(x)/S(x).h$,[1] *i.e.* if we have no information about a relevant to $\phi(a)$ except $S(a)$.

Or alternatively we might express our definition as follows: Consider a particular instance a, where the object of our inquiry is the probability of $\phi(a)$ relative to evidence h. Let us discard that part of our knowledge $h(a)$ which is irrelevant to $\phi(a)$, leaving us with relevant knowledge $h'(a)$. Let the class of instances a_1, a_2, etc., which satisfy $h'(x)$ be designated by S. Then, relative to evidence h, a is a random member of the class or series S for the characteristic ϕ.

Let us denote the proposition 'x is, on evidence h, a random member of S for characteristic ϕ' by $R(x, S, \phi, h)$; then our statistical generalisation is of the form $\phi(x)/R(x, S, \phi, h).h=p$.

If $R(a, S, \phi, h)$ holds, then, on evidence h, S is the appropriate statistical series to which to refer a for the purposes of the characteristic ϕ.

It is not always the case that the evidence indicates any series at all as 'appropriate' in the above sense. In particular,

[1] The use of variables in probability, as has been pointed out on p. 58, is very dangerous. It might therefore be better to enunciate the above: a is a random member of S for characteristic ϕ, if $\phi(a)/S(a).h=\phi(b)/S(b).h$ where $S(b).h$ contains no information about b, except that b is a member of S

if evidence h indicates S as the appropriate series, and evidence h' indicates S′ as the appropriate series, then relative to evidence hh' (assuming these to be not incompatible), it may be the case that no determinate series is indicated as appropriate. In this case the method of statistical induction fails us as a means of determining the probability under inquiry.

6. We can now remove our attention from the individual instance a to the properties of the series S. What sort of evidence is capable of justifying the conclusion that p is the probability that a random member of the series S will have the characteristic ϕ?

In the simplest case, S is a finite series of which we know the truth frequency for the characteristic ϕ, namely f.[1] Then by a straightforward application of the Principle of Indifference we have $p=f$, so that $\phi(x)/\mathrm{R}(x, \mathrm{S}, \phi, h).h=f$.

In another important type S is a series, with an indefinite number of members which, however, group themselves in such a way that for every member of which $\phi(x)$ is true, there corresponds a determinate number of members of which $\phi(x)$ is false. The series, that is to say, contains an indefinite number of atoms, but each atom is made up of a set of molecules of which $\phi(x)$ is true and false respectively in fixed and determinate proportions. If this determinate proportion is known to be f, we have, as before, $p=f$. The typical instance of this type is afforded by games of chance. Every possible state of affairs which might lead to a divergence in one direction is balanced by another probability leading in the opposite direction; and these alternative possibilities are of a kind to which the Principle of Indifference is applicable. Thus for every poise of the dice box which leads to the fall of the six-face, there is a corresponding poise which leads to the fall of each of the other faces; so that if S is the series of possible poises, we may equate p to $\frac{1}{6}$ where ϕ is the fall of the six-face. It is not necessary, in order to obtain this result, to assert that S is a finite series with an actual determinate frequency f for the fall of each face.

So far no inductive element enters in. But in general we do not know the constitution of S for certain, and can only infer it inductively from its resemblance to other series of which we know the constitution. This presents a normal inductive problem—

[1] *I.e.* if f is the proportion of the members of the series for which $\phi(x)$ is true.

the determination by an analysis of the positive and negative analogies as to whether the respects in which S differs or may differ from the other series is or is not relevant in the particular context ϕ; and it involves the same sort of considerations as those discussed in Part III.

There is, however, a further difficulty to be introduced before we have reached the typical statistical problem. In the case now to be considered our actual data do not consist of positive knowledge of the constitutions either of S itself or of other series more or less resembling S, but only of the frequency of the characteristic in actually observed sets of *selections*, great or small, either from S itself or from other series more or less resembling S.

Thus in the most general case our inquiry falls into two parts. We are given the observed frequency in statistical sets *selected* from S_1, S_2, etc., respectively. The first part of our inquiry is the problem of arguing from these observed frequencies to the probable constitutions of S_1, S_2, etc., *i.e.* of determining the values of $\phi(x)/R(x, S_1, \phi, h).h$, etc.; we may call this part the statistical problem. The second part of our inquiry is the problem of arguing from the probable constitutions of S_1, S_2, etc., to the probable constitution of S, where S, S_1, S_2 resemble one another more or less, and we have to determine whether the differences are or are not relevant to our inquiry; we may call this part the inductive problem.

Now if the observed statistical sets are made up of random instances of S_1, S_2, etc., we can argue in certain conditions from the observed frequencies to the probable constitutions of the series, out of which the random selections have been made, by an inverse application of Bernoulli's Theorem on the lines explained in Chapter XXXI. Moreover, if the series S_1, S_2, etc., are finite series and the observed selections cover a great part of their members, we can reach an at least approximate conclusion without raising all the theoretical difficulties or satisfying all the conditions of Chapter XXXI. The commonly received opinions as to the bearing of the observed frequencies in a random sample on the constitution of the universe out of which the sample is drawn, though generally stated too precisely and without sufficient insistence on the assumptions they involve, our actual evidence not warranting in general more than an approximate result, are not, I think, fundamentally erroneous. The most usual error in modern method consists in treating too lightly what I have termed above the *inductive* problem, *i.e.* the problem of passing from the series S_1, S_2, etc., of which we have observed samples, to the series S of which we have not observed samples.

Let us, then, assume that we have ascertained p_1, p_2, etc., with more or less exactness, by examining either all the instances of the series S_1, S_2, etc., or random selections from them, *i.e.* $\phi(x)/R(x, S_1, \phi, h) . h = p_1$, etc. This can be expressed for short by saying that the series S_1, S_2, etc., are subject to probable-frequencies p_1, p_2, etc., for the characteristic ϕ. Our problem is to infer from this the probable-frequency p of the unexamined series S. The class characteristics of the series S_1, S_2, etc., will be partly the same and partly different. Using the terminology of Part III. we may term the class characteristics which are common to all of them the Positive Analogy, and the class characteristics which are not common to all of them the Negative Analogy.

Now, if the observed or inferred probable-frequencies of the series S_1, S_2, are to form the basis of a statistical induction, they must show a *stable* value; that is to say, either we must have $p_1 = p_2 =$ etc., or at least p_1, p_2, etc., must be stably grouped about their mean value. Our next task, therefore, must be to discover whether the probable-frequencies p_1, p_2, etc., display a significant stability. It is the great merit of Lexis that he was the first to investigate the problem of stability and to attempt its measurement. For, until a *primâ facie* case has been established for the existence of a stable probable-frequency, we have but a flimsy basis for any statistical induction at all; indeed we are limited to the class of case where the instance under inquiry is a member of identically the *same* series as that from which our samples were drawn, *i.e.* where $S = S_1$, which in social and scientific inquiries is seldom the case.

What is the meaning of the assertion that p_1, p_2, etc., are *stably* grouped about their mean value? The answer is not simple and not perfectly precise. We could propound various formulae for the measurement of stability and dispersion, respectively, and the problem of translating the conception of stability, which is not quantitatively precise, into a numerical formula involves an arbitrary or approximative element. For practical purposes, however, I doubt if it is possible to improve on Lexis's measure of stability Q, the mathematical definition of which has been given above on p. 399. Lexis describes the stability as subnormal, normal, or supernormal according as Q is less than, equal to, or greater than 1. This is too precise, and it is better perhaps to say that the stability about the mean is normal if the dispersion is such as would not be improbable *à priori*, if we had assumed that the members of S_1, S_2, etc., were obtained by random selection out of a single universe U, that it is subnormal if the dispersion is less than one would have expected on the same hypothesis, and that it is supernormal if the dispersion is greater than one would have expected.

Let us suppose that we find that on this definition p_1, p_2, etc., are stable about p, and let us postpone consideration of the cases of subnormal or supernormal dispersion. This is equivalent to saying that the frequencies of S_1, S_2, etc., are within limits which we should expect *à priori*, if we had knowledge relative to which their members were chosen at random from a universe U of which the frequency was p for the characteristic under inquiry. We next seek to extend this result to the unexamined series S and to justify anticipations about it on the basis of the members of S also being chosen at random from the universe U. This leads us to the strictly inductive part of our inquiry.

The class characteristics of the several series S_1, S_2, etc., will be partly the same and partly different, those that are the same constituting the positive analogy and those that are different constituting the negative analogy, as stated above. The series S will share part of the positive analogy. The argument for assimilating the properties of S, in relation to the characteristic under inquiry, to the properties of S_1, S_2, etc., in relation to this characteristic depends on the differences between S, S_1, S_2, etc., being *irrelevant* in this particular connection. The method of strengthening this argument seems to me to be the same as the general inductive method discussed in Part III. and to present the same, but not greater, difficulties.

In general this inductive part of our inquiry will be best advanced by classifying the aggregate series of instances with which we are presented in such a way as to analyse most clearly the significant positive and negative analogies, to group them, that is to say, into sub-series S_1, S_2, etc., which show the most marked and definite class characteristics. Our knowledge of the differences between the particular observed instances which constitute our original data will suggest to us one or more principles of classification, such that the members of each sub-series all have in common some set of positive or negative characteristics, not all of which are shared in common by all the members of any of the other sub-series. That is to say, we classify our whole set of instances into a series of series S_1, S_2, etc., which have frequencies f_1, f_2, etc., for the characteristic under inquiry; and then again we classify them by another principle or criterion of classification into a second series of series S_1', S_2', etc., with frequencies f_1', f_2', etc.; and so on, so far as our knowledge of the possible relevant differences between the instances extends; the whole result being then summed up in a statement of the positive and negative analogies of the series of series. If we then find that all the frequencies f_1, f_2, etc., f_1', f_2', etc., are stable about a value p, and if, on the basis of the above positive and negative analogies, we have a normal inductive argument for assimilating the unexamined series S to the examined series S_1, S_2, etc., S_1', S_2', etc., in respect of the characteristic under inquiry, in this case we have, not conclusive grounds, but grounds of some weight for asserting the probability p, that an instance taken at random from S will have the characteristic in question.

Let me recapitulate the two essential stages of the argument. We first find that the observed frequencies in a set of series are such as would have been not improbable *à priori* if, relative to our knowledge, these series had all been made up of random members of the same universe U; and we next argue that the positive and negative analogies of this set of series furnish an inductive argument of some weight for supposing that a further unexamined series S resembles the former series in having a frequency for the characteristic under inquiry such as would have been not improbable *à priori* if, relative to our knowledge, S was also made up of random members of the hypothetical universe U.

7. It is very perplexing to decide how far an argument of this character involves any new and theoretically distinct difficulties or assumptions, beyond those already admitted as inherent in Universal Induction. I believe that the foregoing analysis is along the right lines and that it carries the inquiry a good deal further than it has been carried hitherto. But it is not conclusive, and I must leave to others its more exact elucidation.

There is, however, a little more to be said about the half-felt reasons which, in my judgment, recommend to common sense some at least of the scientific (or semi-scientific) arguments which run along the above lines. In expressing these reasons I shall be content to use language which is not always as precise as it ought to be.

I gave in Chapter XXIV. §§ 7-9 an interpretation of what is meant by an 'objectively chance' occurrence, in the sense in which the results of a game, such as roulette, may be said to be governed by 'objective chance.' This interpretation was as follows: "An event is due to objective chance if in order to predict it, or to prefer it to alternatives, at present equi-probable, with any high degree of probability, it would be necessary to know a great many more facts of existence about it than we actually do know, and if the addition of a wide knowledge of general principles would be little use." The ideal instance of this is the game of chance; but there are other examples afforded by science in which these conditions are fulfilled with more or less perfection. Now the field of statistical induction is the class of phenomena which are due to the combination of two sets of influences, one of them constant and the other liable to vary in accordance with the expectations of objective chance,—Quetelet's 'permanent causes' modified by 'accidental causes.' In social and physical statistics the ultimate alternatives are not as a rule so perfectly fixed, nor the selection from them so purely random, as in the ideal game of chance. But where, for example, we find stability in the statistics of crime, we could explain this by supposing that the population itself is stably constituted, that persons of different temperaments are alive in proportions more or less the same from year to year, that the motives for crime are similar, and that those who come to be influenced by these motives are selected from the population at large in the same kind of way. Thus we have stable causes at work leading to the several alternatives in fixed proportions, and these are modified by random influences. Generally speaking, for large classes of social statistics we have a more or less stable population including different kinds of persons in certain proportions and on the other hand sets of environments; the proportions of the different kinds of persons, the proportions of the different kinds of environments, and the manner of allotting the environments to the persons vary in a *random* manner from year to year (or, it may be, from district to district). In all such cases as these, however, prediction beyond what has been observed is clearly open to sources of error which can be neglected in considering, for example, games of chance;—our so-called 'permanent' causes are always changing a little and are liable at any moment to radical alteration.

Thus the more closely that we find the conditions in scientific examples assimilated to those in games of chance, the more confidently does common sense recommend this method. The rather surprising frequency with which we find apparent stability in human statistics may possibly be explained, therefore, if the biological theory of Mendelism can be established. According to this theory the qualities apparent in any generation of a given race appear in proportions which are determined by methods very closely analogous to those of a game of chance. To take a specific example (I am giving not the correct theory of sex but an artificially simplified form of it), suppose there are two kinds of spermatozoa and two kinds of ova and of the four possible kinds of union two produce males and two females, then if the kinds of spermatozoa and ova exist in equal numbers and their union is determined by random considerations in precisely the same sense in which a game of chance such as roulette depends upon random considerations, we should expect the observed proportions to vary from equality, as indeed they do, in the same manner as variations from equality of red and black occur at roulette.[1] If the sphere of influence of Mendelian considerations is wide, we have both an explanation in part of what we observe and also a large opportunity in future of using with profit the methods of statistical analysis.

This is all familiar. This is the way in which in fact we do think and argue. The inquiry as to how far it is covered by the abstract analysis of the preceding paragraphs, and by what

[1] The fluctuations in the proportion of the sexes which, as is well known, is not in fact one of equality, correspond, as Lexis has shown, to what one would expect in a game of chance with an astonishing exactitude. But it is difficult to find any other example, amongst natural or social phenomena, in which his criteria of stability are by any means as equally well satisfied.

logical principle the use of this analysis can be justified as rational, I have pushed as far as I can. It deserves a profounder study than logicians have given it in the past.

8. Two subsidiary questions remain to be mentioned. The first of these relates to the character of series which, in the terminology of Lexis, show a subnormal or supernormal stability; for I have pressed on to the conclusion of the argument on the assumption that the stabilities are normal. Subnormal stability conceals two types: the one in which there is really no stability at all and the results are in fact chaotic; and the other in which there is mutual dependence between the successive instances of such a kind that they tend to resemble one another so that any divergence from the normal tends to accentuate itself. Supernormal stability corresponds in the other direction to the second of these two types; that is to say, there is mutual dependence of a regulative kind between the successive instances which tends to prevent the frequency from swinging away from its mean value. The case, where the dog was fed with scraps when he looked thin and not fed when he looked fat, illustrated this. The typical example of this type is where balls are drawn from urns, containing black and white balls in certain proportions and *not* replaced; so that every time a black ball is drawn the next ball is more likely than before to be white, and there is a tendency to redress any excess of either colour beyond the proper proportions. Possibly the aggregate annual rainfall may afford a further illustration.

Where there is no stability at all and the frequencies are chaotic, the resulting series can be described as 'non-statistical.' Amongst 'statistical series,' we may term 'independent series' those of which the instances are independent and the stability normal, and 'organic series,' those of which the instances are mutually dependent and the stability abnormal, whether in excess or in defect. 'Organic series' have been incidentally discussed elsewhere in this volume. I shall not pursue them further now, because I do not think that they introduce any new *theoretical* difficulty into the general problem of statistical inference; although the problem of fitting them into the general theoretical scheme is not easy.[1]

[1] The following more precise definitions bring these ideas into line with what has gone before: consider the terms $a_1, a_2 \ldots a_n$ of a series $s(x)$; let 'a_r is g'

9. The second question is concerned with the relation between the Inductive Correlation, which has been the subject-matter of this chapter, and the Correlation Coefficient, or, as I should prefer to call it, the *Quantitative Correlation*, with which recent English statistical theory has chiefly occupied itself. I do not propose to discuss this theory in detail, because I suspect that it is much more concerned, at any rate in its present form, with statistical description than with statistical induction. The transition from defining the 'correlation coefficient' as an algebraical expression to its employment for purposes of inference is very far from clear even in the work of the best and most systematic writers on the subject, such as Mr. Yule and Professor Bowley.

In the notation employed in the earlier part of this chapter I have classified each examined instance a according as it did or did not possess the characteristic ϕ, *i.e.* satisfy the propositional function $\phi(x)$, or, in other words, according as $\phi(a)$ was true or false. Thus only two possible alternatives were contemplated, and ϕ was not considered as a quantitative characteristic which the instance could satisfy in greater or less degree. Equally the common element in all the instances, required to constitute them as instances for the purpose of our statistical generalisation (or, as I have sometimes put it, required to satisfy the *condition* of the generalisation), was regarded as definite and unique and not capable of quantitative variation. That is to say, all the instances satisfied a function $\psi(x)$, and the question was, what proportion

---

$\equiv g_r$ and let $g_r/h = p_r$, where h is our data. Then, if $g_r/g_s \ldots \bar{g}_t \ldots h = p_r$ for all values of $r, s, \ldots, t \ldots$, the terms of the series are *independent* relative to h. If $p_1 = p_2 = \ldots = p$ the terms are *uniform*. If the terms are both independent and uniform, the series may be called an *independent Bernoullian series*, subject to a *Bernoullian probability* p. If the terms are independent but not uniform, the series may be called an *independent compound series*, subject to a *compounded probability* $1/n\Sigma p_r$. If the terms are not independent, the series is an *organic series*.

The same terminology can then be applied to the series $S_1, S_2, \ldots S_n$, regarded as members of the series of series $S(x)$. Let the frequencies of the series for the characteristic under inquiry be $x_1, x_2, \ldots x_n$, and let $x_1/h = \theta_1(x_1)$, *i.e.* $\theta_1(x_1)$ is the probability of a frequency x_1 in the first series. Then if $x_r/x_s \ldots h = \theta_r(x_r)$ for all values of r, s, etc., the frequencies are *independent*; and if $\theta_1(x) = \theta_2(x_2) = \ldots \theta(x)$, the frequencies are *stable*. If the frequencies are stable and independent, the series of series may be called *Gaussian*. If the frequencies are stable and independent, and if in addition each individual series is subject to a Bernoullian probability, the probable dispersion of the frequency is normal and symmetrical. If the individual series are organic, the dispersion of the frequencies may be normal, subnormal, or supernormal. If the series of series is Gaussian, and the individual series Bernoullian, we have the type of the perfect statistical series.

of them also satisfied the function $\phi(x)$. A typical example was that of sex-ratio,—$\psi(x)$ being the birth of a child and $\phi(x)$ its sex, where there is no question of *degree* in either $\psi(x)$ or $\phi(x)$.

It might be the case, however, that the characteristics under examination were capable of degree or quantitative variation; for example $\psi(x)$ might be the age of the mother and $\phi(x)$ the weight of the child at birth. In this case we should have a series $\psi_1(x)$, $\psi_2(x)$, etc., corresponding to the various age-periods of the mothers, and a series $\phi_1(x)$, $\phi_2(x)$, etc., corresponding to the various weights of the children. Now if we concentrated our attention on $\psi_1(x)$ and $\phi_1(x)$ alone, *i.e.* on mothers of a particular age and the proportions of their children which had a particular weight at birth, we have a one-dimensional problem of the same kind as before; out of all the instances which satisfy $\psi_1(x)$ a certain proportion satisfy $\phi_1(x)$ also. But clearly we can push our observations further and we can take note what proportion of the instances which satisfy $\psi_1(x)$ satisfy $\phi_2(x)$, $\phi_3(x)$, and so on, respectively; and then we can do the same as regards the instances which satisfy $\psi_2(x)$, $\psi_3(x)$, etc. The total results of this two-dimensional set of observations can then be tabulated in what is called a twofold correlation table. Thus if f_{rs} is the proportion of instances satisfying $\psi_s(x)$ which also satisfy $\phi_r(x)$ we have a table as follows:

	$\psi_1(x)$	$\psi_2(x)$	$\psi_3(x)$	...
$\phi_1(x)$	f_{11}	f_{12}	f_{13}	
$\phi_2(x)$	f_{21}	f_{22}	f_{23}	
$\phi_3(x)$	f_{31}	f_{32}	f_{33}	
⋮	⋮	⋮	⋮	

We could, further, increase the complexity and completeness of our observations to any required degree. For example we might take account also of $\theta(x)$, the age of the father, and construct a threefold table where f_{rst} is the proportion of instances satisfying $\phi_r(x)$, $\psi_s(x)$, $\theta_t(x)$; and so on up to an n-fold table.

Clearly it is not necessary for the construction of tables of this kind that $\phi(x)$ and $\psi(x)$ should stand for degrees of the same quantitative characteristic; they might be any set of exclusive alternatives; for example, $\psi(x)$ might be the colour of the baby's eyes, and $\phi(x)$ its Christian name.

But in order that the correlation table may be of any practical interest for the purposes of inference, it is necessary—and this, I think, is one of the critical assumptions of correlation—that $\phi_1(x)$, $\phi_2(x)$ . . . and also $\phi_1(x)$, $\phi_2(x)$ . . . should be arranged in an order that is *significant*, *i.e.* such that we have some *à priori* reason for expecting some connection to exist between the *order* of the ϕ's and the *order* of the ϕ's. The point of this will be illustrated by concentrating our attention on the simplest type of case where $\phi(x)$ and $\phi(x)$ are quantitative characteristics arranged in order of magnitude. Now suppose it were the case that the younger mothers tended to bear heavier babies, then, if $\phi_1(x)$ $\phi_2(x)$ are the ages increasing upwards and $\phi_1(x)$ $\phi_2(x)$ the weights diminishing downwards, f_{11} would probably be the greatest of the f_{r1}'s and, generally speaking, f_{r1} would be greater than $f_{r+1,1}$; also f_{22} might be the greatest of the f_{r2}'s, and so on; so that the frequencies lying on the diagonal of the table would be the greatest and the frequencies would tend to be less the farther they lay from the diagonal. If we had some reason *à priori* (*i.e.* based on our pre-existing knowledge), if only a slight one, for supposing that there might be some connection between the age of the mother and the weight of the baby, then, if in a particular set of instances the frequencies were grouped about the diagonal as suggested above, this might be taken as affording some inductive support for the hypothesis.

Now the theory of correlation, as it is expounded in the text-books, is almost entirely concerned with measuring how nearly the observed frequencies are grouped about the diagonal of the table (though the complete theory is not, of course, so restricted as this). The 'coefficient of correlation' is an algebraical formula which may be regarded as measuring this phenomenon in a way that is sufficiently satisfactory for all ordinary purposes. If it is defined thus, it is simply a statistical description of a particular set of observations arranged in a particular order. How can we make use of this coefficient for the purposes of inference?

Dr. Bowley faces this problem a little more definitely than do most statistical writers. Mr. Yule warns the student that the problem exists,[1] but he does not himself attack it systematically or do more than apply common sense to particular problems. So much greater emphasis, however, has been laid hitherto on the mathematical complications, that many statistical students hazily float from defining the correlation coefficient as a statistical description to employing it as a measure of the probability of a statistical generalisation as to the association between quantitative variations of $\phi(x)$ and $\psi(x)$ respectively. If, for example, it is found in a particular set of observations of mothers' ages and babies' weights that the frequencies are closely ranged about the diagonal, this is considered a sufficiently good reason for attributing probability to a generalisation as to the 'correlation' (*i.e.* tendency to quantitative correspondence) between the age of the mother and the weight of the baby.

Dr. Bowley's line of thought is as follows. He begins by defining the correlation coefficient r merely as a statistical description (*Elements of Statistics*, p. 354). He then shows (p. 355), as an illustration of the nature of r, that if x and y are two variable quantities which depend (more strictly, *are known* to depend) on other variables U, V, W in such a way that

$$X_t = {}_1U_t + {}_2U_t + \ldots + {}_pU_t + {}_1V_t + {}_2V_t + \ldots + {}_qV_t$$
$$Y_t = {}_1U_t + {}_2U_t + \ldots + {}_pU_t + {}_1W_t + {}_2W_t + \ldots {}_qW_t$$

where ${}_1U_t$, ${}_2U_t \ldots {}_1V_t$, ${}_2V_t \ldots {}_1W_t$, ${}_2W_t \ldots$ are selected at random each from an independent group of quantities (more strictly, are *relative to our data*, random members of independent groups); then, if we know *à priori* certain statistical coefficients descriptive of the constitution of these groups, the value of r will probably tend towards a certain value. So far we are on fairly safe, but not very fruitful, ground. We have no basis for arguing backwards from the observed value of r; but, provided we have rather extensive and peculiar knowledge *à priori* as to how X_t and Y_t are constituted, then we have calculable expectations as to the limits within which the value

[1] *Introduction to the Theory of Statistics*, p. 191: "The coefficient of correlation, like an average or a measure of dispersion, only exhibits in a summary and comprehensible form one particular aspect of the facts on which it is based, and the real difficulties arise in the interpretation of the coefficient when obtained."

of r, namely the correlation coefficient between X and Y, will probably turn out to lie, when we have observed it.

Dr. Bowley's next move is more dubious. If the constitutions of the independent groups are similar in a certain statistical respect (*i.e.* if they have the same standard deviations), then, Dr. Bowley concludes, $r=\frac{p}{p+q}$, which "expressed in words shows that the correlation coefficient tends to be the ratio of the number of causes common in the genesis of two variables to the whole number of independent causes on which each depends." By this time the student's mind, unless anchored by a more than ordinary scepticism, will have been well launched into a vague, fallacious sea.

Neglecting, however, the *dictum* just quoted, we find that the second stage of the argument consists in showing that, *if* we have a certain sort of knowledge *à priori* as to how our variables are constituted, then the various possible values for the coefficients of correlation, which would be yielded by actual sets of observations made in prescribed conditions, will have, *à priori*, and before the observations have been made, calculable probabilities, certain ranges of values being probable and others improbable.

As a rule, however, we are not arguing from knowledge about the variables to anticipations about their correlation coefficient; but the other way round, that is from observations of their correlation coefficients to theories about the nature of the variables. Dr. Bowley perceives that this involves a third stage of the argument, and appeals accordingly (p. 409) to "the difficult and elusive theory of inverse probability." He apprehends the difficulty but he does not pursue it; and, like Mr. Yule, he really falls back for practical purposes on the criteria of common sense, an expedient well enough in his case, but not a universal safeguard.

The general argument from inverse probability to which Dr. Bowley makes his vague appeal is doubtless on the following lines: If there is no causal connection between the two sets of quantities, then a close grouping of the frequencies about the diagonal would be *à priori* improbable (and the greater the number of the individual observations, the greater the improbability since, if the quantities are independent, there is, then, all the more opportunity for 'averaging out'); therefore, inversely, if the frequencies do group themselves about the diagonal, we have a presumption in favour of a causal connection between the two sets of quantities.

But if the reader recalls our discussion of the principle of inverse probability, he will remember that this conclusion cannot be reached unless *à priori*, and quite apart from the observations in question, we have some reason for thinking that there may be such a causal connection between the quantities. The argument can only strengthen a pre-existing presumption; it cannot create one. And in the absence of reasons peculiar to the particular inquiry, we have no choice but to fall back on the general methods and the general presumptions of induotion.

It is apparent that, where the correlation argument seems plausible, some tacit assumption must have slipped in, if we return to the case where our correlation table relates to the weights of the babies and their Christian names. Either by accident or because we had arranged the order of the Christian names to suit, it might happen with a particular set of observations, even a fairly numerous set, that the correlation coefficient was large. Yet on that evidence alone we should hardly assert a generalisation connecting the weights of babies with their Christian names.

The truth is that sensible investigators only employ the correlation coefficient to test or confirm conclusions at which they have arrived on other grounds. But that does not validate the crude way in which the argument is sometimes presented, or prevent it from misleading the unwary,—since not all investigators are sensible.

If we abandon the method of inverse probability in favour of the less precise but better founded processes of induction, 'quantitative correlation,' as I should like to term this particular branch of statistical induction, is more complicated than, but not theoretically distinct from, the kind of arguments which have occupied the earlier paragraphs of this chapter. The character of the additional complication can be described by saying that we are presented with a two-dimensional problem instead of a one-dimensional problem. The mere existence of a particular correlation coefficient as descriptive of a group of observations, even of a large group, is not in itself a more conclusive or significant argument than the mere existence of a particular frequency coefficient would be. Of course if we have a considerable body of pre-existing knowledge relevant to the particular inquiry, the calculation of a small number of correlation coefficients may be crucial. But otherwise we must proceed as in the case of frequency coefficients; that is to say we must have before us, in order to found a satisfactory argument, many sets of observations, of which the correlation coefficients display a significant stability in the midst of variation in the non-essential class characteristics (*i.e.* those class characteristics which our generalisation proposes to neglect) of the different sets of observations.

10. I am now at the conclusion of an inquiry in which, beginning with fundamental questions of logic, I have endeavoured to push forward to the analysis of some of the actual arguments which impress us as rational in the progress of knowledge and the practice of empirical science. In writing a book of this kind the author must, if he is to put his point of view clearly, pretend sometimes to a little more conviction than he feels. He must give his own argument a chance, so to speak, nor be too ready to depress its vitality with a wet cloud of doubt. It is a heavy task to write on these problems; and the reader will perhaps excuse me if I have sometimes pressed on a little faster than the difficulties were overcome, and with decidedly more confidence than I have always felt.

In laying the foundations of the subject of Probability, I have departed a good deal from the conception of it which governed the minds of Laplace and Quetelet and has dominated through their influence the thought of the past century,—though I believe that Leibniz and Hume might have read what I have written with sympathy. But in taking leave of Probability, I should like to say that, in my judgment, the practical usefulness of those modes of inference, here termed Universal and Statistical Induction, on the validity of which the boasted knowledge of modern science depends, can only exist—and I do not now pause to inquire again whether such an argument *must* be circular—if the universe of phenomena does in fact present those peculiar characteristics of atomism and limited variety which appear more and more clearly as the ultimate result to which material science is tending:

> fateare necessest
> materiem quoque *finitis* differre figuris.

The physicists of the nineteenth century have reduced matter to the collisions and arrangements of particles, between which the ultimate qualitative differences are very few; and the Mendelian biologists are deriving the various qualities of men from the collisions and arrangements of chromosomes. In both cases the analogy with the perfect game of chance is really present; and the validity of some current modes of inference may depend on the assumption that it is to material of this kind that we are applying them. Here, though I have complained sometimes at their want of logic, I am in fundamental sympathy with the deep underlying conceptions of the statistical theory of the day. If the contemporary doctrines of Biology and Physics remain tenable, we may have a remarkable, if undeserved, justification of some of the methods of the traditional Calculus of Probabilities. Professors of probability have been often and justly derided for arguing as if nature were an urn containing black and white balls in fixed proportions. Quetelet once declared in so many words—" l'urne que nous interrogeons, c'est la nature." But again in the history of science the methods of astrology may prove useful to the astronomer; and it may turn out to be true—reversing Quetelet's expression—that " La nature que nous interrogeons, c'est une urne."

BIBLIOGRAPHY

BIBLIOGRAPHY

INTRODUCTION

There is no opinion, however absurd or incredible, which has not been maintained by some one of our philosophers.—DESCARTES.

THE following Bibliography does not pretend to be complete, but it contains a much longer list of what has been written about Probability than can be found elsewhere. I have hesitated a little before burdening this volume with the titles of many works, so few of which are still valuable. But I was myself much hampered, when first I embarked on the study of this subject, by the absence of guide-posts to the scattered but extensive literature of the subject; and a list which I drew up for my own convenience, without much attention to bibliographical nicety or to exact uniformity in the style of entry, may be useful to others.

It is rather an arbitrary matter to decide what to include and what to exclude. Probability overlaps many other topics, and some of the most important references to it are to be found in books, the main topic of which is something else. On the other hand it would be absurd to include every casual reference; and no useful purpose would have been served by cataloguing the very numerous volumes dealing with Insurance, Games of Chance, Statistics, Errors of Observation, and Least Squares, which treat in detail these various applications of the Theory of Probability. It has been a matter of some difficulty, therefore, to know precisely where to draw the line. Where the main subject of a book or paper is Probability proper, I have included it, nearly regardless of my own view as to its importance, and have not attempted to act as censor; but where Probability is not the main subject or where an application of Probability is concerned, the chief interest of which is solely in the application itself, I have only included the entry where I think it important, intrinsically or historically or from the celebrity of the author. In particular, the existence of Professor Mansfield Merriman's very extensive bibliography, published in the *Transactions of the Connecticut Academy* for 1877, has made it possible to deal very lightly (and to the extent of but few entries) with the inordinately large literature of Least Squares. This list comprises 408 titles of writings relating to the Method of Least Squares and the theory of accidental errors of observation, and is sufficiently exhaustive so far as relates to memoirs on this topic published before 1877.

Of bibliographical sources for Probability proper, Todhunter's *History of the Mathematical Theory of Probability* and Laurent's *Calcul des probabilités* are alone important. Of *mathematical* works published before the time of Laplace, Todhunter's list, and also his commentary and analysis, are complete and exact,—a work of true learning, beyond criticism. The bibliographical catalogue at the conclusion of Laurent's *Calcul* (published in 1873) is the longest list published hitherto of general works on Probability. But it is unduly swollen by the inclusion of numerous items on Insurance and Errors of Observation, the bearing of which on Probability is very slight;[1] it is chiefly mathematical in bias; and it is now nearly fifty years old.

I have not read all these books myself, but I have read more of them than it would be good for any one to read again. There are here enumerated many dead treatises and ghostly memoirs. The list is too long, and I have not always successfully resisted the impulse to add to it in the spirit of a collector. There are not above a hundred of these which it would be worth while to preserve,—if only it were securely ascertained which these hundred are. At present a bibliographer takes pride in numerous entries; but he would be a more useful fellow, and the labours of research would be lightened, if he could practise deletion and bring into existence an accredited *Index Expurgatorius*. But this can only be accomplished by the slow mills of the collective judgment of

[1] Laurent's list contains 310 titles, of which I have excluded 174 from my list as being insufficiently relevant.

the learned; and I have already indicated my own favourite authors in copious footnotes to the main body of the text.

The list is long; yet there is, perhaps, no subject of equal importance and of equal fascination to men's minds on which so little has been written. It is now fifty-five years since Dr. Venn, still an accustomed figure in the streets and courts of Cambridge, first published his *Logic of Chance*; yet amongst systematic works in the English language on the logical foundations of Probability my Treatise is next to his in chronological order.

The student will find many famous names here recorded. The subject has preserved its mystery, and has thus attracted the notice, profound or, more often, casual, of most speculative minds. Leibniz, Pascal, Arnauld, Huygens, Spinoza, Jacques and Daniel Bernoulli, Hume, D'Alembert, Condorcet, Euler, Laplace, Poisson, Cournot, Quetelet, Gauss, Mill, Boole, Tchebychef, Lexis, and Poincaré, to name those only who are dead, are catalogued below.

ABBOTT, T. K. "On the Probability of Testimony and Arguments." Phil. Mag. (4), vol. 27, 1864.

ADRAIN, R. "Research concerning the Probabilities of the Errors which happen in making Observations." The Analyst or Math. Museum, vol. 1, pp. 93-109, 1808.

[This paper, which contains the first deduction of the normal law of error, was partly reprinted by Abbé with historical notes in Amer. Journ. Sci. vol. i. pp. 411-415, 1871.]

AMMON, O. "Some Social Applications of the Doctrine of Probability." Journ. Pol. Econ. vol. 7, 1899.

AMPÈRE. Considérations sur la théorie mathématique du jeu. Pp. 63. 4to. Lyon, 1802.

ANCILLON. "Doutes sur les bases du calcul des probabilités." Mém. Ac. Berlin, pp. 3-32, 1794-5.

ARBUTHNOT, J. Of the Laws of Chance, or a Method of Calculation of the Hazards of Game plainly Demonstrated. 16mo. London, 1692.

[Contains a translation of Huygens, De ratiociniis in ludo aleae.]

4th edition revised by John Hans. By whom is added a demonstration of the gain of the banker in any circumstance of the game call'd Pharaon, etc. Sm. 8vo. London, 1738.

[For a full account of this book and discussion of the authorship, see Todhunter's History, pp. 48-53.]

"An Argument for Divine Providence, taken from the constant Regularity observ'd in the Births of both Sexes." Phil. Trans. vol. 27, pp. 186-190 (1710-12).

[Argues that the excess of male births is so invariable, that we may conclude that it is not an even chance whether a male or female be born.]

ARISTOTLE. Anal. Prior. ii. 27, 70a 3.
Rhet. i. 2, 1357 a 34. [See Zeller's *Aristotle* for further references.]
ARNAULD. (The Port Royal Logic.) La Logique ou l'Art de penser. 12mo. Paris, 1662. Another ed. C. Jourdain, Hachette, 1846. Transl. into Eng. with introduction by T. S. Baynes. London, 1851. xlvii + 430. See especially pp. 351-370.

BABBAGE, C. An Examination of some Questions connected with Games of Chance. 4to. 25 pp. Trans. R. Soc. Edin., 1820.
BACHELIER, LOUIS. Calcul des probabilités. Tome i. 4to. Pp. vii + 517. Paris, 1912.
Le Jeu, la chance, et le hasard. Pp. 320. Paris, 1914.
[BAILEY, SAMUEL.] Essays on the pursuit of truth, on the progress of knowledge and on the fundamental principle of all evidence and expectation. Pp. xii + 302. London, 1829.
BALDWIN. Dictionary of Philosophy. Bibliographical volumes; *s.v.* "Probability."
BANIOL, A. "Le Hasard." Revue Internationale de Sociologie. Pp. 16. 1912.
BARBEYRAC. Traité du jeu. 1st ed. 1709. 2nd ed. 1744.
[Todhunter states (p. 196) that Barbeyrac is said to have published a discourse "Sur la nature du sort."]
BAYES, THOMAS. An Essay towards solving a Problem in the Doctrine of Chances. Phil. Trans. vol. liii. pp. 370-418, 1763. A demonstration, etc. Phil. Trans. vol. liv. pp. 296-325, 1764.
[Both the above were communicated by the Rev. Richard Price, and the second is partly due to him.]
German transl. Versuch zur Lösung eines Problems der Wahrscheinlichkeitsrechnung. Herausgegeben von H. E. Timerding. Sm. 8vo. Leipzig, 1908. Pp. 57.
BÉGUELIN. "Sur les suites ou séquences dans le loterie de Gênes." Hist. de l'Acad. Pp. 231-280. Berlin, 1765.
"Sur l'usage du principe de la raison suffisante dans le calcul des probabilités." Hist. de l'Acad. Pp. 382-412. Berlin, 1767. (Publ. 1769.)
BELLAVITIS. "Osservazioni sulla theoria delle probabilità." Atti del Instituto Veneto di Scienze, Lettere, ed Arti, Venice, 1857.
BENARD. "Note sur une question de probabilités." Journal de l'École royale politechnique. Vol. 15, Paris, 1855.
BENTHAM, J. Rationale of Judicial Evidence.
See Introductory View, chap. xii., and Bk. i. chaps. v., vi., vii.
BERNOULLI, DANIEL. "Specimen theoriae novae de mensura sortis." Comm. Acad. Sci. Imp. Pet. vol. v. pp. 175-192, 1738.
Germ. transl. 1896, by A. Pringsheim: Die Grundlage der modernen Wertlehre. Versuch einer neuen Theorie der Wertbestimmung von Glücksfällen (Einleitung von Ludvig Fick). Pp. 60. Leipzig, 1896.
"Recueil des pièces qui ont remporté le prix de l'Académie Royale des Sciences." 1734. iii. pp. 95-144.
[On "La cause physique de l'inclinaison des plans des orbites des planètes par rapport au plan de l'équateur de la révolution du soleil autour de son axe."]
"Essai d'une nouvelle analyse de la mortalité causée par la petite vérole." Hist. de l'Acad. pp. 1-45. Paris, 1760.
De usu algorithmi infinitesimalis in arte conjectandi specimen. Novi Comm. Petrop., 1766. xii. pp. 87-98. A 2nd memoir. Petrop., 1766. xii. pp. 99-126. See a criticism by Trembley, Mém. de l'Acad., Berlin, 1799.

BERNOULLI, DANIEL.—*continued.*

Disquisitiones analytiquae de novo problemate conjecturali. Novi Comm. Petrop. xiv. pp. 1-25, 1769. A 2nd memoir, Petrop. xiv. pp. 26-45, 1769.

"Dijudicatio maxime probabilis plurium observationum discrepantium atque verisimillima inductio inde formanda." Acta Acad., pp. 3-23. Petrop., 1777. Crit. by Euler, pp. 24-33.

BERNOULLI, JAC. Ars conjectandi, opus posthumum. Pp. ii + 306 + 35. Sm. 4to, Basileae, 1713.

[Published by N. Bernoulli eight years after Jac. Bernoulli's death.]

Part I. Reprint with notes and additions of Huygens, De ratiociniis in ludo aleae.

Part II. Doctrina de permutationibus et combinationibus.

Part III. Explicans usum praecedentis doctrinae in variis sortitionibus et ludis aleae. [Twenty-four problems.]

Part IV. Tradens usum et applicationem praecedentis doctrinae in civilibus, moralibus et oeconomicis.

Tractatus de seriebus infinitis. [Not connected with the subject of Probability.]

Lettre à un amy, sur les partis du jeu de paume.

[The most important sections, including Bernoulli's Theorem, are in Part IV. For a very full account of the whole volume see Todhunter's *History*, chap. vii.]

Engl. Transl. of Part II. only, vide *Maseres.*

Fr. transl. of Part I. only, vide *Vastel.*

Germ. transl.: Wahrscheinlichkeitsrechnung. 4 Teile mit dem Anhange: Brief an einem Freund über das Ballspiel, übers. u. hrsg. v. R. Haussner. 2 vols. Sm. 8vo. 1899.

[See also *Leibniz.*]

BERNOULLI, JOHN. De alea, sive arte conjectandi, problemata quaedam. Collected ed. vol. iv. pp. 28-33. 1742.

BERNOULLI, JOHN (grandson). "Sur les suites ou séquences dans la loterie de Gênes." Hist. de l'Acad, pp. 234-253. Berlin, 1769.

"Mémoire sur un problème de la doctrine du hasard." Hist. de l'Acad., pp. 384-408. Berlin, 1768.

BERNOULLI, NICHOLAS. Specimina artis conjectandi, ad quaestiones juris applicatae. Basel, 1709. Repr. Act. Erud. Suppl., pp. 159-170, 1711.

BERTRAND, J. Calcul des probabilités. Pp. lvii + 332. Paris, 1889.

"Sur l'application du calcul des probabilités à la théorie des jugements." Comptes rendus, 1887.

"Les Lois du hasard." Rev. des Deux Mondes, p. 758. Avril 1884.

BESSEL. "Untersuchung über die Wahrscheinlichkeit der Beobachtungsfehler." Astr. Nachrichten, vol. xv. pp. 369-404, 1838.

Also Abhandl. von Bessel, vol. ii. pp. 372-391. Leipzig, 1875.

BICQUILLEY, C. F. DE. Du calcul des probabilités. 164 pp., 1783. 2nd ed. 1805.

Germ. transl. by C. F. Rüdiger. Leipzig, 1788.

BIENAYMÉ, J. "Sur un principe que Poisson avait cru découvrir et qu'il avait appelé loi des grands nombres." Comptes rendus de l'Acad. des Sciences morales, 1855.

[Reprinted in Journal de la Soc. de Statistiques de Paris, pp. 199-204, 1876.]

"Probabilité de la constance des causes conclue des effets observés." Procès-verbaux de la Soc. Philomathique, 1840.

"Sur la probabilité des résultats moyens des observations, etc." Sav. Étrangers. v., 1838.

BIENAYMÉ, J.—*continued.*
"Théorème sur la probabilité des résultats moyens des observations." Procès-verbaux de la Soc. Philomathique, 1839.
"Considérations à l'appui de la découverte de Laplace sur la loi de probabilité dans la méthode des moindres carrés." Comptes rendus des séances de l'Académie des Sciences, vol. xxxvii., 1853.
[Reprinted in Journal de Liouville, 2nd series, vol. xii., 1867, pp. 158-176.]
"Remarques sur les différences qui distinguent l'interpolation de Cauchy de la méthode des moindres carrés." Comptes rendus, 1853.
"Probabilité des erreurs dans la méthode des moindres carrés." Journ. Liouville, vol. xvii., 1852.

BINET. "Recherches sur une question de probabilité" (Poisson's Theorem). Comptes rendus, 1844.

BLASCHKE, E. Vorlesungen über mathematische Statistik. Pp. viii + 268. Leipzig, 1906.

BOBEK, K. J. Lehrbuch der Wahrscheinlichkeitsrechnung. Nach System Kleyer. Pp. 296. Stuttgart, 1891.

BOHLMANN, G. "Die Grundbegriffe der Wahrscheinlichkeitsrechnung in ihrer Anwendung auf die Lebensversicherung." Atti del IV Congr. intern. dei matematici, Rome, 1909.

BOOLE, G. Investigations of Laws of Thought on which are founded the Mathematical Theories of Logic and Probabilities. Pp. ix + 424. London, 1854.
"Proposed Questions in the Theory of Probabilities." Cambridge and Dublin Math. Journal, 1852.
"On the Theory of Probabilities, and in particular on Michell's Problem of the Distribution of the Fixed Stars." Phil. Mag., 1851.
"On a General Method in the Theory of Probabilities." Phil. Mag., 1852.
"On the Solution of a Question in the Theory of Probabilities." Phil. Mag., 1854.
"Reply to some Observations published by Mr. Wilbraham in the Phil. Mag. vii. p. 465, on Boole's 'Laws of Thought.'" Phil. Mag., 1854.
"Further Observations in reply to Mr. Wilbraham." Phil. Mag., 1854.
"On the Conditions by which the Solutions of Questions in the Theory of Probabilities are limited." Phil. Mag., 1854.
"On certain Propositions in Algebra connected with the Theory of Probabilities." Phil. Mag., 1855.
"On the Application of the Theory of Probabilities to the Question of the Combination of Testimonies or Judgments." Edin. Phil. Trans. vol. xxi. pp. 597-652, 1857.
"On the Theory of Probabilities." Roy. Soc. Proc. vol. xii. pp. 179-184, 1862-1863.

BORCHARDT, B. Einführung in die Wahrscheinlichkeitslehre. vi + 86. Berlin, 1889.

BORDONI, A. Sulle probabilità. 4to. Giorn. dell' I. R. Instit. Lombardo di Scienze. T. iv. Nuova Serie. Milano, 1852.

BOREL, E. Éléments de la théorie des probabilités. 8vo, pp. vii + 191. Paris, 1909. 2nd ed. 1910.
Le Hasard. Pp. iv + 312. Paris, 1914.
"Le Calcul des probabilités et la méthode des majorités." L'Année psychologique, vol. 14, pp. 125-151. Paris, 1908.
"Les Probabilités dénombrables et leurs applications arithmétiques." Rendiconti del Circolo matematico di Palermo, 1909.
"Le Calcul des probabilités et la mentalité individualiste." Revue du Mois, vol. 6, pp. 641-650, 1908.

BOREL, E.—*continued.*
"La Valeur practique du calcul des probabilités." Revue du Mois, vol. 1, pp. 424-437, 1906.
"Les Probabilités et M. le Dantec." Revue du Mois, vol. 12, pp. 77-91, 1911.

BORTKIEWICZ, L. VON. Das Gesetz der kleinen Zahlen. 8vo, pp. viii + 52, Leipzig, 1898.
"Anwendungen der Wahrscheinlichkeitsrechnung auf Statistik." Encyklopädie der mathematischen Wissenschaften, Band 1, Heft 6.
"Wahrscheinlichkeitstheorie und Erfahrung." Zeitschrift für Philosophie und philosophische Kritik, vol. 121, pp. 71-81. Leipzig, 1903.
[With reference to Marbe, Brömse, and Grimsehl, *q.v.*]
"Kritische Betrachtungen zur theoretischen Statistik." Jahrb. f. Nationalök. u. Stat. (3), vol. 8, pp. 641-680, 1894; vol. 10, pp. 321-360, 1895; vol. 11, pp. 671-705, 1896.
"Die erkenntnistheoretischen Grundlagen der Wahrscheinlichkeitsrechnung." Jahrb. f. Nationalök. u. Stat. (3), vol. 17, pp. 230-244, 1899.
[Criticised by Stumpf., *q.v.*, who is answered by Bortkiewicz, *loc. cit.*, vol. 18, pp. 239-242, 1899.]
"Zur Verteidigung des Gesetzes der kleinen Zahlen." Jahrb. f. Nationalök. u. Stat. (3), vol. 39, pp. 218-236, 1910.
[The literature of this topic is not fully dealt with in this Bibliography, but very full references to it will be found in the above article.]
"Über den Präzisionsgrad des Divergenzkoeffizientes." Mitteil. des Verbandes der österr. und ungar. Versicherungstechniker, vol. 5.
"Realismus und Formalismus in der mathematischen Statistik." Allg. Stat. Archiv, vol. ix. pp. 225-256. Munich, 1915.
Die Iterationen: ein Beitrag zur Wahrscheinlichkeitstheorie. Pp. xii + 205. Berlin, 1917.
Die radioaktive Strahlung als Gegenstand wahrscheinlichkeitstheoretischer Untersuchungen. Pp. 84. Berlin, 1913.
"Wahrscheinlichkeitstheoretische Untersuchungen über die Knabenquote bei Zwillings Gebieten." Sitzungsber. der Berliner Math. Ges., vol. xvii. pp. 8-14, 1918.
Homogeneität und Stabilität in der Statistik. Pp. 81. (Extracted from the Skandinavisk Aktuarietidskrift.) Uppsala, 1918.

BOSTWICK, A. E. "The Theory of Probabilities." Science, iii., 1896, p. 66.

BOUTROUX, PIERRE. "Les Origines du calcul des probabilités." Revue du Mois, vol. 5, pp. 641-654, 1908.

BOWLEY, A. L. Elements of Statistics. Pp. xi + 459. 4th ed. London, 1920.

BRADLEY, F. H. The Principles of Logic. Bk. i. chap. 8, §§ 32-63, pp. 201-20. London, 1883.

BRAVAIS. "Analyse mathématique sur les probabilités des erreurs de situation d'un point." Mém. Sav. vol. 9, pp. 255-332, Paris, 1846.

BRENDEL. Wahrscheinlichkeitsrechnung mit Einschluss der Anwendungen. Göttingen, 1907.

BROAD, C. D. "The Relation between Induction and Probability." Mind, vol. xxvii. (1918). Pp. 389-404, and vol. xxix. (1920) pp. 11-45.

BROMSE, H. Untersuchungen zur Wahrscheinlichkeitslehre. (Mit besonderer Beziehung auf Marbes Schrift (*q.v.*).)
Zeitschrift für Philosophie und philosophische Kritik. Band 118. Leipzig, 1901. Pp. 145-153.
(See also Marbe, Grimsehl, and v. Bortkiewicz.)

BRUNN, Dr. HERMANN. "Über ein Paradoxon der Wahrscheinlichkeitsrechnung." Sitzungsberichte der philos.-philol. Klasse der K. bayrische Akademie, pp. 692-712, 1892.

BRUNS, H. Wahrscheinlichkeitsrechnung und Kollektivmasslehre. 8vo. Pp. viii + 310 + 18. Leipzig, 1906.

"Das Gruppenschema für zufällige Ereignisse." Abhandl. d. Leipz. Ges. d. Wissensch. vol. xxix. pp. 579-628, 1906.

BRYANT, SOPHIE. "On the Failure of the Attempt to deduce inductive Principles from the Mathematical Theory of Probabilities." Phil. Mag. S. 5, No. 109, Suppl. vol. 17.

BUFFON. "Essai d'arithmétique morale." Supplément à l'Histoire Naturelle, vol. 4, 103 pp. 4to. 1777. Hist. Ac. Par. pp. 43-45, 1733.

BUNYAKOVSKI. Osnovaniya, etc. (Principles of the Mathematical Theory of Probabilities.) Petersburg, 1846.

BURBURY, S. H. "On the Law of Probability for a System of correlated variables." Phil. Mag. (6), vol. 17, pp. 1-28, 1909.

CAMPBELL, R. "On a Test for ascertaining whether an observed Degree of Uniformity, or the reverse, in tables of Statistics is to be looked upon as remarkable." Phil. Mag., 1859.

"On the Stability of Results based upon average Calculations." Journ. Inst. Act. vol. 9, p. 216.

A popular Introduction to the Theory of Probabilities. Pp. 16, Edinburgh, 1865.

CANTELLI, F. P. "Sulla applicazione delle probabilità parziali alla statistica." Giornale di Matematica finanziaria, vol. i. (1919), pp. 30-44.

CANTOR, G. Historische Notizen über die Wahrscheinlichkeitsrechnung. 4to. 8 pp. Halle, 1874.

CANTOR, M. Politische Arithmetik oder die Arithmetik des täglichen Lebens. Pp. x + 155. Leipzig, 1898, 2nd ed. 1903.

CANZ, E. C. Tractatio synoptica de probabilitate juridica sive de praesumtione. 4to. Tübingen, 1751.

CARAMUEL, JOHN. Kybeia, quae combinatoriae genus est, de alea, et ludis fortunae serio disputans. 1670. [Includes a reprint of Huygens, which is attributed to Longomontanus.]

CARDAN. De ludo aleae. fo., 15 pp. 1663. [Cardan ob. 1576.]

CARVELLO, E. Le Calcul des probabilités et ses applications. 8vo. Pp. ix + 169. Paris, 1912.

CASTELNUOVO, GUIDO. Calcolo delle probabilità. Large 8vo. Pp. xxiii + 373. Rome, 1919.

CATALAN, E. "Solution d'un problème de probabilité, relatif au jeu de rencontre." Journ. Liouville, vol. ii., 1837.

"Deux problèmes de probabilités." Journ. Liouville, vol. vi.

Problèmes et théorèmes de probabilités. 4to. 1884.

CAUCHY. Sur le système de valeurs qu'il faut attribuer à divers éléments déterminés par un grand nombre d'observations. 4to. Paris, 1814.

CAYLEY, A. "On a Question in the Theory of Probabilities." Phil. Mag., 1853.

CESÀRO, E. "Considerazioni sul concetto di probabilità." Periodico di Matematica, vi., 1891.

CHARLIER, C. V. L. Researches into the Theory of Probability. Publ. in Engl. in Meddelanden from Lund's Astronom. Observatorium, Series ii., No. 24. 4to. 51 pp. Lund, 1906.

"Contributions to the Mathematical Theory of Statistics," Arkiv för matematik, astronomi och fysik, vols. 7, 8, 9, *passim.*

Vorlesungen über die Grundzüge der mathematischen Statistik. Sm. 4to. Pp. 125. Lund, 1920.

Charpentier, T. V. "Sur la nécessité d'instituer la logique du probable." Comptes rendus de l'Acad. des Sciences morales, vol. i. p. 103, 1875.
"La Logique du probable." Rev. phil. vol. vi. pp. 23-38, 146-163, 1878.

Chrystal, G. On some Fundamental Principles in the Theory of Probability. London, 1891.

Clark, Samuel. The Laws of Chance: or a Mathematical Investigation of the Probability arising from any proposed Circumstance of Play, etc. Pp. ii + 204, 1758.

Cohen, J. Chance: A Comparison of 4 Facts with the Theory of Probabilities. Pp. 47. London, 1905.

Condorcet, Marquis de. Essai sur l'application de l'analyse à la probabilité des décisions rendues à la pluralité des voix. 4to. Pp. cxci + 304. Paris, 1785. Another edition, 1804.
"Sur les événements futurs." Acad. des Sc., 1803.
Memoir on Probabilities in six parts:
1. "Réflexions sur la règle générale qui prescrit de prendre pour valeur d'un événement incertain la probabilité de cet événement, multipliée par la valeur de l'événement en lui-même." Hist. de l'Acad. pp. 707-728. Paris, 1781.
2. "Application de l'analyse à cette question: Déterminer la probabilité qu'un arrangement régulier est l'effet d'une intention de le produire." Hist. de l'Acad., Paris, 1781. With Part i.
3. Sur l'évaluation des droits éventuels. 1782, pp. 674-691.
4. Réflexions sur la méthode de déterminer la probabilité des événements futurs, d'après l'observation des événements passés. 1783, pp. 539-559.
5. Sur la probabilité des faits extraordinaires. 1783, with Part 4.
6. Application des principes de l'article précédent à quelques questions de critique. 1784, pp. 454-468.

Coover, J. Experiments in Psychical Research at Leland Stanford Junior University. Pp. 641. Stanford University, California, 1917.
[See Psychical Research and Statistical Method by F. Y. Edgeworth, Stat. JL., vol. lxxxii. (1919), p. 222.]

Corbaux, F. Essais métaphysiques et mathématiques sur le hasard. 8vo. Paris, 1812.

Costa. Probabilité du tir. 8vo. Paris, 1825.
"Question de probabilité applicable aux décisions rendues par les jurés." Liouv. J. (1), vii., 1842.

Courcy, Alph. de. Essai sur les lois du hasard suivi d'étendus sur les assurances. 8vo. Paris, 1862.

Cournot, A. Revue de Métaphysique et de Morale, May 1905. Numéro spécialement consacré à Cournot. See especially:
F. Faure: "Les Idées de Cournot sur la statistique," pp. 395-411.
D. Parodi: "Le Criticisme de Cournot," pp. 451-484.
F. Mentré: "Les Racines historiques du probabilisme rationnel de Cournot," pp. 485-508.
Art. "Probabilités." Dictionnaire de Franck.
"Sur la probabilité des jugements et la statistique." Journal de Liouville, t. iii. p. 257.
"Mémoire sur les applications du calcul des chances à la statistique judiciaire." Liouv. J. (1) iii., 1838.
Exposition de la théorie des chances et des probabilités. Pp. viii + 448. Paris, 1843.
German translation by C. H. Schnuse. 8vo. Braunschweig, 1849.

Couturat, L. La Logique de Leibniz d'après des documents inédits. Pp. xiv. + 608. Paris, 1901.

COUTURAT, L.—*continued.*
[See especially chap. vi. for references to Leibniz's views on Probability.]
Opuscules et fragments inédits de Leibniz. Paris, 1903.
CRAIG. Theologiae Christianae principia mathematica. 4to. London, 1699. Reprinted Leipzig, 1755.
[CRAIG (?).] "A Calculation of the Credibility of Human Testimony." Phil. Trans. vol. xxi. pp. 359-365, 1699.
[Also attributed to HALLEY.]
CRAKANTHORPE, R. Logica. 1st ed. London, 1622. 2nd ed. London, 1641 (auctior et emendatior). 3rd ed. Oxon., 1677.
[Book v. "De syllogismo probabili."]
CROFTON, M. W. "On the Theory of Local Probability, applied to Straight Lines drawn at random in a Plane." Phil. Trans. vol. 158, pp. 181-199, 1869.
[Summarised in Proc. Lond. Math. Soc. vol. 2, pp. 55-57, 1868.]
"Probability." Encycl. Brit. 9th ed., 1885.
"Geometrical Theorems relating to Mean Values." Proc. Lond. Math. Soc. vol. 8, pp. 304-309, 1877.
CZUBER, E. Zum Gesetz der grossen Zahlen. Prag, 1889.
Geometrische Wahrscheinlichkeiten und Mittelwerte. Pp. vii + 244. Leipzig, 1884.
Theorie der Beobachtungsfehler. Pp. xiv + 418. Leipzig, 1891.
Die Entwicklung der Wahrscheinlichkeitstheorie und ihrer Anwendungen. Pp. viii + 279. Leipzig, 1899.
Wahrscheinlichkeitsrechnung und ihre Anwendung auf Fehlerausgleichung, Statistik und Lebensversicherung. Leipzig, 1903.
Ditto. 2 vols, 8vo. x + 410 + x + 470. Leipzig, 1908-10. Second edition, revised and enlarged. Vol. i. Warscheinlichkeitstheorie, Fehlerausgleichung, Kollektivmasslehre, 1908. Vol. ii. Mathematische Statistik, mathematische Grundlagen der Lebensversicherung, 1910.

D'ALEMBERT. Opuscules mathématiques: Paris, 1761–1780.
[Réflexions sur le calcul des probabilités, ii. pp. 1-25, 1761.
Sur l'application du c. des p. à l'inoculation, ii. pp. 26-95.
Sur le calcul des probabilités, etc., iv. pp. 73-105; iv. pp. 283-341; v. pp. 228-231; v. pp. 508-510; vii. pp. 39-60.]
Mélanges de littérature, d'histoire et de philosophie. Amsterdam, 1770.
[Doutes et questions sur le calcul des probabilités, vol. v. pp. 223-246.
Réflexions sur l'inoculation. Vol. v. (These two papers were reprinted in the first volume of D'Alembert's collected works published at Paris in 1821 (pp. 451-514).)]
Articles in Encyclopédie ou Dictionnaire raisonné:
"Croix ou Pile," 1754.
"Gageure," 1757.
Article in Encyclopédie méthodique: "Cartes."
D'ANIÈRES. "Réflexions sur les jeux de hasard." Mém. de l'Acad. pp. 391-398. Berlin, 1784.
DANTEC, FÉLIX LE. "Le Hasard et la question d'échelle." Revue du Mois, vol. 4, pp. 257-288, 1907.
Le Chaos et l'harmonie universelle. Paris, 1911.
DARBISHIRE, A. D. Some Talks illustrating Statistical Correlation. (Reprinted from Memoirs of the Manchester Literary and Philosophical Society.) 21 pp. and plates. 8vo. 1907.
DARBON, A. Le Concept du hasard dans la philosophie de Cournot. Étude critique. Pp. 60. Paris, 1911.
DAVENPORT, C. B. Statistical Methods. 1904.

De Moivre, A. "De mensura sortis, seu, de probabilitate eventuum in ludis a casu fortuito pendentibus." Phil. Trans. vol. xxvii. pp. 213-264, 1711.

Doctrine of Chances, or A Method of Calculating the Probabilities of Events in Play. 1st ed. 4to. Pp. xiv + 175. 1718. 2nd ed. Large 4to. Pp. xiv + 258. 1738. 3rd ed. Large 4to. Pp. xii + 348. 1756.

La dottrina d. azzardi applic. ai problemi d. probabilità di vita, di pensi, ecc., trad. da R. Gaeta e G. Fontana. Milan, 1776.

Miscellanea analytica de seriebus et quadraturis. 4to. Pp. 250 + 22. London, 1730.

De Morgan, A. Essay on Probabilities and their Application to Life Contingencies and Insurance Offices. 1838.

Formal Logic: or the Calculus of Inference Necessary and Probable. 1847.

Theory of Probabilities. 4to. 1849.

[From the Encyclopaedia Metropolitana.]

On the Structure of the Syllogism and on the Application of the Theory of Probabilities to Questions of Argument and Authority. 4to. Camb. Phil. Soc. pp. 393-405, 1847 (read Nov. 9, 1846).

On the Symbols of Logic, the Theory of the Syllogism, and in particular of the Copula, and the Application of the Theory of Probabilities to some Questions of Evidence. 4to. Camb. Phil. Soc. vol. ix. pp. 116-125, 1851.

De Witt, John. De vardye van de lif-renten na proportie van de los-renten. La Haye, 1671.

English transl.: Contributions to the History of Insurance, by Frederick Hendriks in the Assurance Magazine, vol. 2, p. 231 (1852).

[For an abstract see N. Struyck, Inleiding tot het algemeine geography, etc. 4to. Amsterdam, 1740. P. 345.]

Dedekind, R. Bemerkungen zu einer Aufgabe der Wahrscheinlichkeitsrechnung. Pp. 268-271. Crelle J. vol. l., 1855.

Degen, C. F. Tabularum ad faciliorem probabilitatis computationem utilem Enneas. Kiobenhavn, 1824.

Diderot. Art. "Probabilité" in the Encyclopédie.

Didion, J. Calcul des probabilités appliqué au tir des projectiles. 8vo. 1858.

Dodson, James. Mathematical Repository. 3 vols. 1753. Vol. ii. pp. 82-136.

Donkin, W. F. "Sur la théorie de la combinaison des observations." Liouv. J. (1), vol. xv. 1850.

"On Certain Questions relating to the Theory of Probabilities." Phil. Mag., May 1851.

Dormoy, E. Théorie mathématique des assurances sur la vie. 2 vols. Paris, 1878.

Drobisch, A. "Über die nach der Wahrscheinlichkeitsrechnung zu erwartende Dauer der Ehen." Berichte über die Verhandlungen der Königl. Sächsischen Gesellschaft der Wissenschaften mathem.-physik. 1880.

Drobisch, M. W. Neue Darstellung der Logik. 2nd ed. Leipzig, 1851. 3rd ed. 1863. 4th ed. 1875. 5th ed. 1887.

[Probability, pp. 181-209, §§ 145-157 (references to 4th ed).]

Edgeworth, F. Y. "Calculus of Probability applied to Psychical Research." Proceedings of Soc. for Psych. Res. Parts viii. and x.

"On the Method of ascertaining a Change in the Value of Gold." Roy. Stat. Soc. J. xlvi. pp. 714-718. 1883.

"Law of Error." Phil. Mag. (5) vol. xvi. pp. 300-309, 1883.

"Method of least Squares." Phil. Mag. (5) vol. xvi. pp. 360-375, 1883.

"Physical Basis of Probability." Phil. Mag. vol. xvi. pp. 433-435, 1883.

EDGEWORTH, F. Y.—*continued.*
"Chance and Law." Hermathene (Dublin), 1884.
"On the Reduction of Observations." Phil. Mag. (5) vol. xvii. pp. 135-141, 1884.
"Philosophy of Chance." Mind, April 1884.
"*A priori* Probabilities." Phil. Mag. (5) vol. xviii. pp. 209-210, 1884.
"On Methods of Statistics." Stat. Journ. Jub. vol. pp. 181-217, 1885.
[Criticised by Bortkiewicz and defended by Edgeworth, Jahrb. f. nat. Ök. u. Stat. (3), vol. 10, pp. 343-347; vol. 11, pp. 274-277, 701-705, 1896.]
"Observations and Statistics." Phil. Soc. 1885.
"Law of Error and Elimination of Chance." Phil. Mag., 1886, vol. xxi. pp. 308-324.
"Problems in Probabilities." Phil. Mag., 1886, vol. xxii. pp. 371-384, and 1890, vol. xxx. pp. 171-188.
Metretike: or the Method of Measuring Probability and Utility. 8vo. 1887.
"On Discordant Observations." Phil. Mag. (5) vol. xxiii. pp. 1887.
"The Empirical Proof of the Law of Error." Phil. Mag. (5) vol. xxiv. pp. 330-342, 1887.
"The Element of Chance in Competitive Examinations." Roy. Stat. Soc. Journ. liii. pp. 460-475 and 644-663, 1890.
"The Law of Error and Correlated Averages." Phil. Mag. (5) vol. xxxv. pp. 63-64, 1893.
"Statistical Correlation between Social Phenomena." Roy. Stat. Soc. Journ. lvi. pp. 670-675, 1893.
"The Asymmetrical Probability-Curve." 1896. Phil. Mag. vol. xli. pp. 90-99.
"Miscellaneous Applications of the Calculus of Probabilities." Roy. Stat. Soc. Journ. lx. pp. 681-698, 1897; lxi. pp. 119-131 and 534-544, 1898.
"Law of Error." Phil. Trans. vol. xx.
"The Generalised Law of Error." Stat. Journ. vol. lxix., 1906.
"On the Probable Errors of Frequency-Constants." Stat. Journ. vol. lxxi. pp. 381-397, 499-512, 651-678, 1908; and vol. lxxii. pp. 81-90, 1909.
"On the Application of the Calculus of Probabilities to Statistics." Bulletin xviii. of the International Statistical Institute, Paris, 1910, 32 pp.
"Applications of Probabilities to Economics." Economic Journal, vol. xx. pp. 284-304, 441-465, 1910.
"Probability." Encyclopaedia Britannica, 11th ed. vol. 22, pp. 376-403, 1911.
"On the Application of Probabilities to the Movement of Gas-Molecules." Phil. Mag., vol. xl., pp. 249-272, 1920.
"Molecular Statistics." Roy Stat. Soc. Journ., vol. lxxxiv. pp. 71-89, 1921.

EGGENBERGER, J. "Beiträge zur Darstellung des bernoullischen Theorems." Berner Mitth. vol. 50 (1894); and Zeitschr. f. Math. u. Ph. 45 (1900), p. 43.

ELDERTON, W. P. Frequency-Curves and Correlation. 8vo. London, 1907. xiii + 172.
[Contains a useful list of papers on Correlation, p. 163.]

ELLIS, R. L. "On the Foundations of the Theory of Probability." 4to. Camb. Phil. Soc. vol. viii., 1843.
[Reprinted in "Mathematical and other Writings," 1863.]
"On a Question in the Theory of Probabilities." Camb. Math. Journ. No. xxi. vol. iv., 1844.
[Reprinted in "Mathematical and other Writings," 1863.]

Ellis, R. L.—*continued.*
"On the Method of Least Squares." Trans. Camb. Phil. Soc. vol. viii., 1844.
[Reprinted in "Mathematical and other Writings," 1863.]
"Remarks on an alleged Proof of the 'Method of Least Squares.'" Phil. Mag. (3) vol. xxxvii., 1850.
[Reprinted in "Mathematical and other Writings," 1863.]
"Remarks on the Fundamental Principle of the Theory of Probabilities." Trans. Camb. Phil. Soc. vol. ix., 1854.
[Reprinted in "Mathematical and other Writings," 1863.]
Elsas, A. "Kritische Betrachtungen über die Wahrscheinlichkeitsrechnung." Philos. Monatssch. vol. xxv. pp. 557-584, 1889.
Emerson, William. Miscellanies, 1776. [See espec. pp. 1-48.]
Encke, J. F. Methode der kleinsten Quadrate. Fehler theoret. Untersuchungen. Berlin, 1888.
Engel, G. "Über Möglichkeit und Wirklichkeit." Philos. Monatssch. vol. v. pp. 241-271, 1875.
Ermakoff, W. P. Wahrscheinlichkeitslehre (in Russian).
Euler. "Calcul de la probabilité dans le jeu de rencontre." Hist. Ac. Berl. (1751), pp. 255-270, 1753.
"Sur l'avantage du banquier au jeu de pharaon." Hist. Ac. Berl. (1764), pp. 144-164, 1766.
"Sur la probabilité des séquences dans la loterie génoise." Hist. Ac. Berl. (1765), pp. 191-230, 1767.
"Solution d'une question très difficile dans le calcul des probabilités." Hist. Ac. Berl. (1769), pp. 285-302, 1771.
"Solutio quarundam quaestionum difficiliorum in calculo probabilium." Opuscula analytica, vol. ii. pp. 331-346, 1785.
"Solutio quaestionis ad calculum probabilitatis pertinentis: Quantum duo conjuges persolvere debeant, ut suis haeredibus post utriusque mortem certa argenti summa persolvatur." Opuscula analytica, vol. ii., pp. 315-330, 1785.
"Wahrscheinlichkeitsrechnung." Opera omnia, ser. 1, A, vol. iv. Leipzig.

Fahlbeck. "La Régularité dans les choses humaines, ou les types statistiques et leurs variations." Journ. Soc. Stat. de Paris, pp. 188-200, 1900.
Fechner, G. Th. Kollektivmasslehre. (Edited by G. F. Lipps.) 1897.
Fick, A. Philosophischer Versuch über die Wahrscheinlichkeiten. Pp. 46. Würzburg, 1883.
Fisher, A. The Mathematical Theory of Probabilities. Translated from the Danish. Pp. xx+171. New York, 1915.
Forbes, J. D. "On the alleged Evidence for a Physical Connexion between Stars forming Binary or Multiple Groups, deduced from the Doctrine of Chances." Phil. Mag., Dec. 1850. (See also Phil. Mag., Aug. 1849.)
Forncey. The Logic of Probabilities. Transl. from the French. 8vo. London, n.d. (? 1760.)
Förster, W. Wahrheit und Wahrscheinlichkeit. Pp. 40. Berlin, 1875.
Fries, J. J. Versuch einer Kritik der Principien der Wahrscheinlichkeitsrechnung. Braunschweig, 1842.
Frömmichen. Über Lehre der Wahrscheinlichkeit. 4to. Braunschweig, 1773.
Fuss, N. "Recherches sur un problème du calcul des probabilités." Act. Ac. Petr. (1779), pars posterior, pp. 81-92, 1783.
"Supplément au mémoire sur un problème du calcul des probabilités." Act. Ac. Petr. (1780), pars posterior, pp. 91-96, 1784.

GALILEO, G. "Considerazioni sopra il giuoci dei dadi." Opere, vol. iii. pp. 119-121, 1718. Also, Opere, vol. xiv. pp. 293-296. Firenze, 1855.

"Lettere intorno le stima di un cavallo." Opere, vol. xiv. pp. 231-284. Firenze, 1855.

GALLOWAY, T. A Treatise on Probability. 8vo. Edinburgh, 1839. (From the 7th edition of the Encyclopaedia Britannica.)

GALTON, F. "Correlations and their Measurement." Proc. Roy. Soc., vol. xlv. pp. 136-145.

Probability, the Foundation of Eugenics. Herbert Spencer Lecture, 1907. (Reprinted—Essays in Eugenics. 8vo. ii + 109 pp. London, 1909.)

GARDON, C. Antipathies des 90 nombres, probabilités, et observations comparatives, sur les loteries de France et de Bruxelles. 8vo. Paris, 1801.

Traité élémentaire des probabilités, etc. Paris, 1805.

L'investigateur des chances . . . pour obtenir souvent des succès aux loteries impériales de France. Paris.

GARVE, C. De nonnullis quae pertinent ad logicam probabilium. 4to. Halae, 1766.

GATAKER, T. On the Nature and Use of Lots. 4to. 1619.

GAUSS, C. F. Theoria motus corporum coelestium. 4to. Hamburg, 1809.

"Theoria combinationis observationum erroribus minimis obnoxiae." Comm. Soc. Göttingen, vol. v. pp. 33-90. 1823.

Méthode des moindres carrés. Traduit en français par J. Bertrand. 8vo. 1855.

[A translation of part of the above.]

Wahrscheinlichkeitsrechnung. Werke, vol. iv. pp. 1-53. 4to. Göttingen, 1873.

GEISENHEIMER, L. Über Wahrscheinlichkeitsrechnung. 8vo. Berlin, 1880.

GILMAN, B. I. "Operations in Relative Number with Applications to Theory of Probability." Johns Hopkins Studies in Logic, 1883.

GLADSTONE, W. E. "Probability as a Guide to Conduct." Nineteenth Cent. vol. v. pp. 908-934, 1879; and in "Gleanings," vol. ii. pp. 153-200.

GLAISHER, J. W. L. "On the Rejection of Discordant Observations." Monthly Notices R. Astr. S. vol. xxiii., 1873.

"On the Law of Facility of Errors of Observation, and on the Method of Least Squares." Mem. R. Astr. S. vol. xxxix., 1872.

GOLDSCHMIDT, L. "Wahrscheinlichkeit und Versicherung." Bull. du Comité permanent des Congrès Internationaux d'Actuaires, 1897.

Die Wahrscheinlichkeitsrechnung: Versuch einer Kritik. Pp. 279. Hamb., 1897.

[Cf. Zeitschr. f. Philos. u. phil. Kr., cxiv., pp. 116-119.]

GONZALEZ, T. Fundamentum theologiae moralis, id est tractatus theologicus de recto usu opinionum probabilium. 4to. Dillingen, 1689. Naples, 1694.

[An abridgement entitled: Synopsis tract. theol. de recto usu opin. prob., concinnata a theologo quodam Soc. Jesu: cui accessit logistica probabilitatum. 3rd ed. 8vo. Venice, 1696. See Migne, Theol. Cur. Compl., vol. xi., p. 1397.]

GOURAND, CH. Histoire du calcul des probabilités depuis ses origines jusqu'à nos jours. 8vo. Paris, 1848, 148 pp.

[His history seems to be a portion of a very extensive essay in 3 folio volumes containing 1929 pp., written when he was very young, in competition for a prize proposed by the Fr. Acad. on a subject entitled "Théorie de la certitude"; see Séances et Travaux de l'Académie des Sciences morales et politiques, vol. x. pp. 372, 382, vol. xi. p. 137. See TODHUNTER.]

GRAVESANDE, W. J. 'S. Introductio ad philosophiam, metaphysicam et logicam continens. 8vo. Venetiis, 1737.

GRAVESANDE, W. J. 'S —*continued.*
Œuvres philosophiques et mathématiques. 4to. Amsterdam, 1774, 2 vols. 4to. ii. pp. 82-93, 221-248.
GRELLINGS, K. "Die philosophischen Grundlagen der Wahrscheinlichkeitsrechnung." Abhandlungen der Friesschen Schule, N.F., vol. iii., 1910.
GRIMSEHL, E. "Untersuchungen zur Wahrscheinlichkeitslehre. (Mit besonderer Beziehung auf Marbes Schrift (*q.v.*).)" Zeitschrift für Philosophie und philosophische Kritik. Band 118, pp. 154-167. Leipzig, 1901.
[See also BRÖMSE, MARBE, and V. BORTKIEWICZ.]
GROLOUS. "Sur une question de probabilité appliquée à la théorie des nombres." Journal de l'Institut, 1872.
GROSCHIUS, J. A. Logica probabilium in artium practicarum subsidium adornata. Sm. 8vo. Halae, 1764. Pp. xvi + 352.
GRÜNBAUM, H. Isolierte und reine Gruppen und die Marbesche Zahl "p." Würzburg, 1904.
GUIBERT, A. "Solution d'une question relative à la probabilité des jugements rendus à une majorité quelconque." Liouv. J. (1) vol. iii., 1838.

HACK. Wahrscheinlichkeitsrechnung. Leipzig, 1911.
HAGEN, G. F. Meditationes philosophicae de methodo mathematico. Norimbergae, 1734.
Fortsetzung einiger aus der Mathematic abgenommenen Regeln, nach welchen sich der menschliche Verstand bei Erfindung der Wahrheiten richtet. Halle, 1737.
HAGEN, G. Grundzüge der Wahrscheinlichkeitsrechnung. Berlin, 1837. (2nd ed. 1867, 3rd ed. 1882.)
Der constante wahrscheinliche Fehler: Nachtrag zur 3ten Auflage der Grundzüge der Wahrscheinlichkeitsrechnung. 38 pp. Berlin, 1884.
HALLEY. See CRAIG.
HANS, JOHN. See J. ARBUTHNOT.
HANSDORFF, F. "Beiträge zur Wahrscheinlichkeitsrechnung." Leipz. Ber., vol. 53, pp. 152-178, 1901.
"Das Risiko bei Zufallsspielen." Leipz. Ber., vol. 49, pp. 497-548, 1897.
HANSEN, P. A. "Über die Anwendung der Wahrscheinlichkeitsrechnung auf geodätische Vermessungen." Astr. N. vol. ix. 1831.
HARTMANN, E. von. "Die Grundlage der Wahrscheinlichkeitsurteils." Vierteljahrsschr. f. wiss. Phil. u. Soz., vol. xxviii., 1904.
HAUTESERVE, GAUTHIER D'. Traité élémentaire sur les probabilités. Paris, 1834.
Application de l'algèbre élémentaire au calcul des probabilités. Paris, 1840.
HÉLIE. Mémoire sur la probabilité du tir. 8vo. 1854.
HELM. "Eine Anwendung der Theorie des Tauschwerthes auf die Wahrscheinlichkeitsrechnung." Zeitschr. f. Math. u. Phys., vol. 38, pp. 374-376. Leipzig, 1893.
"Die Wahrscheinlichkeitslehre als Theorie der Kollektiv-begriffe." Annalen der Naturphilosophie, vol. 1.
HENRY, CHARLES. La Loi des petits nombres. Recherches sur le sens de l'écart probable dans les chances simples à la roulette, au trente-et-quarante etc., en général dans les phénomènes dépendant de causes purement accidentales. 72 pp. 8vo. Paris, 1908.
HERSCHEL, W. "On the Theory of Probabilities." Journal of Actuaries, 1869.
"Quetelet on Probabilities." Edin. Rev., 1850.
[Reprinted in Quetelet's Physique Sociale, vol. i. pp. 1-89, 1869.]
"On an Application of the Rule of Succession." Edin. Rev., 1850.

HERZ, N. Wahrscheinlichkeits- und Ausgleichungsrechnung. Pp. iv + 381. Leipzig, 1900.

HIBBEN, J. G. Inductive Logic. London, 1896.
[See chaps. xv., xvi.]

HOBHOUSE, L. T. Theory of Knowledge.
[See Part II., chaps. x., xi.]

HOYLE. An Essay towards making the Doctrine of Chances easy to those who understand vulgar Arithmetic only. Pp. viii + 73, 1754, 1758, 1764.

HUBERDT, A. Die Principien der Wahrscheinlichkeitsrechnung. 4to. Berlin, 1845.

HUME, DAVID. Treatise on Human Nature. 1st ed. 1739.
[See especially Part III.]
An Enquiry concerning Human Understanding.
[See specially Section vi.]
Essays, Part I., XIV. On the Rise and Progress of the Arts and Sciences, pp. 115, 116, 1742.

HUYGENS, CH. "De ratiociniis in ludo aleae." Schooten's Exercitat. math. pp. 519-534. 4to. Lugd. Bat., 1657.
[Written by Huygens in Dutch and translated into Latin by Schooten.]
Engl. transl. by W. Browne. Sm. 8vo, pp. 24. London, 1714.
[See also JAC. BERNOULLI, ARBUTHNOT (Engl. Transl.), and VASTEL (Fr. Transl.).]

JAHN, G. A. Die Wahrscheinlichkeitsrechnung und ihre Anwendung auf das wissenschaftliche und praktische Leben. Leipzig, 1839.

JANET. La Morale. Paris, 1874. [See Bk. iii. chap. 3 for Probabilism.]
Engl. transl. The Theory of Morals. New York, 1883, pp. 292-308.

JEVONS, W. S. Principles of Science. 2 vols. 1874.

JORDAN, C. "De quelques formules de probabilité (sur les causes)." Comptes rendus, 1867.

JOURDAIN, P. E. B. "Causality, Induction, and Probability." Mind, vol. xxviii. pp. 162-179, 1919.

KAHLE, L. M. Elementa logicae probabilium methodo mathematica, in usu scientiarum et vitae adornata. Pp. 10 + xxii + 245. Sm. 8vo. Halae, 1735.

KANNER, M. "Allgemeine Probleme der Wahrscheinlichkeitsrechnung und ihre Anwendung auf Fragen der Statistik." Journ. des Collegiums für Lebens-Versicherungs-Wissenschaft. Berlin, 1870.

KAUFMANN, AL. Theorie und Methoden der Statistik. [Translated from the Russian.] Pp. xii + 540. Tübingen, 1913.

KEPLER, J. "De stella nova in pede serpentarii." 1606. See J. Kepler's Astr. Op. Omn. edidit Frisch, ii. pp. 714-716.

KIRCHMANN, J. H. VON. Über die Wahrscheinlichkeit. Pp. 60. Leipzig, 1878.

KNAPP. "Quetelet als Theoretiker." Jahrb. f. nat. Ök. und Stat. (New Series), vol. xviii.

KOZÁK, JOSEF. Grundlehren der Wahrscheinlichkeitsrechnung als Vorstufe für das Studium der Fehlerausgleichung, Schiesstheorie, und Statistik. Vienna, 1912.
Theorie des Schiesswesens auf Grundlage der Wahrscheinlichkeitsrechnung und Fehlertheorie. Vienna, 1908.

KRIES, J. VON. Die Principien der Wahrscheinlichkeitsrechnung. Eine logische Untersuchung. Pp. 298. 8vo. Freiburg, 1886.
[See also LEXIS, MEINONG and SIGWART.]

LACROIX, S. F. Traité élémentaire du calcul des probabilités. Pp. viii + 299. 8vo. Paris, 1816.
[2nde éd., revue et augmentée, 1822 ; 4th ed. 1864.]
[Translated into German : E. S. Unger, Erfurt, 1818.]

LAGRANGE. "Mémoire sur l'utilité de la méthode de prendre le milieu entre les résultats de plusieurs observations, dans lequel on examine les avantages de cette méthode par le calcul des probabilités, et où l'on résout différents problèmes relatifs à cette matière." Misc. Taurinensia, vol. 5, pp. 167-232, 1770–1773. Œuvres complètes, vol. 2, Paris, 1867–1877.
"Recherches sur les suites recurrentes . . . et sur l'usage de ces équations dans la théorie des hasards." Nouv. Mém. Ac. Berl. (1775), pp. 183-272, 1777. Œuvres complètes, vol. 4. Paris, 1867–1877.

LAISANT, C. A. Algèbre. Théorie des nombres, probabilités, géométrie de situation. Paris, 1895.

LAMBERT, J. H. "Examen d'une espèce de superstition ramenée au calcul des probabilités." Nouv. Mém. Ac. Berl., 1771, pp. 411-420.

LÄMMEL, R. Untersuchungen über die Ermittelung von Wahrscheinlichkeiten. (Inaug.-Dissert.) Pp. 80. Zürich, 1904.

LAMPE, E. "Über eine Aufgabe aus der Wahrscheinlichkeitsrechnung." Grun. Arch., vol. 70, 1884.

LANGE, F. A. Logische Studien.

LAPLACE. Essai philosophique sur les probabilités. (Printed as introduction to Théorie analytique des probabilités, from 2nd ed. of the latter onwards.) 4to. Paris, 1814.
German translation by Tönnies. Heidelberg, 1819. German translation by N. Schwaiger. Leipzig, 1886.
A Philosophical Essay on Probabilities: transl. from the 6th French ed. by E. W. Truscott and F. L. Emory. 8vo. New York, 1902, 196 pp.
Théorie analytique des probabilités.
1st ed. 4to. Paris, 1812. 1st and 2nd Suppl., 1812–1820. 2nd ed. 4to. cxi + 506 + 2, Paris, 1814. 3rd Suppl. 1820. 3rd ed. Paris, 1820. 4th Suppl. after 1820. Œuvres complètes, vol. 7, pp. cxcv + 691, Paris, 1847. Œuvres complètes, vol. 7, pp. 832, Paris, 1886.
"Recherches sur l'intégration des équations différentielles aux différences finies, et sur leur usage dans la théorie des hasards." Mém. prés. à l'Acad. des Sc., pp. 113-163, 1773.
"Mémoire sur les suites récurro-récurrentes et sur leurs usages dans la théorie des hasards." Mém. prés. à l'Acad. des Sc., vol. 6, pp. 353-371, 1774.
"Mémoire sur la probabilité des causes par les événements." Mém. prés. à l'Acad. des Sc., vol. 6, pp. 621-656, 1774.
"Mémoire sur les probabilités." Mém. prés. à l'Acad. des Sc., pp. 227-332, 1780.
"Mémoire sur les approximations des formules qui sont fonctions de très grands nombres, et sur leurs applications aux probabilités." Mém. de l'Inst., pp. 353-415, 539-565, 1810.
"Mémoire sur les intégrales définies, et leur application aux probabilités." Mém. de l'Inst., pp. 279-347, 1810.
[The above memoirs are reprinted in Œuvres complètes, vols. 8, 9, and 12, Paris, 1891–1898.]
Sur l'application du calcul des probabilités appliqué à la philosophie naturelle. Conn. des temps. Œuvres complètes, vol. 13. Paris, 1904.
"Applications du calcul des probabilités aux observations et spécialement aux opérations du nivellement." Annales de Chimie. Œuvres complètes, vol. 14, Paris, 1913.

LA PLACETTE, J. Traité des jeux de hasard. 18mo. 1714.

LAURENT, H. Traité du calcul des probabilités. Paris, 1873.
[A la fin une liste des principaux ouvrages (320) ou mémoires publiés sur le calcul des probabilités.]
"Application du calcul des probabilités à la vérification des répartitions." Journ. des Actuaires français, vol. i.
"Sur le théorème de J. Bernoulli." Journ. des Actuaires français, vol. i.

LECHALAS, G. "Le Hasard." Rev. Néo-scolastique, 1903.
"A propos de Cournot: hasard et déterminisme." Rev. de Mét. et de Mor., 1906.

LEGENDRE. "Méthode des moindres carrés." Mém. de l'Inst., 1810, 1811.
Nouvelles méthodes pour la détermination des orbites des comètes. Paris, 1805-6.

LEHR. "Zur Frage der Wahrscheinlichkeit von weiblichen Geburten und von Totgeburten." Zeitschrift f. des ges. Staatsw., vol. 45, p. 172, and p. 524, 1889.

LEIBNIZ. Nouveaux Essais. Liv. ii. chap. xxi.; liv. iv. chaps. ii. § 14, xv., xvi., xviii., xx.
Opera omnia, ed. Dutens, v. 17, 22, 28, 29, 203, 206; vi. pt. i., 271, 304, 36, 217; iv. pt. iii. 264.
Correspondence between Leibnitz and Jac. Bernoulli. L.'s Gesammelte Werke (ed. Pertz and Gerhardt), vol. 3, pp. 71-97, *passim*. Halle, 1855.
[These letters were written between 1703 and 1705.]
See also s.v. COUTURAT.

LEMOINE, E. "Solution d'un problème sur les probabilités." Bulletin de la Soc. math. de Paris, 1873.
Questions de probabilités et valeurs relatives des pièces du jeu des échecs. 8vo. 1880.
"Quelques questions de probabilités résolues géométriquement." Bull. de la Soc. math. de France, 1883.
"Divers problèmes de probabilité." Ass. française pour l'Avancement des Sciences, 1885.

LEXIS, W. Abhandlungen zur Theorie der Bevölkerungs- und Moral-statistik. Pp. 253. Jena, 1903.
Zur Theorie der Massenerscheinungen in der menschlichen Gesellschaft. Pp. 95. Freiburg, 1877.
"Über die Wahrscheinlichkeitsrechnung und deren Anwendung auf die Statistik. Jahrb. f. nat. Ök. u. Stat. (2), vol. 13, pp. 433-450, 1886.
[Contains a review of v. Kries's "Principien."]
"Über die Theorie der Stabilität statistischer Reihen." Jahrb. f. nat. Ök. u. Stat. (1), vol. 32, p. 604, 1879.
[Reprinted in Abhandlungen.]
"Das Geschlechtsverhältnis der Geborenen und die Wahrscheinlichkeitsrechnung." Jahrb. f. nat. Ök. u. Stat. (1), vol. 27, p. 209, 1876.
[Reprinted in Abhandlungen.]
Einleitung in die Theorie der Bevölkerungsstatistik. Strassburg, 1875.

LIAGRE, J. B. J. Calcul des probabilités et théorie des erreurs avec des applications aux sciences d'observation en général et à la géodésie en particulier. 416 pp. Brussels, 1852. 2nd ed. 8vo. 1879.
"Sur la probabilité d'une cause d'erreur régulière, etc." Bull. de l'Acad. de Belgique, 1855.

LIAPOUNOFF, A. "Sur une proposition de la théorie des probabilités." Bull. de l'Acad. des Sc. de Saint-Pét., v. série, vol. xiii.
"Nouvelle Forme du théorème sur la limite de probabilité." Mém. de l'Acad. des Sc. de Saint-Pét., viii. série, vol. xiii. (1901).

LIEBERMEISTER, C. "Über Wahrscheinlichkeitsrechnung in Anwendung auf therapeutische Statistik." Sammlung klinische Vorträge, Nr. 110. 1877.

LILIENFELD, J. "Versuch einer strengen Fassung des Begriffs der mathematischen Wahrscheinlichkeit." Zeitschr. f. Philos. u. phil. Kr., vol. cxx, pp. 58-65, 1902.

LIPPS, G. F. Kollectivmasslehre. 1897.

LITTROW, J. J. Die Wahrscheinlichkeitsrechnung in ihrer Anwendung auf das wissenschaftliche und praktische Leben. 8vo. Wien, 1833.

LOBATCHEWSKY, N. J. "Probabilité des résultats moyens tirés d'observations répétées." Crelle J. 1824.

Reprinted. Liouv. J. vol. 24. 1842.

LOTTIN, J. Le Calcul des probabilités et les régularités statistiques. 32 pp. 8vo. Louvain, 1910. (Originally published in the Revue Néo-scolastique, Feb. 1910.)

Quetelet, statisticien et sociologue. Louvain, 1912. Pp. xxx + 564. [Contains a very full discussion of Quetelet's Work on Probability.]

LOTZE, H. Logik. 1st ed. 1874, 2nd ed. 1880.

Engl. transl. by B. Bosanquet. Oxford, 1884.

[See Bk. ii. chap. ix.: Determination of Single Facts and Calculus of Chances.]

LOURIÉ, S. Die Prinzipien der Wahrscheinlichkeitsrechnung. Tübingen, 1910.

LUBBOCK, J. W., and DRINKWATER. Treatise on Probability. [Library of Useful Knowledge.]

[Often wrongly ascribed to DE MORGAN.]

MACALISTER, DONALD. The Law of the Geometric Mean. Phil. Trans., 1879.

MCCOLL, HUGH. Symbolic Logic. 1906. [Especially chaps. xvii., xviii.]

The Calculus of Equivalent Statements. Proc. Lond. Math. Soc. Six papers.

[See particularly 1877, vol. ix. pp. 9-20; 1880, xi. 113-121, 4th paper; 1897, xxviii. p. 556, 6th paper.]

"Growth and Use of a Symbolical Language." Memoirs Manchester Lit. Phil. Soc. series iii. vol. 7, 1881.

"Symbolical or Abbreviated Language with an Application to Mathematical Probability." Math. Questions, vol. 28, pp. 20-23.

Various Papers in Mathematical Questions from the Journal of Education, vols. 29, 33, etc.

"A Note on Prof. C. S. Peirce's Probability Notation of 1867." Proc. Lond. Math. Soc. vol. xii. p. 102.

MACFARLANE, ALEXANDER. Principles of the Algebra of Logic.

[See especially chaps. ii., iii., v., xx., xxi., xxii., xxiii., and the examples.]

Various Papers in Mathematical Questions from the Journal of Education, vols. 32, 36, etc.

MACMAHON, P. A. "On the Probability that the Successful Candidate at an Election by Ballot may never at any time have fewer Votes than the one who is unsuccessful, etc." Phil. Trans. (A), vol. 209, pp. 153-175, 1909.

MALDIDIER, JULES. "Le Hasard." Rev. Philos. xliii., 1897, pp. 561-588.

MALFATTI, G. F. "Esame critico di un problema di probabilità del Sig. Daniele Bernoulli, e soluzione d'un altro problema analogo al bernulliano." Memorie di Matematica e Fisica della Società Italiana, vol. 1, pp. 768-824, 1782.

MALLET. "Sur le calcul des probabilités." Act. Helv. Basileae, 1772, vii. pp. 133-163.

MANSIONS, P. "Sur la portée objective du calcul des probabilités." Bulletin de l'Académie de Belgique (Classe des sciences), pp. 1235-1294, 1903.

MARBE, DR. KARL. Naturphilosophische Untersuchungen zur Wahrscheinlichkeitslehre. 50 pp. Leipzig, 1899.

Die Gleichförmigkeit in der Welt. Munich, 1916.

MARKOFF, A. A. "Über die Wahrscheinlichkeit *à posteriori*" (in Russian). Mitteilungen der Charkowv Math. Gesell. 2 Serie, vol. iii. 1900.

"Untersuchung eines wichtigen Falles abhängiger Proben" (in Russian). Abh. der K. Russ. Ak. d. W., 1907.

"Über einige Fälle der Theoreme vom Grenzwert der mathematischen Hoffnungen und vom Grenzwert der Wahrscheinlichkeiten" (in Russian). Abh. der K. Russ. Ak. d. W., 1907.

"Erweiterung des Gesetzes der grossen Zahlen auf von einander abhängige Grössen" (in Russian). Mitt. d. phys.-math. Ges. Kazan, 1907.

"Über einige Fälle des Theorems vom Grenzwert der Wahrscheinlichkeiten" (in Russian). Abh. der K. Russ. Ak. d. W., 1908.

"Erweiterung gewisser Sätze der Wahrscheinlichkeitsrechnung auf eine Summe verketteter Grössen" (in Russian). Abh. der K. Russ. Ak. d. W., 1908.

"Untersuchung des allgemeinen Falles verketteter Ereignisse" (in Russian). Abh. der K. Russ. Ak. d. W., 1910.

"Über einen Fall von Versuchen, die eine komplizierte zusammenhängendes Kette bilden," and "Über zusammenhängende Grössen, die keine echte Kette bilden" (both in Russian). Bull. de l'Acad. des Sciences. Petersburg, 1911

Wahrscheinlichkeitsrechnung. Transl. from 2nd Russian edition by H. Liebmann. Leipzig, 1912. Pp. vii + 318.

Démonstration du second théorème—limite du calcul des probabilités par la méthode des moments. Saint-Pétersbourg, 1913. Pp. 66.

[Supplement to the 3rd Russian edition of Wahrscheinlichkeitsrechnung, in honour of the bicentenary of the Law of Great Numbers, with a Portrait of Jacques Bernoulli.]

MASARYK, T. G. David Hume's Skepsis und die Wahrscheinlichkeitsrechnung. Wien, 1884.

MASERES, F. The Doctrine of Permutations and Combinations, being an Essential and Fundamental Part of the Doctrine of Chances: As it is delivered by Mr. James Bernoulli, in his excellent Treatise on the Doctrine of Chances, intitled, Ars conjectandi . . . 8vo. London, 1795.

MEINONG, A. Review of Von Kries's "Die Principien der Wahrscheinlichkeitsrechnung." Göttingische Gelehrte Anzeigen, vol. 2, pp. 56-75, 1890.

Über Möglichkeit und Wahrscheinlichkeit: Beiträge zur Gegenstandstheorie und Erkenntnistheorie. Pp. xvi. + 760. Leipzig, 1915.

MEISSNER (OTTO). Wahrscheinlichkeitsrechnung: I. Grundlehren; II. Anwendungen. Leipzig, 1912; 2nd ed., 1919. Pp. 56 + 52.

[An elementary primer.]

MENDELSSOHN, MOSES. Philos. Schriften, 2 Tle. 12mo. Pp. xxii + 278 + 283. Berlin, 1771. (*Vide* especially vol. ii. pp. 243-283, entitled "Ueber die Wahrscheinlichkeit.")

MENTRÈ, F. "Rôle du hasard dans les inventions et découvertes." Rev. de Phil., 1904.

"Les Racines historiques du probabilisme rationnel de Cournot." Rev. de Métaphysique et de Morale, pp. 485-508, May 1905.

Cournot et la renaissance du probabilisme au xixe siècle. Paris, 1908.

MERRIMAN, M. A Text-book of the Method of Least Squares. New York, 1884. Pp. vii + 198. 6th ed., 1894.

"List of Writings relating to the Method of Least Squares, with Historical and Critical Notes." Trans. Connecticut Acad. vol. 4, pp. 151-232, 1877.

MERTZ. Die Wahrscheinlichkeitsrechnung und ihre Anwendung, etc. Frankfort, 1854.

MESSINA, I. "Intorno a un nuovo teorema di calcolo delle probabilità." 20 pp. 4to. Giornale di Matematiche di Battaglini, vol. lvi. (1918). Naples.

MESSINA, I.—*continued.*
[Described Stat. Jl. vol. lxxxii. (1919), p. 612.]
"Su di un nuovo teorema di calcolo delle probabilità, sul teorema di Bernoulli e sui postulati empirici per la loro applicazione." Boll. del Lavoro et della Presidenza, vol. xxxiii. (1920).

MEYER, A. Essai sur une exposition nouvelle de la théorie analytique des probabilités *à posteriori*. 4to. Pp. 122. Liége, 1857.
Cours de calcul des probabilités fait à l'université de Liége de 1849 à 1857. Publié sur les mss. de l'auteur par F. Folie. Bruxelles, 1874.
Vorlesungen über Wahrscheinlichkeitsrechnung. (Translation of the above by E. Czuber.) Pp. xii + 554. Leipzig, 1879.

MICHELL. "An Inquiry into the Probable Parallax and Magnitude of the Fixed Stars, from the Quantity of Light which they afford us, and the particular Circumstances of their Situation." Phil. Trans. vol. 57, pp. 234-264, 1767.

MILHAUD, G. "Le Hasard chez Aristote et chez Cournot." Revue de Méta. et de Mor. vol. x. pp. 667-681, 1902.

MILL, J. S. System of Logic. Bk. iii. chaps. 18, 23.

MONDÉSIR. "Solution d'une question qui se présente dans le calcul des probabilités." Liouville Journ. vol. ii.

MONRO, C. J. "Note on the Inversion of Bernoulli's Theorem in Probabilities." Proc. Lond. Math. Soc. vol. 5, pp. 74-78 and 145, 1874.

MONTESSUS, R. DE. Leçons élémentaires sur le calcul des probabilités. Pp. 191. Paris, 1908. (Reviewed Stat. Journ., 1909, p. 113.)
"Le Hasard." Rev. du Mois, March 1907.

MONTESSUS, R. DE, and LECHALAS, G. "Un Paradoxe du calcul des probabilités." Nouv. Ann. iv. (3), 1903.

MONTMORT, P. DE. Essai d'analyse sur les jeux de hasard. 4to. Pp. xxiv + 189. Paris, 1708.
Ditto. 4to. pp. 414. Paris, 1714. (The 2nd ed. is increased by a treatise on Combinations, and the correspondence between M. and Nicholas Bernoulli.)

MONTUCLA, J. T. Histoire des mathématiques. 4 vols. 4to. Paris, 1799–1802.
Vol. iii. pp. 380-426.

NEWCOMB, SIMON. A Statistical Inquiry into the Probability of Causes of the Production of Sex in Human Offspring. (Published by the Carnegie Institution of Washington.) Pp. 34. 8vo. Washington, 1904.

NICOLE, F. "Examen et résolution de quelques questions sur les jeux." Hist. Ac. Par. pp. 45-56, 331-344, 1730.

NIEUPORT, C. F. DE. Un peu detort ou amusemens d'un sexagenaire. 8vo. Bruxelles, 1818. Containing "Conversations sur la théorie des probabilités."

NITSCHE, A. "Die Dimensionen der Wahrscheinlichkeit und die Evidenz der Ungewissheit." Vierteljahrsschr. f. wissensch. Philos. vol. 16, pp. 20-35, 1892.

NIXON, J. W. "An Experimental Test of the Normal Law of Error." Stat. Journ. vol. 76, pp. 702-706, 1913.

OETTINGER, L. Die Wahrscheinlichkeitslehre. 4to. Berlin, 1852.
[Reprinted from Crelle, J., vols. 26, 30, 34, 36, under the title, Untersuchungen über Wahrscheinlichkeitsrechnung.]

OSTROGRADSKY. "Probabilité des jugements." Acad. de St-Pétersbourg, 1834.
"Sur la probabilité des hypothèses." Mélanges math. et astr., 1859.

PAGANO, F. Logica dei probabili. Napoli, 1806.
PARISOT, S. A. Traité du calcul conjectural ou l'art de raisonner sur les choses futures et inconnues. 4to. Paris, 1810.
PASCAL, B. "Letters to Fermat." Varia opera mathematica D. Petri de Fermat. pp. 179-188, Toulouse, 1678.
Œuvres, vol. 4, pp. 360-388, Paris, 1819.
PATAVIO. Probabilismus methodo mathematico demonstratus. 1840.
PAULHAN, FR. "L'erreur et la sélection." Rev. Philos. vol. viii. pp. 72-86, 179-190, 290-306, 1879.
PEABODY, A. P. "Religious Aspect of the Logic of Chance and Probability." Princeton Rev. vol. v. pp. 303-320, 1880.
PEARSON, K. "On a Form of Spurious Correlation which may arise when Indices are used, etc." Proc. Roy. Soc. vol. lx. pp. 489-498.
"On the Criterion that a given System of Deviations from the Probable in the case of a Correlated System of Variables is such that it can be reasonably supposed to have arisen from Random Sampling." Phil. Mag. (5), vol. 50, pp. 157-160, 1900.
"On some Applications of the Theory of Chance to Racial Differentiation." Phil. Mag. (6), vol. 1, pp. 110-124, 1901.
Contributions to the Mathematical Theory of Evolution.
[The main interest of the twelve elaborate memoirs published in the Phil. Trans. under the above title is in every case statistical. References are given below to those of them which have most reference to the theory of Probability and in which Professor Pearson's general theory is mainly developed.]
II. "Skew Variation in Homogeneous Material." Phil. Trans. (A), vol. 186, Part i. pp. 343-414, 1895.
III. "Regression, Heredity, and Panmixia." Phil. Trans. (A), vol. 187, pp. 253-318, 1897.
IV. "On the Probable Errors of Frequency Constants and on the Influence of Random Selection on Variation and Correlation." Phil. Trans. (A), vol. 191, pp. 229-311, 1898. (With L. N. G. Filon.)
VII. "On the Correlation of Characters not quantitatively measurable." Phil. Trans. (A), vol. 195, pp. 1-47, 1901.
"Mathematical Contributions to the Theory of Evolution." Roy. Stat. Soc. Journ. lvi., 1893, pp. 675-679; lix., 1896, pp. 398-402; lx., 1897, pp. 440-449.
"On the Mathematical Theory of Errors of Judgment, with special reference to the Personal Equation." Phil. Trans. (A), vol. 198, pp. 235-299, 1902.
On the Theory of Contingency and its relation to Association and Normal Correlation. Pp. 35. 4to. London, 1904.
On the General Theory of Skew Correlation and Non-linear Regression. Pp. 54. 4to. London, 1905.
On further Methods of determining Correlation. London, 1907. (Reviewed by G. U. Yale Journ. Roy. Stat. Soc., Dec. 1907.)
"On the Influence of Past Experience on Future Expectation." Phil. Mag. (6), vol. 13, pp. 365-378, 1907.
"The Fundamental Problem of Practical Statistics." Biometrika, vol. xiii. pp. 1-16, 1920.
[On Inverse Probability.]
"Notes on the History of Correlation." Biometrika, vol. xiii. pp. 25-45, 1920.
"The Chances of Death" and other essays. 2 vols. 8vo, London, 1897.
The Grammar of Science. London, 1892.

PEIRCE, C. S. "A Theory of Probable Inference." Johns Hopkins "Studies in Logic," 1883.

"On an Improvement in Boole's Calculus of Logic." Proc. Amer. Acad. Arts and Sci. vol. vii. pp. 250-261, 1867. Pp. 62. Camb., 1870.

PEROZZO. "Nuove applicazioni del calcolo delle probabilità allo studio dei fenomeni statistici." Proceedings of Academia dei Lincei, 1881–82.

Germ. transl. by O. Elb. Neue Anwendungen der Wahrscheinlichkeitsrechnung in der Statistik. Pp. 33. 4to. Dresden, 1883.

PIÈRON, H. "Essai sur le hasard. La Psychologie d'un concept." Rev. de Méta. et de Mor. vol. x. pp. 682-693, 1902.

PINARD, H. "Sur la Convergence des Probabilités." Rev. Néo-Schol. de Phil. No. 84 (1919) and No. 85 (1920).

PINCHERLE, S. "Il calcolo delle probabilità e l' intuizione." Scientia, vol. xix. pp. 417-426, 1916.

PIZZETTI, P. I fondamenti matematici per la critica dei risultati sperimentali. Atti della R. Univ., Genova, 1892.

PLAATS, J. D. VAN DER. Over de toepassing der waarschijnlijkheidsrekening op medische statistick. 1895.

PLANA, G. "Mémoire sur divers problèmes de probabilité." Mémoires de l'Académie de Turin for 1811–12, vol. xx. pp. 355-408, 1813.

POINCARÈ, H. Calcul des probabilités. Pp. 274. Paris, 1896.

2nd edition (with additions). Pp. 333. Paris, 1912.

Science et hypothèse. Paris.

Engl. transl., London, 1905.

Science et méthode. Paris. (Includes a chapter on "Le Hasard.")

Eng. transl. (by F. Maitland). Pp. 288. London, 1914.

"Le Hasard." Rev. du Mois, March 1907.

POISSON, S. D. Recherches sur la probabilité des jugements en matière criminelle et en matière civile, précédées des règles générales du calcul des probabilités. 4to. Pp. ix + 415. Paris, 1837.

Lehrbuch der Wahrscheinlichkeitsrechnung. German translation of the above by H. Schnuse. Braunschweig. 8vo. 1841.

"Sur la probabilité des résultats moyens des observations." Conn. des Temps. Pp. 273-302, 1827. Pp. 3-22, 1832.

"Formules relatives aux probabilités qui dépendent de très grand nombres." Compt. Rend., Acad. Paris, vol. 2, pp. 603–613, 1836.

"Sur le jeu de trente et quarante." Annal. de Gergonne, xv.

"Solution d'un problème de probabilité." Liouv. J. (1), vol. 2, 1837.

"Mémoire sur la proportion des naissances des filles et des garçons." Mém. Acad. Paris, vol. 9, pp. 239-308, 1830.

PONDRA et HOSSARD. Question de probabilité résolue par la géométrie. 8vo. Paris, 1819.

PORETZKI, PLATON. S. "Solution of the general Problem of the Theory of Probability by means of Mathematical Logic." (In Russian.) Bull. of the physico-mathematical Academy of Kasan, 1887.

PREVOST, P. "Sur les principes de la théorie des gains fortuits." Nouv. Mém. Pp. 430-472. Berlin, 1780.

PREVOST, P., and LHUILIER, S. A. "Sur les probabilités." Mém. Ac. Berl. (1796), pp. 117-142, 1799.

"Sur l'art d'estimer la probabilité des causes par les effets." Mém. Ac. Berl. (1796), pp. 3-24, 1799.

"Remarques sur l'utilité et l'étendue du principe par lequel on estime la probabilité des causes." Mém. Ac. Berl. (1796), pp. 25-41, 1799.

Note on last. Mém. Ac. Berl. (1797), p. 152, 1800.

"Mémoire sur l'application du calcul des probabilités à la valeur du témoignage." Mém. Ac. Berl. (1797), pp. 120-151, 1800.

PRICE, R. See BAYES.

PRINGSHEIM, A. See DANIEL BERNOULLI.

"Weiteres zur Geschichte des Petersburger Problems." Grunert, Archiv, 77, 1881.

PROCTOR, R. A. Chance and Luck. A Discussion of the Laws of Luck, Coincidences, Wagers, Lotteries, and the Fallacies of Gambling; with Notes on Poker and Martingales. Pp. vii + 263. London, 1887.

PROTIMALETHES. Miracle *versus* Nature: being an Application of Certain Propositions in the Theory of Chances to the Christian Miracles. 8vo. Cambridge, 1847.

QUETELET, A. Instructions populaires sur le calcul des probabilités. 12mo. Bruxelles, 1828.

Engl. transl.: Popular Instructions on the Calculation of Probabilities, transl. with notes by R. Beamish. 1839.

Dutch transl. by H. Strootman. Breda, 1834.

Lettres sur la théorie des probabilités appliquée aux sciences, morales et politiques. Bruxelles, 1846.

Engl. transl.: Letters on the Theory of Probabilities as applied to the Moral and Political Sciences, transl. by O. G. Downes. 8vo. 1849.

"Sur la possibilité de mesurer l'influence des causes qui modifient les élémens sociaux." Corresp. mathém. et phys. vol. vii. pp. 321-346. Bruxelles, 1832.

"Sur la constance qu'on observe dans le nombre des crimes qui se commettent." Corresp. mathém. et phys. vol. vi. pp. 214-217. Brussels, 1830.

"Théorie des probabilités." (In the Encycl. populaire.) Brussels, 1853.

"Sur le calcul des probabilités appliqué à la science de l'homme." Bull. de l'Acad. roy. vol. xxvi. pp. 19-32. Brussels, 1873.

[For a full bibliography and discussion of Quetelet's writings on these topics see Lottin's Quetelet.]

RAYLEIGH, LORD. "On James Bernouilli's Theorem in Probabilities." Phil. Mag. (5), vol. 47, pp. 246-251, 1899.

REGNAULT. Calcul des chances et philosophie de la bourse. 8vo. Paris, 1863.

RENOUVIER, CH. L'Homme: la raison, la passion, la liberté, la certitude, la probabilité morale. 8vo. 1859.

REVEL, P. CAMILLE. Esquisse d'un système de la nature fondé sur la loi du hasard. 1890. 2nd ed. (corrigée), 1892.

Le Hasard, sa loi et ses conséquences dans les sciences et en philosophie. Paris, 1905. 2nd ed. (corrigée et augmentée)., Pp. 249. Paris, 1909.

RIZZETTI, J. "Ludorum scientia, sive artis conjectandi elementa ad alias applicata." Act. Erud. Suppl. vol. 9, pp. 215-229, 295-307. Leipzig, 1729.

ROBERTS, HON. FRANCIS. "An Arithmetical Paradox concerning the Chances of Lotteries." Phil. Trans. vol. xvii. pp. 677-681, 1693.

ROGER. "Solution d'un problème de probabilité." Liouv. J. (1), vol. 17, 1852.

ROUSE, W. Doctrine of Chances, or the Theory of Gaming made easy to every Person—Lotteries, Cards, Horse-Racing, Dice, etc. 1814.

RUDIGER, ANDREAS. De sensu falsi et veri libri iv. [Lib. i. cap. xii. et lib. iii.] Editio Altera. 4to. Lipsiae, 1722.

RUFFINI. Critical Reflexions on the Essai philosophique of Laplace (in Italian). Modena, 1821.

SABUDSKI-EBERHARD. Die Wahrscheinlichkeitsrechnung, ihre Anwendung auf das Schiessen und auf die Theorie des Einschiessens. Stuttgart, 1906.

SAWITSCH, A. Die Anwendung der Wahrscheinlichkeitstheorie auf die Berechnung der Beobachtungen und geodätischen Messungen oder die Methode der kleinsten Quadrate. (Translated into German from the Russian by Lais.) Leipzig, 1863.

SCHELL, W. Über Wahrscheinlichkeit. 8vo.

SCHNUSE, H. Vid. POISSON.

SCHWEIGGER, F. Berechnung der Wahrscheinlichkeit beim Würfeln.

SCOTT, JOHN. The Doctrine of Chance: the Arithmetic of Gambling. 56 pp. 8vo. 1908.

SEGUERI, PAOLO. Lettere sulla materia del probabile. 12mo. Colonia, 1732.

SEXTUS EMPIRICUS. Works.

SHELDON, W. H. "Chance." Journal of Phil., Psych., and Sci. Meth., vol. ix. pp. 281-290. 1912.

SHEPPARD, W. F. "On the Application of the Theory of Error to Cases of Normal Distribution and Normal Correlation." Phil. Trans. A. vol. 192, pp. 101-167, 1899.

"On the Calculation of the most Probable Values of the Frequency Constants for Data arranged according to Equidistant Divisions of a Scale." Proc. Lond. Math. Soc. vol. xxix. pp. 353-380.

"Normal Correlation." Camb. Phil. Soc. vol. xix.

"Normal Distribution and Correlation. Roy. Soc. Trans., 1898.

SIGWART, C. Review of von Kries in Vierteljahrsschr. für Wiss. Phil. xiv. p. 90.

Logik. Tübingen, 1878.

2nd ed. Freiburg, i. B., 1893. English ed., 1895.

Vol. ii. Part 3, chap. 3, § 85, Die Wahrscheinlichkeitsrechnung; 5, § 102. Die Wahrscheinlichkeit auf statischem Boden.

References in English ed.:

Probability, vol. ii. pp. 216-230, 261-271 (errors of observation), 303-309 (induction), 504-507 (statistics).

SIMMONS, T. C. "A New Theorem in Probability." Proc. Lond. Math. Soc. vol. 26, pp. 290-323, 1895.

"Sur la probabilité des événements composés." Ass. Franc. pour l'Avancement des Sciences. 1896.

SIMON. "Exposition des principes du calcul des probabilités." Journ. des Actuaires français, i.

SIMPSON, T. "A Letter to the Right Honourable George, Earl of Macclesfield, President of the Royal Society, as to the Advantage of taking the Mean of a Number of Observations in Practical Astronomy." Phil. Trans. vol. xlix. pp. 82-93, 1755.

"An Attempt to show the Advantage arising by taking the Mean of a Number of Observations in Practical Astronomy." (Miscellaneous tracts on some curious subjects, pp. 64-75). London, 4to, 1757.

[A reprint of the above with some new matter. The probability, assuming positive and negative errors to be equally likely, that the mean is nearer to the truth than a single observation taken at random, is here investigated for the first time.]

Treatise on the Nature and Laws of Chance. 4to. London, 1740.

Another edition. 8vo. 1792.

SOREL, G. "Le Calcul des probabilités et l'expérience." Rev. Philos. vol. xxiii. pp. 50-66, 1887.

SPEHR, F. W. Vollständiger Lehrbegriff der reinen Combinationslehre mit Anwendungen derselben auf Analysis und Wahrscheinlichkeitsrechnung. 2. wohlfeile Ausg. 4to. Braunschweig, 1840.

SPINOZA. "Letter to Jan van der Meer." Opera ed. Van Vloten and Land, vol. ii. pp. 145-149, Ep. 38 (in Latin and Dutch).

See also Spinoza's Briefwechsel in J. H. v. Kirchmann's Philos. Bibliothek, vol. xlvi. pp. 145-147.

SPRAGUE, T. B. On Probability and Chance and their Connexion with the Business of Insurance. 8vo. 1892.

STAMKART, F. J. Over de waarschijnlijkheidsrekening. 8vo.

STERZINGER, O. Zur Logik und Naturphilosophie der Wahrscheinlichkeitslehre. Leipzig, 1911.

STEWART, DUGALD. "On the Calculus of Probabilities, in reference to the Preceding Argument for the Existence of God, from Final Causes." Philosophy of the Moral Powers, vol. ii. pp. 108-119. (Sir W. Hamilton's ed., Edin., 1860.) 1st ed., 1828.

STIEDA, L. Über die Anwendung der Wahrscheinlichkeitsrechnung in der anthropologischen Statistik. Arch. f. Anthrop., 1882.

2nd ed. 8vo. Braunschweig, 1892.

STREETER, T. E. The Elements of the Theory of Probabilities. 31 pp. 8vo. 1908.

STRUVE. Catalogus novus stellarum duplicium et multiplicium. Dorpati, 1827, pp. xxxvii-xlviii.

STUMPF, C. "Bemerkung zur Wahrscheinlichkeitslehre." Jahrb. f. national. Ök. u. Stat. (3), vol. 17, pp. 671, 672, 1899; vol. 18, p. 243, 1899.

[In criticism of Bortkiewicz, *q.v.*]

STUMPF, K. "Über den Begriff der mathematischen Wahrscheinlichkeit." Ber. bayr. Ak. (Phil. Cl.), pp. 37-120, 1892.

"Uber die Anwendung des mathematischen Wahrscheinlichkeitsbegriffes auf Teile eines Continuums." Ber. bayr. Ak. (Phil. Cl.), pp. 681-691, 1892.

SUPPANTSCHITSCH. Einführung in die Wahrscheinlichkeitsrechnung. Leipzig.

TAIT, P. G. "Law of Frequency of Error." Edin. Phil. Trans. vol. 4, 1865.

On a Question of Arrangement and Probabilities. 1873.

TCHEBYCHEF, P. L. Essai d'analyse élémentaire de la théorie des probabilités. 4to. Moscow, 1845 (in Russian, degree thesis). Pp. ii + 61 + iii.

"Démonstration élémentaire d'une proposition générale de la théorie des probabilités." Crelle J. vol. 33, pp. 259-267, 1846.

"Des valeurs moyennes." Liouv. J. (2), vol. 12, pp. 177-184, 1867. (Extrait du Recueil des Sciences mathématiques, vol. ii.)

"Sur deux théorèmes relatifs aux probabilités." Petersb. Abh. vol. 55, 1887. (In Russian.) French translation by J. Lyon: Act. Math. Petr. vol. 14, pp. 305-315, 1891.

Œuvres. 2 vols. 4to. St-Pétersbourg, 1907.

(The three memoirs preceding are here reprinted in French.)

TERROT, BISHOP. "Summation of a Compound Series and its Application to a Problem in Probabilities." Edin. Phil. Trans., 1853, vol. xx. pp. 541-545.

"On the Possibility of combining two or more Probabilities of the same Event, so as to form one Definite Probability." Edin. Phil. Trans., 1856, vol. xxi. pp. 369-376.

THIELE, T. N. Theory of Observations. Pp. 6 + 143. 4to. London, 1903.

THOMSON, ARCHBISHOP. Laws of Thought. § 124, Syllogisms of Chance (13 pp.).

THUBEUF. Élémens et principes de la royale arithmétique aux jettons, etc. 12mo. Paris, 1661.

TIMERDING. Die Analyse des Zufalls. Pp. ix + 168. Braunschweig, 1915.

TODHUNTER, I. "On the Method of Least Squares." Camb. Phil. Trans. vol. ii.

A History of the Mathematical Theory of Probability from the Time of Pascal to that of Laplace. Lge. 8vo. pp. xvi + 624, Camb. and Lond., 1865.

TOZER, J. On the Measure of the Force of Testimony in Cases of Legal Evidence. 4to. Camb. Phil. Soc. vol. viii. Part II. 16 pp. (read Nov. 27, 1843). 1844.

TREMBLEY. "Observations sur le calcul d'un jeu de hasard." Mém. Ac. Berl. (1802), pp. 86-102.

"Recherches sur une question relative au calcul des probabilités." Mém. Ac. Berl. (1794-5), pp. 69-108, 1799.

(On Euler's memoir, "Solutio quarundam quaestionum difficiliorum in calculo probabilitatum.")

"De probabilitate causarum ab effectibus oriunda." Comm. Soc. Reg. Gott. (1795-8), vol. 13, pp. 64-119, 1799.

"Observations sur la méthode de prendre les milieux entre les observations." Mém. Ac. Berl. (1801), pp. 29-58, 1804.

"Disquisitio elementaris circa calculum probabilium." Comm. Soc. Reg. Gott. (1793-4), vol. 12, pp. 99-136, 1796.

TSCHUPROW, A. A. "Die Aufgaben der Theorie der Statistik." Jahrb. f. gesetzg. Verwalt. u. Volkswirtsch. vol. 29, pp. 421-480, 1905.

"Zur Theorie der Stabilität statistischer Reihen." Skandinavisk Aktuarietidskrift, pp. 199-256, 1918; pp. 80-133, 1919.

TWARDOWSKI, K. "Über sogenannte relative Wahrheiten." Arch. f. syst. Philos. vol. viii. pp. 439-447, 1902.

URBAN, F. M. "Über den Begriff der mathematischen Wahrscheinlichkeit." Vierteljahrsschr. f. wiss. Phil. und Soz., vol. x. (N.S.), 1911.

VASTEL, L. G. F. L'Art de conjecturer. Traduit du latin de J. Bernoulli, avec observations, éclaircissemens et additions. Caen, 1801.

[Translation of Part I. only of Bernoulli's Ars Conjectandi (*q.v.*) containing a commentary on and reprint of Huygens, De ratiociniis in ludo aleae.]

VENN, J. The Logic of Chance. 1866. 2nd ed., 1876. 3rd ed., 1888.

"The Foundations of Chance." Princeton Rev. vol. 2, pp. 471-510, 1872.

"On the Nature and Uses of Averages." Stat. Journ. vol. 54, pp. 429-448, 1891.

WAGNER, A. Die Gesetzmässigkeit in den scheinbar willkürlichen Handlungen des Menschen. Hamburg, 1864.

"Wahrscheinlichkeitsrechnung und Lebensversicherung." Zeitschr. f. d. ges. Versicherungswissenschaft. Berlin, 1906.

WARING, E. (M.D. Lucasian Prof.) On the Principles of translating Algebraic Quantities into Probable Relations and Annuities, etc. Pp. 59. Cambridge, 1792.

An Essay on the Principles of Human Knowledge. Pp. 244. Cambridge, 1794.

WELTON, J. Manual of Logic. (Probability, vol. ii. pp. 165-185.) London, 1896.

WESTERGAARD. Grundzüge der Theorie der Statistik.

WHITAKER, LUCY. "On the Poisson Law of Small Numbers." Biometrika, vol. x. 1914.

WHITTAKER (E. T.). "On Some Disputed Questions of Probability." Transactions of the Faculty of Actuaries in Scotland, vol. viii. (1920), pp. 163-206.

[Problems of Inverse Probability including the Law of Succession. This paper is followed by others on the same subject by various writers.]

WHITWORTH, W. A. Choice and Chance, An Elementary Treatise on Permutations, Combinations, and Probability, with 300 Exercises. 1867. 2nd ed., 1870. 3rd ed. pp. viii + 244. Cambridge, 1878.

Expectations of Parts into which a Magnitude is divided at Random. 1898.

WICKSELL, S. D. "Some Theorems in the Theory of Probabilities." Skandinavisk Aktuarietidskrift, p. 196 (1910).

WIJNNE, H. A. De leer der waarschijnlijkheid in hare toepassing op het dagelijksche leven. 1862.

WILBRAHAM, H. "On the Theory of Chances developed in Prof. Boole's 'Laws of Thought.'" Phil. Mag., 1854.

WILD, A. Die Grundsätze der Wahrscheinlichkeitsrechnung und ihre Anwendungen. München, 1862.

WINDELBAND, W. Die Lehren vom Zufall. Berlin, 1870.

WOLF, A. "The Philosophy of Probability." Proc. Arist. Soc. vol. xiii. pp. 29. London, 1913.

WOLF, R. "Über eine neue Serie von Würfelversuchen." Viertelјs. Naturforsch. Gesellsch. in Zürich, vol. 26, pp. 126-136 and 201-224, 1881; vol. 27, pp. 241-262, 1882; vol. 28, pp. 118-124, 1883.

"Neue Serie von Würfelsversuchen." *Ibid.* vol. 38, pp. 10-32, 1893.

"Versuche zur Vergleichung der Erfahrungswahrscheinlichkeit mit der mathematischen Wahrscheinlichkeit." Mitth. d. Naturforsch. Gesellsch., Bern, 1849-1851, 1853.

WOLFF, CHRISTIAN. Philosophia rationalis sive logica. Leipzig, 1732.

WOODWARD, R. S. Higher Mathematics, chap. x. pp. 467, 507. "Probability and Theory of Error." New York, 1900.

Probability and Theory of Errors. New York, 1906.

WYROUBOFF, G. "Le Certain et le probable." La Philos. posit. p. 165, 1867.

YOUNG, J. R. Elementary Treatise on Algebra, Theoretical and Practical, with an Appendix on Probabilities and Life Annuities. 4th ed. enlarged, post 8vo. 1844.

YOUNG, REV. M. "On the Force of Testimony in establishing Facts contrary to Analogy." Trans. Roy. Ir. Acad. vol. vii. pp. 79-118, 1800.

YOUNG, T. "Remarks on the Probabilities of Error in Physical Observations, etc." Phil. Trans., 1819.

YULE, G. U. "On the Theory of Correlation." Journ. Stat. Soc. vol. lx. p. 812, 1897.

"On the Association of Attributes in Statistics." Phil. Trans. (A), vol. 194, pp. 257-319, 1900.

"On the Theory of Consistence of Logical Class-frequencies." Phil. Trans. (A), vol. 197, pp. 91-132, 1901.

An Introduction to the Theory of Statistics. Pp. xiii. + 376. London, 1911.

YULE and GALTON. "The Median." Stat. Journ. pp. 392-398, 1896.

INDEX[1]

Acquaintance, direct, 12
Addition, of probabilities, 37, 135
  definition of, 120
  Theorem of, 104, 121, 144
  and measurement, 158
Analogy, principle of, 68
  and induction, 218, 222
  negative, 219, 223, 233, 415
  positive, 220, 223, 415
  and generalisation, 223
  logical foundation of, 258
  and Bacon, 268
  and Leibniz, 272
  and Jevons, 273
  and statistics, 391, 407, 415 f.
Ancillon, 5 *n.*, 82
Apprehension, direct, and ethical judgment, 316
Argument, 13
Aristotle, 80, 92
  and induction, 274
Arithmetic mean (or average), 205
  and laws of error, 197
  Laplace on, 206
  Gauss on, 206
Astronomers and Least Squares, 210
Asymmetry, and Bernoulli's Theorem, 358 f.
Atomic Uniformity, 249
Averages, 205 f.
  weighting of, 211
  and discordant observations, 214
Axioms, 135 f.
  non-self-evident, 299

Bachelier, 347 *n.*
  and statistical frequency, 349 *n.*, 351
  and Rule of Succession, 376 *n.*
Bacon, 265 f.
  tables of, 269
  and limited variety, 270
Bayes, and Inverse Probability, 174
  Theorem of, 379
Belief, rational, 4 f., 10, 16, 307
  degrees of, 11
Bentham, measurement of Probability, 20
Bernoulli, Daniel, and Inverse Probability, 174
  and planets, 293 *n.*, 294
  and Petersburg Paradox, 316, 317
Bernoulli, Jac., 15 *n.*, 41, 76, 81, 83, 86, 368, 369
  weight of evidence, 313
  second axiom of, 322
  and regular frequency, 333
  and statistical series, 392
Bernoulli's Theorem, 109, 314, 319 *n.*, 333, 337 f.
  and asymmetry, 358 f.
  empirical verification of, 361 f.
  Inverse of, 368 f., 385 f.,
Bertrand, 48 *n.*, 49
  on multiplication, 136
  and Maxwell, 172 *n.*
  and independence, 173
  and Law of Error, 208 *n.*
  and chance, 284
  and Petersburg Paradox, 317
  and Bernoulli's Theorem, 339
  and Rule of Succession, 382
Bicquilley and testimony, 184 *n.*
Bobek and Rule of Succession, 383
Bode's Law, 304
Boole, 43 *n.*, 50 *n.*, 84, 294 *n.*
  and German logicians, 87
  and relation of Probability, 90
  and symbolic probability, 155
  and approximation, 161
  and independence, 167
  and Whately, 179
  and combination of premisses, 179

[1] This Index does not cover the Bibliography.

Boole (*contd.*)—
 and testimony, 180
 and Challenge Problem, 187
 and Cournot, 284 *n.*
 and Rule of Succession, 382
Borel, 47 *n.*, 48
Bortkiewicz, von, and great numbers, 333 *n.*
 and Marbe, 365 *n.*
 method of, 384
 and Lexis, 393 f.
 and Law of Small Numbers, 401 f.
 and Quetelet, 402
Boscovitch and Least Squares, 210
Bowley, 421, 423, 424 f.
Bradley, 319 *n.*
 and relativity of Probability, 91
 and Bernoulli's Theorem, 341 *n.*
Broad, C. D., 257 *n.*
Brömse and Marbe, 365 *n.*
Brünn and lotteries, 364
Bruns and Marbe, 365 *n.*
Buffon, 317, 322
 and coin-tossing, 362
Butler, Bishop, 79, 80, 309, 310
 and risk, 321

Calculus of Probability, 83 *n.*, 149, 164, 303, 428
 and Psychical Research, 302
 and Sociology, 335
'*Casual*,' 288
Causality, 263, 276
 and independence, 164
'*Cause*,' 275
Cause, final, 297
Cayley, and tradition, 185
 and Challenge Problem, 187
Certainty, 10, 127, 128
 and truth, 15
 Kahle and, 90 *n.*
 definition of, 120
 relation of, 134
 and Bacon, 267
 and Leibniz, 272
Chance, objective, 281, 286 f., 295, 418
 Conturat on, 283
 Poincaré on, 284, 289
 Condorcet on, 284
 definition of, 287
 and planets, 293
 and binary stars, 295
Coefficient of Credibility, 183
 of Correlation, 421 f.
Combination of premisses, 149, 178
Comte, and '*seven*,' 246
 and statistics, 335
Condorcet, 83 *n.*, 317
 and testimony, 180
 and chance, 282, 284
 and ethics, 313, 316
 and gambling, 319
Conduct and Probability, 307
Consistence and group theory, 124
Contradiction, 143
Coover, J., 298 *n.*
Correlation, 329, 390
 and statistical frequency, 330
 Quantitative, 391, 426
 Inductive, 406
 coefficient, 421 f.
Cournot, and frequency theory, 92
 and independence, 166
 on testimony, 180
 and causality, 275
 and chance, 282, 283
Couturat, 272 *n.*, 311 *n.*
Craig and tradition, 184
Cramer and Petersburg Paradox, 318
Crofton, 47 *n.*
Cumulative Formula, 150
 Johnson and, 121
Czuber, 47 *n.*, 78, 82, 86, 339 *n.*, 345 *n.*, 347
 and symbolic probability, 156
 and '*cause*,' 275 *n.*
 and risk, 315 *n.*
 and Bernoulli's Theorem, 340 *n.*
 and statistical frequency, 351, 394
 and Tchebycheff's Theorem, 353 *n.*, 355 *n.*
 and verification of Bernoulli, 362 *n.*
 and lotteries, 364
 and Marbe, 365
 and Inverse of Bernoulli's Theorem, 370 *n.*
 and Rule of Succession, 376 *n.*, 382

D'Alembert, 82, 170 *n.*, 321, 365 *n.*, 369
 and chance, 282
 and planets, 293
 and mathematical expectation, 314
 and ethics, 316
 and Petersburg Paradox, 317
 and Marbe, 365
Darbon, A., and Cournot, 284
Darwin, 108
 and Lyell, 161
 and Mill, 265
Dedekind and '*Challenge Problem*,' 187 *n.*

Definitions, 134 f.
summary of, 120
de la Placette, Jean, and chance, 283
De Morgan, 21, 74, 83
and inference, 139
and independence, 168
and Inverse Probability, 178
and combination of premisses, 179
and tradition, 184 *n.*
and planets, 293
pupil of, 362
and Inverse of Bernoulli's Theorem, 370 *n.*
and Rule of Succession, 375, 382
De Witt and arithmetic averages, 206
Dice-tossing, 361 f.
Diderot on testimony, 183
Discordant observations, rejection of, 213
Donkin, W. F., 20
and Inverse Probability, 176
Dormoy, 394

Edgeworth, 29 *n.*, 84, 85, 362 *n.*, 379, 400
use of '*Probability*,' 96 *n.*
and randomness, 290
and Psychical Research, 298 *n.*
and ethics, 316
and German statisticians, 394
Eggenberger, 340 *n.*
Ellis, Leslie, 84, 85
and frequency theory, 92
and Least Squares, 207 *n.*, 209
and Bacon, 265 *n.*, 266 *n.*, 269 *n.*, 271 *n.*, 274 *n.*
and Bernoulli's Theorem, 341
Empirical School, 85, 86
Epistemology, 302
and inductive hypothesis, 261
Equiprobability, 41, 63, 65
Equivalence, definition of, 120, 134
axiom of, 135
principle of, 141
Error, probable, 329
Ethics, 307 f.
Euler and Least Squares, 210
Event, probability of, 5
Evidence, and measurement of Probability, 7, 35
relevant and irrelevant, 53, 54
independent and complementary, 55
external, 57
addition of, 66, 68
weight of, 71
and Induction, 221

Excluded Middle, Law of, 143
Experience and the Principle of Indifference, 100

Fechner, and median, 201
and law of sensation, 208
and lotteries, 364
Fermat, formula of, 242
Forbes, J. D., 20 *n.*, 21, 294 *n.*
Frazer, Sir J., 245
Frequency curves, 199
and statistics, 328
Frequency, statistical, 330
Frequency theory, 92 f.
and randomness, 290
and Bernoulli's Theorem, 344
and Rule of Succession, 378
Fresnel and simplicity, 206
Fries, 15 *n.*

Galton, 321
and Fechner's law, 208
Gambling, 319
Gauss, and laws of error, 196 *n.*, 198
and arithmetic mean, 206
and Least Squares, 210
Generalisation, 389
definition of, 222
from statistics, 328
Generator properties, 253
plurality of, 254, 256, 257
Geometrical probability, 47, 62
German logicians, 87
Gibbon, 29, 322, 333
Gilman, B. I., and symbolic probability, 156
Goldschmidt, 29 *n.*
Goodness, organic nature of, 310
Graunt, 392 *n.*
Great Numbers, Law of, 82, 330, 333 f.
Greville, Fulke, 466
Grimsehl, 248 *n.*
and Marbe, 365 *n.*
Groups, of propositions, 117, 124
definition of, 120, 125
real and hypothetical, 129
Grünbaum and Marbe, 365 *n.*

Hagen, and error, 207
and discordant observations, 214 *n.*
Halley and mortality statistics, 332
Herodotus, 307
Herschell and binary stars, 294
Houdin, 364 *n.*
Hudson, W. H., and animism, 247 *n.*
Hume, 52, 70, 80, 81, 82, 83, 239, 427
and testimony, 182

Hume (*contd.*)—
 and Induction, 218, 233, 265, 272
 and analogy, 222, 224
 and chance, 282
Huyghens, 82
 and '*six*,' 247
Hypothesis, 7
Hypothetical entities, 299

Implication, 124
Impossibility, 15
 definition of, 120
 relation of, 134
Inconsistency, definition of, 120
Independence, for knowledge, 107, 165
 definition of, 120, 138
 Theorem of, 121, 146
 of events, 164
 and law of error, 195
 and measurement, 204
 and averages, 212
 and discordant observations, 214
 and chance, 283
Index numbers, 211
'*Induction*,' 274
Induction, 97
 Principle of, 68
 and frequency theory, 98, 99, 107
 and Logic, 217
 pure, 218
 universal, 220, 406, 417
 validity of, 221
 and statistics, 327 f.
 statistical, 406 f.
Inductive correlation, 220, 257, 258, 392, 397, 406
Inductive hypothesis, 260, 264
Inductive method, 260
Inference, 129
 necessary, 120, 139
 hypothetical and assertoric, 130
 statistical, 327 f.
Insurance, 22, 285, 404
Intuition *versus* experience, 86
 and ethical judgment, 312
Inverse Probability, 149, 174
 and Venn, 100
 and frequency theory, 106
 Theorem of, 121
 and statistics, 369, 370 *n.*
 and Bowley, 425
Irrelevance, 255
 judgments of, 54
 definition of, 55, 120, 138
 Theorem of, 121, 146

James, W., and spirits, 301
Jesuits, 308
Jevons, 244 *n.*
 and equiprobability, 42 *n.*
 and Inverse Probability, 178
 and index numbers, 212
 and Induction, 222, 238, 243, 265, 273, 274
 and analogy, 246
 and coin-tossing, 362
 and Rule of Succession, 382
Johnson, W. E., 116
 and propositions, 11 *n.*
 and added evidence, 68
 and cumulative formula, 121, 150, 153, 155
 and groups, 124
 and testimony, 183
Judgments, 54
 of preference and relevance, 65
 direct, 70
 disjunctive, 77

Kahle and the Probability relation, 90
Kant, 333
 and Hume, 272
Kapteyn, Prof. J. C., and law of error, 199
Knowledge, 10
 kinds of, 3, 4
 direct and indirect, 12, 262
 incomplete and proper, 13
 of logical relations, 14
 probable and vague, 17
 relativity of, 17
 vague and distinct, 53
 homologic and ontologic, 276, 288
 and ignorance, 281
 and chance, 289
Kries, von, 42, 44 *n.*, 45 *n.*, 46 *n.*, 50, 67 *n.*, 84
 and equiprobability, 87
 and Principle of Indifference, 172
 and independence, 173
 and Inverse Probability, 176
 and knowledge, 276
 and Cournot, 284 *n.*
 and School of Lexis, 394

Lacroix, 184 *n.*
Lambert and Least Squares, 210
Lämmel, 47 *n.*
 and symbolic probability, 156
Laplace, 15 *n.*, 28 *n.*, 31, 82, 83, 84, 318, 427
 school of, 44, 51, 86, 358, 365

Laplace (*contd.*)—
and relation of Probability, 91
and independence, 170
and Inverse Probability, 175, 178
and testimony, 180, 182
and doctrine of averages, 202
and arithmetic mean, 206
and Least Squares, 210
and Induction, 220, 239, 265, 273
and chance, 282
and planets, 293 *n.*
and Quetelet, 334
and Bernoulli's Theorem, 340, 341, 370
and Rule of Succession, 351 *n.*, 359 *n.*, 368
and birth proportions, 364
and unknown probabilities, 370
and Bayes' Theorem, 380
and statistical series, 392
Laurent and gambling, 319
Law, 311 *n.*
Law of error, 194 f.
and arithmetic mean, 197
and geometric mean, 198
and median, 200
and mode, 203
normal law, 199, 202, 205
Lexis and, 398
Least Squares and Venn, 206
method of, 202, 205, 206, 209
Lee and tradition, 184 *n.*
Legendre and Least Squares, 210
Leibniz, 24 *n.*, 308, 368, 392, 427
and arithmetic average, 206
and Induction, 272
Lexis, and asymmetry of statistical frequency, 359 *n.*
and Marbe, 365 *n.*
method of, 384, 393 f., 397 f.
and Edgeworth, 401
and statistical stability, 415, 419 *n.*
Locke, 76, 80, 82, 83, 308, 323
on tradition, 184
and weight of evidence, 313
Logic, academic, 3
of probability, 8
of implication, 58
and Induction, 217, 245
and initial probability, 299
Logical priority, 129
Lotteries, 333 *n.*, 361, 364 f.
published results of, 363
Lotze, 89
and Rule of Succession, 382
Lucretius, 427
M'Alister, Sir Donald, and laws of error, 198
Macaulay and Bacon, 266
McColl, and symbolic probability, 155
and Boole, 167 *n.*
and Inverse Probability, 176
and '*Challenge Problem*,' 188 *n.*
Macfarlane, and independence, 169 *n.*
and tradition, 185
and '*Challenge Problem*,' 187 *n.*
Maclaurin, Theorem of, 207
Marbe, Dr. Karl, and roulette, 365
Marginal utility, 318
Markoff, A. A., 177 *n.*
and Inverse Probability, 176
and Tchebycheff's Theorem, 357
Mathematical Expectation, 311, 315, 316
Mathematicians, and probability, 84
and cumulative formula, 152
and laws of error, 207
and ethics, 316
Maxim, Sir Hiram, 364 *n.*
Maxwell, 172 *n.*
and theory of gases, 172
Mayer and Least Squares, 210
Means and laws of error, 194 f.
Measurement of Probability, 34, 158, 311
and frequency theory, 94
and induction, 259, 388
and psychical research, 302
and ethics, 311
Median and laws of error, 200
Meinong, 78
Meissner, Otto, and dice-throwing, 363
Memory, 14
Mendelism and statistics, 335, 419, 428
Merriman, Mansfield, and Least Squares, 209
Metaphysics and certainty, 239
Method of Difference, 246
Michell, 302
and Inverse Probability, 174
and binary stars, 294
Middle Term, Fallacy of, 68, 155
Mill, and inductive correlations, 220
and induction, 265 f.
and plurality of causes, 267 *n.*
and probability, 268 *n.*
and pure induction, 269
methods of, 270
and limited variety, 271
Modality and probability, 16 *n.*

Modality (*contd.*)—
Venn and, 98
Mode, and law of error, 203
asymmetry about, 361
Monte Carlo, 364
Moore, G. E., 19, 240 *n.*, 309
Morgan, *vide* De Morgan
Multiplication, 135
definition of, 120
theorems of, 121, 148, 342
of instances, 233 f.
Munro, 370 *n.*

Necessary connection, law of, 251
Newton, and induction, 244
and '*seven*,' 247
and Bacon, 265
Nitsche, A., 45 *n.*, 50 *n.*, 78, 172 *n.*

Occurrences, remarkable, 302

Pascal, 82
Pearson, Karl, 84, 351 *n.*
and frequency theory, 100
and arithmetic mean, 208
and stars, 297
and asymmetry, 347, 359 *n.*
and generalised Probability curves, 347
and roulette, 364
and Rule of Succession, 379, 382
Peirce, 50 *n.*, 304
and randomness, 290
Petersburg Paradox, 316
psychology of, 318
and Buffon, 362
Peterson and tradition, 184
Physics and initial probability, 299
Planets, movements of, 293
Playfair, Dr. Lyon, 305
Plurality of causes and Mill, 267
Poetry and statistics, 401
Poincaré, Henri, 48, 84
and independence, 173
and chance, 284, 289
Poisson, 51 *n.*, 362 *n.*
on testimony, 180
and least errors, 207
and Petersburg Paradox, 317
and gambling, 319
and great numbers, 333, 336
Theorem of, 344
and statistical frequency, 348
and Tchebycheff, 357
and inverse of Bernoulli's Theorem, 370

Poretzki, Platon S., and symbolic probability, 157
Port Royal logic, 70, 80, 321
and probabilism, 308
Prediction, value of, 305
Price and Bayes, 174 *n.*
Primitive people and rational belief, 245
Principle of compelling reason, 86
Principle of Indifference, 42, 81, 83 f., 87, 104, 107, 171
analysis of, 53
modification of, 55, 58
and induction, 99
and measurement, 160
and Psychical Research, 302
and ethics, 310
and statistics, 367
and Laplace, 372, 374
and Rule of Succession, 377
Principle of Non-Sufficient Reason, 41, 85
Principle of superposition of small effects, 249
Probabilism, 308
'*Probability*,' 8
Venn's use of, 95
Edgeworth's use of, 96 *n.*
Probability, and relevant knowledge, 4
objective relation of, 5, 8, 281
mathematical, 6
dependent on evidence, 7
philosophical definition of, 8
three senses of, 11
measurement of, 20 f., 37
and law, 24
and similarity, 28, 36
comparison of, 34, 66, 160
series of, 35, 38
'*geometrical*,' 47, 48, 62
and rational belief, 97
and statistical frequency, 98
and truth frequency, 101 f., 337 f.
Inverse, 106, 149
and truth, 116, 322
negative, 139
finite, 237
and randomness, 291
and planetary orbits, 293
and binary stars, 294
and star drifts, 296
and final causes, 297
and spirits, 300, 301
and telepathy, 300
and ethics, 307

Probability (*contd.*)—
  from statistics, 367 f.
  '*unknown*,' and Laplace, 372
Probability relation, 4, 8, 13, 134
  intuition of, 52
Probable error, 74
Proctor, 364 *n.*
Proposition, characterisation of, 3, 4
  primary and secondary, 11, 13
  knowledge of, 12
  self-evident, 17
  classes of, 101 f.
  groups of, 117, 124
  sub-groups of, 126, 129
  disjunction and conjunction of, 134
  synthetic, 263
  existential, 276
Propositional function, 56
  and induction, 222
  and randomness, 291
Psychical Research, 278 f.
Psychology and probability, 52
Pythagoras and '*seven*,' 246

Quetelet, 333 *n.*, 334, 335, 401, 418, 427, 428
  and arithmetic mean, 208
  and balls, 362
  and statistical stability, 393

Randomness, 281, 290, 412
  Pearson's use of, 297
Relation, of probability, 6
  of '*between*,' 35, 39
Relativity, of knowledge, 17
  of probabilities, 102
  doctrine of, and the Law of Uniformity, 248 *n.*
Relevance, judgments of, 54
  and frequency theory, 104
  theorems of, 147
Remarkableness, 302
Requirement, 129
Risk, 315
  and ethics, 313
  and Petersburg Paradox, 319
  '*moral*,' 320, 322
  '*physical*,' 322
Roulette, 361, 364
  published results of, 363 *n.*
Rule of Succession, 359 *n.*, 368, 372, 374
  proof of, 375
  and frequency theory, 378
  and Pearson, 380 *n.*
Russell, Bertrand, 19, 115, 124 *n.*, 126
Russell, Bertrand (*contd.*)—
  and inference, 117
  and implication, 124

Schematisation, 67
Schröder and symbolic probability, 157
Selection, random, 292
Series of probabilities, 35, 38
  and frequency theory, 93
  independent, 283, 420
  organic, 399, 420
  Gaussian, 421 *n.*
Sigwart, 88
  and inverse probability, 178
  and induction, 273
Simmons and asymmetry in Bernoulli's Theorem, 359
Simpson and Least Squares, 210
Small Numbers, Law of, 401 f.
Society for Psychical Research, 298 *n.*
Space, 255
  and uniformity, 226
  irrelevance of, 301
Spedding and Ellis and Bacon, 265 *n.*, 266 *n.*
Spielräume, doctrine of, 88
Spinoza, 116 *n.*, 282 *n.*
Spirits, probability of, 300
Star drifts, 296
Stars, binary, 294
Statistical frequency, theory of, 93 f.
  generalisation of, 101
  criticism of, 103
  stability of, 336, 392-415
  fluctuation of, 392
Statistical inference, 327 f.
  induction, 406 f.
Statistics, and prediction, 306
  descriptive and inductive, 327
Stumpf, 44 *n.*, 50 *n.*, 172 *n.*
Sub-analogies, 223, 229
Sub-groups of propositions, 126, 129
Succession, Law of, 82
  See Rule of
Süssmilch and regular frequencies, 333

Taylor, Jeremy, 308 *n.*
Tchebycheff, Theorem of, 353, 355
  and Poisson's Theorem, 357
Telepathy, probability of, 300
Terrot, Bishop, 43 *n.*
  and Whately, 179 *n.*
  and combination of premisses, 179
Testimony, theory of, 180

Time, 255
  and uniformity, 226
  irrelevance of, 301
Todhunter, 294 *n.*, 318 *n.*, 370 *n.*
  and Bayes, 175
  and Craig, 184
  and Petersburg Paradox, 316
  and Bernoulli's Theorem, 340 *n.*
Truth and probability, 116 *n.*, 322
Truth frequency, 101, 406
Tschuprow, 358, 399 *n.*
  and statistical frequency, 348, 394 *n.*
  method of, 384

Uniformity of Nature, Law of, 226, 248, 255, 263, 276
  and Mill, 270
Universal Causation, Law of, 248
Universal Induction and statistical methods, 389, 406-417
Universe of reference, 117, 129, 130
Unknown probabilities, 372, 373, 375

Variables in Probability, 58, 123, 412 *n.*
Variety, 234
  and induction, 219
  limitation of, 258, 260, 427
Venn, 84, 106 *n.*, 294 *n.*
Venn (*contd.*)—
  and experience, 85
  and Bernoulli, 86, 341
  and frequency theory, 93 f.
  and inverse probability, 100
  and Least Squares, 206 *n.*
  and induction, 273
  and chance, 288
  and '*random*,' 290
  and Rule of Succession, 372, 378, 382

Weight, of evidence, 312
  and ethics, 315
Weighting of averages, 211
Weldon and dice, 362
Whately and combination of premisses, 178
Whitehead, and frequency theory, 101
  and invalid inference, 329 *n.*
Whittaker, E. T., and Rule of Succession, 376 *n.*
Wilbraham, H., and Boole, 167 *n.*
Wolf and dice, 362

Yule, 349 *n.*, 361 *n.*
  and approximation, 161
  and independence, 166
  and '*statistics*,' 327
  and coin-tossing, 346 *n.*, 361 *n.*
  and correlation, 421, 424

O False and treacherous Probability,
Enemy of truth, and friend to wickednesse;
With whose bleare eyes Opinion learnes to see,
Truth's feeble party here, and barrennesse.

THE END